Water Resources Development and Management

More information about this series at http://www.springer.com/series/7009

Reinhard F. Hüttl • Oliver Bens
Christine Bismuth • Sebastian Hoechstetter
Editors

Society - Water - Technology

A Critical Appraisal of Major Water
Engineering Projects

Editors
Reinhard F. Hüttl
Helmholtz Centre Potsdam - GFZ German
  Research Centre for Geosciences
Telegrafenberg, Potsdam, Germany

Oliver Bens
Helmholtz Centre Potsdam - GFZ German
  Research Centre for Geosciences
Telegrafenberg, Potsdam, Germany

Christine Bismuth
Interdisciplinary Research Group
  Society - Water - Technology
Berlin-Brandenburg Academy of Sciences
  and Humanities
Berlin, Germany

Sebastian Hoechstetter
Helmholtz Centre Potsdam - GFZ German
  Research Centre for Geosciences
Telegrafenberg, Potsdam, Germany

Helmholtz Centre Potsdam - GFZ
  German Research Centre for Geosciences
Telegrafenberg, Potsdam, Germany

This publication presents the results of the Interdisciplinary Research Group "Society – Water – Technology" of the Berlin-Brandenburg Academy of Sciences and Humanities (Forschungsberichte der interdisziplinären Arbeitsgruppen, vol. 34).

It has been funded by the Senatsverwaltung für Wirtschaft, Technologie und Forschung des Landes Berlin, Germany, and the Ministerium für Wissenschaft, Forschung und Kultur des Landes Brandenburg.

ISSN 1614-810X   ISSN 2198-316X   (electronic)
Water Resources Development and Management
ISBN 978-3-319-18970-3   ISBN 978-3-319-18971-0   (eBook)
DOI 10.1007/978-3-319-18971-0

Library of Congress Control Number: 2015952822

Springer Cham Heidelberg New York Dordrecht London

Contents

Contributors

Oliver Bens Helmholtz Centre Potsdam - GFZ German Research Centre for Geosciences, Telegrafenberg, Potsdam, Germany

Emily S. Bernhardt Department of Biology, Duke University, Durham, NC, USA

Christine Bismuth Interdisciplinary Research Group Society - Water - Technology, Berlin-Brandenburg Academy of Sciences and Humanities, Berlin, Germany

Helmholtz Centre Potsdam - GFZ German Research Centre for Geosciences, Telegrafenberg, Potsdam, Germany

Petra Dobner Institute of Political Science and Japanese Studies, Martin-Luther-Universität Halle-Wittenberg, Halle, Germany

Nils Droste Helmholtz Centre for Environmental Research – UFZ, Leipzig, Germany

Rolf Emmermann Helmholtz Centre Potsdam - GFZ German Research Centre for Geosciences, Telegrafenberg, Potsdam, Germany

Hans-Georg Frede Institute of Landscape Ecology and Resource Management (ILR), Justus-Liebig-Universität Gießen, Gießen, Germany

Gerhard Glatzel Universität für Bodenkultur, Wien BOKU, Vienna, Austria

Armin Grunwald Institute for Technology Assessment and Systems Analysis (ITAS), Karlsruhe, Germany

Hermann H. Hahn Heidelberg Academy of Sciences and Humanities, HAW, Heidelberg, Germany

Ahmad Hamidov Department of Agricultural Economics, Humboldt-Universität zu Berlin, Berlin, Germany

Bernd Hansjürgens Department of Economics, Helmholtz Centre for Environmental Research – UFZ, Leipzig, Germany

Bernd Hillemeier Technische Universität Berlin, Berlin, Germany

Sebastian Hoechstetter Helmholtz Centre Potsdam - GFZ German Research Centre for Geosciences, Telegrafenberg, Potsdam, Germany

Reinhard F. Hüttl Helmholtz Centre Potsdam - GFZ German Research Centre for Geosciences, Telegrafenberg, Potsdam, Germany

Shavkat Kenjabaev Scientific-Information Center of the Interstate Coordination Water Commission (SIC ICWC), Tashkent, Uzbekistan

Anna Koska Leibniz-Institute of Freshwater Ecology and Inland Fisheries (IGB), Berlin, Germany

Hermann Kreutzmann Department of Geography, Freie Universität Berlin, Berlin, Germany

Hans-Joachim Kümpel Federal Institute for Geosciences and Natural Resources (BGR), Hanover, Germany

Abdallah I. Husein Malkawi Jordan University of Science and Technology, Irbid, Jordan

Axel Meyer Universität Konstanz, Konstanz, Germany

Timothy Moss Leibniz Institute for Regional Development and Structural Planning (IRS), Erkner, Germany

Helmar Schubert Karlsruhe Institute for Technology (KIT), Karlsruhe, Germany

Herbert Sukopp Technische Universität Berlin, Berlin, Germany

Klement Tockner Leibniz-Institute of Freshwater Ecology and Inland Fisheries (IGB), Berlin, Germany

Department of Biology, Chemistry and Pharmacy, Freie Universität Berlin, Germany

Yacov Tsur The Hebrew University of Jerusalem, Rehovot, Israel

Ira Yaari Robert H. Smith Faculty of Agriculture, Food and Environment, The Hebrew University of Jerusalem, Rehovot, Israel

Ugur Yaramanci Leibniz Institute for Applied Geophysics (LIAG), Hannover, Germany

Valerie Yorke NCCR Trade Regulation/World Trade Institute, University of Bern, Bern, Switzerland

Christiane Zarfl Center for Applied Geoscience, Eberhard Karls Universität Tübingen, Tübingen, Germany

Abbreviations

ADB	Asian Development Bank
AIS	Administrative irrigation system
AWC	Aqaba Water Company
BAC	Big Andijan Canal
BAIS	Basin administrative irrigation system
BFC	Big Fergana Canal
BGR	Bundesanstalt für Geowissenschaften und Rohstoffe
BMP	Best management practice
BNC	Big Namangana Canal
BOT	Build-operate-transfer
BWO	Basin water management organization
CA	Central Asia
Ca $(HCO_3)^2$	Calcium hydrogen carbonate
$CaSO_4$	Calcium sulfate
CDN	Collector-drainage network
CDW	Collector-drainage water
CWC	Canal water committees
DAAD	German Academic Exchange Service
EEA	European Environment Agency
E-flows	Environmental flows
FAO	Food and Agriculture Organization of the United Nations
FOEME	Friends of the Earth Middle East
FV	Fergana Valley
GDP	Gross domestic product
GIZ	Gesellschaft für Internationale Zusammenarbeit
GSI	Geological Survey of Israel
GTZ	German Technical Cooperation
GWL	Ground water level
GWP	Global Water Partnership
HGME	Hydrogeologic melioration expeditions
HRH	His Royal Highness

IBT	Interbasin water-transfer projects
ICBS	Israel Central Bureau for Statistics
ICG	International Crisis Group
ICOLD	International Commission on Large Dams
ICWC	Interstate Commission for Water Coordination of Central Asia
ILR	Indian Rivers Linking Project
IPCC	Intergovernmental Panel on Climate Change
IPCRI	Israel/Palestine Centre for Research and Information
ISA	Irrigation systems authority
ISF	Irrigation service fee
IUCN	International Union for Conservation of Nature
IWA	Israel Water Authority
IWMI	International Water Management Institute
IWRM	Integrated water resource management
JMGP	Jordan Ministry of Government Performance
JRSP	Jordan Red Sea Project
JVA	Jordan Valley Authority
JWC	Joint Water Committee
JWS	Jordan's water strategy
m^3	Cubic meters
maSL	Meters above sea level
MAWR RUz	Ministry of Agriculture and Water Management of the Republic of Uzbekistan
mbSL	Meters below sea level
MCA	Main canal authority
MDG	Millennium Development Goal
$MgSO_4$	Magnesium sulfate
MoU	Memorandum of understanding
MWEP	Major water engineering project
MWI	Jordan Ministry of Water and Irrigation
NaCl	Sodium chloride
NAWAPA	North American Water and Power Alliance
NFC	North Fergana Canal
NGO	Nongovernmental organization
NIS	New Israeli Shekel
NRW	Non-revenue water
O&M	Operation and maintenance
OECD	Organization for Economic Co-operation and Development
PPP	Public private partnership
RSDS	Red Sea–Dead Sea
SDC	Swiss Agency for Development and Cooperation
SFC	South Fergana Canal
SFG	Strategic Foresight Group
SIC-ICWC	Scientific-Information Center of Interstate Commission for Water Coordination of Central Asia

TEEB	The economics of ecosystems and biodiversity
TEV	Total economic value
TSR	Transboundary small river
UCWU	Union of Canal Water Users
UN	United Nations
UNDP	United Nations Development Programme
UNECE	United Nations Economic Commission for Europe
UNEP	United Nations Environment Programme
UNESCO	United Nations Educational, Scientific and Cultural Organization
UN-ESCWA	United Nations Economic and Social Commission for Western Asia
USAID	US Agency for International Development
USD	US Dollar
VAT	value-added tax
WAJ	Water Authority of Jordan
WCD	World Commission on Dams
WCED	World Commission on Environment and Development
WFD	European Water Framework Directive
WMO	Water management organization
WRG	Water resources group
WTP	Willingness to pay
WUA	Water users association
WUG	Water users group
WWF	World Wildlife Fund
WWTP	Wastewater treatment plant
YWC	Yarmouk Water Company

List of Figures

List of Tables

Part I
Context and Objectives

Chapter 1
Introduction: A Critical Appraisal of Major Water Engineering Projects and the Need for Interdisciplinary Approaches

Reinhard F. Hüttl, Oliver Bens, Christine Bismuth, Sebastian Hoechstetter, Hans-Georg Frede, and Hans-Joachim Kümpel

1.1 Ecological Challenges, Social and Economic Opportunities: The Multiple Facets of Major Water Engineering Projects

Water touches every aspect of human existence on planet earth. While this notion may be regarded as a triviality, it nevertheless has highly complex consequences. Water as a georesource is subject to many pressures due to its multiple functions that go far beyond its role as the fundamental basis of organic life. For instance, antagonistic social, economic and ecological demands meet to form the "water-energy-food" nexus. Rising population numbers, changing lifestyles and climate change have substantial impacts on water resources and aquatic ecosystems. Furthermore, water is a factor in peace among nations: water can be both a source

R.F. Hüttl (✉) • O. Bens • S. Hoechstetter
Helmholtz Centre Potsdam - GFZ German Research Centre for Geoscience,
Telegrafenberg, 14473 Potsdam, Germany
e-mail: huettl@gfz-potsdam.de

C. Bismuth
Interdisciplinary Research Group Society - Water - Technology,
Berlin-Brandenburg Academy of Sciences and Humanities,
Jägerstraße 22/23, 10117 Berlin, Germany

Helmholtz Centre Potsdam - GFZ German Research Centre for Geosciences,
Telegrafenberg, 14473 Potsdam, Germany

H.-G. Frede
Institute of Landscape Ecology and Resource Management (ILR),
Justus-Liebig-Universität Gießen, Heinrich-Buff-Ring 26-32, 35392 Gießen, Germany

H.-J. Kümpel
Federal Institute for Geosciences and Natural Resources (BGR),
Stilleweg 2, 30655 Hannover, Germany

© The Author(s) 2016
R.F. Hüttl et al. (eds.), *Society - Water - Technology*, Water Resources
Development and Management, DOI 10.1007/978-3-319-18971-0_1

of controversy and of cooperation. The "hidden core" of many international conflicts can be regarded as disputes over the access to water.

As a consequence, in many regions of the world, major water engineering projects (MWEPs) such as dams, hydropower plants, large scale irrigation schemes in agriculture, channels for navigation, drinking water transfer connection, etc. are considered as a suitable means for covering the demand for energy and water. In the recent years, rising needs for energy and food have even led to some sort of "renaissance" of MWEPs. New actors – investors, emerging nations and multi-industry companies – have appeared on the scene and economic interests have in many cases impeded a thorough public debate about possible alternatives. The plans for the construction of the Nicaragua Canal as an alternative to the Panama Canal serve as an illustration of such a controversial "megaproject".

Taken all together, MWEPs have significantly shaped societies, economies and ecosystems. This poses crucial questions to scientists in their role as advisors to decision makers: by what means can the future use of water be organised in the most efficient, but also most sustainable way? What are the consequences of further large-scale interventions? And, do the long-term effects of MWEPs limit the number and scope of future options for action and decision-making? Reflecting on these fundamental questions is the main motivation of the book at hand.

1.2   Interdisciplinary Research on Water Resources

Looking back, it has often been "technocratic thinking" that has paved the way for MWEPs. Expected influences on ecosystems, societies and economies have been oversimplified or misjudged in the planning stages. Only retrospectively, in many cases, it has turned out that these interventions had taken place within highly complex and coupled systems. The outcome of supposedly simple and determinate water management measures had proven to be unpredictable in advance, with some of the impacts affecting societies decades or even centuries later. The desire for immediate and simple on-hand solutions has a tendency to blur people's vision and rational thinking when it comes to assessing burdens on future generations. The awareness that technical measures alone are not sufficient for meeting present and future challenges in resource management is a relatively young one, and it still needs to be backed up by scientific, ethical, social and political reflections.

This is where the need for interdisciplinary approaches and research designs comes into play. Interdisciplinary research in its true sense is not a goal as such but must be tailored to address many-faceted research questions in their full complexity and to shed a new light on old problems. In this respect, water resource management and the imbalance between mankind's technological capabilities and its systemic intelligence are an ideal field of interdisciplinary research: we deal with coupled and complex systems, we have to find immediate answers concerning the adaptation and mitigation of the effects of climate change, and we have to integrate social, economic and ecological aspects into long-term water management schemes. Undoubtedly, a task such as this would demand too much from the knowledge provided by one single scientific discipline alone.

1.3 The Interdisciplinary Research Tradition of the Berlin-Brandenburg Academy of Sciences and Humanities

The Berlin-Brandenburg Academy of Sciences and Humanities provides an optimal stage for such integrated and interdisciplinary research endeavours. Since 1994, more than 70 Interdisciplinary Research Groups (IRG) have been established there, becoming an out of the ordinary work concept within German research landscape. The general objective of the IRGs is to seize on topics of high scientific and societal relevance, to conduct research on questions of future importance and to initiate a dialogue between science and society. In all of these projects, close relations between members of the Academy and researchers from different national and international research institutions have developed. Another important objective of the Academy is to provide knowledge-based advice for the society and for politics.

In 2011, based on an initiative of Reinhard Hüttl and Oliver Bens, the IRG *Society – Water – Technology* was established for a term of 3 years. The idea of this IRG, whose members and associates form the group of authors of this book, fits perfectly into this tradition.

1.4 Aims and Working Structure of the Interdisciplinary Research Group *Society – Water – Technology*

The overarching aim of the IRG was to comprehensively describe the framework conditions of major water engineering projects and to analyse their impact on the ecology, economy and society. Based on its analyses and findings, policy recommendations are formulated and research gaps are pointed out. In order to accomplish this, it was necessary to identify those mechanisms that are necessary for the functional continuity of MWEPs – i.e. for their successful and socially acceptable construction – their continuing operation and their adaptation to the targeted conceptual implementation of strategies. The analysis of MWEPs and their impacts and benefits has considered possible future development trends on the basis of present situations and also on the basis of historical reviews. This was accomplished by closely scrutinising two representative case studies: the Fergana Valley in Central Asia and the Lower Jordan Basin in the Middle East.

In the Fergana Valley, the main object of research were the impacts of the construction of a huge irrigation system during the former Soviet period and its effects on present societies and economies as well as on ecosystems along the downstream parts of the big rivers. Thus, the Fergana Valley case study was essential for the development of a common interdisciplinary research methodology and for studying the concept of "path dependencies".

The second case study in the Lower Jordan Basin dealt with the planned Red Sea – Dead Sea Conveyance Project to halt the further decline of the Dead Sea water level resulting from extensive water extraction from the Jordan River. This case study allowed the IRG to assess and to evaluate the planning processes of transboundary projects. It provided valuable information on the strengths and weaknesses of international water law as well as on the relations between different riparian states and the prerequisites for successful cooperation.

The IRG was organised according to these case studies in two different subgroups or "clusters". Overarching themes were discussed in the general assembly. The research instruments of the IRG consisted of profound literature research, thematic workshops, study visits to the Fergana Valley and the Lower Jordan Basin, public symposiums and thematic lectures. In addition, two expertise reports on *Water Ethics* (Armin Grunwald) and on *International Water Law* (Ute Mager) were conducted within the scope of the research group.

A workshop with an international line-up on *The Red Sea – Dead Sea Water Conveyance Project: An opportunity for regional cooperation and improved water management in the Jordan River Basin* was held in Potsdam, Germany, in December 2013. Speakers included Stefan Geyer (Helmholtz Centre for Environmental Research UFZ, Halle), Heinz Hötzl (Karlsruhe Institute of Technology KIT, Karlsruhe), Harald Kunstmann (Karlsruhe Institute of Technology KIT, Garmisch-Partenkirchen), Muath Abou Sadah (Palestine Hydrology Group, Ramallah, Palestinian Territories), Elias Salameh (University of Jordan, Amman, Jordan), Uri Shani (Hebrew University of Jerusalem, Israel), Gerhard Smiatek (Karlsruhe Institute of Technology KIT, Garmisch-Partenkirchen), Katja Tielbörger (Tübingen University, GLOWA Jordan River), Yacov Tsur (Hebrew University of Jerusalem, Israel) and Valerie Yorke (NCCR Trade Regulation, World Trade Institute, University of Bern, Switzerland).

The "academy lectures", a unique event format for disseminating research results to the public in layman's terms, have dealt with *Water Ethics – Reflections on Handling Resource Conflicts* (Armin Grunwald), *Major Water Engineering Projects and their Impact on Ecology and Water Cycles – Opportunities and Risks* (Axel Meyer and Klement Tockner), *The Fergana Valley in Central Asia – Conflicts Regarding the Georesource Water* (Hermann Kreutzmann) and *The Economic Value of Water Habitats – Approaches and Evaluating Experiences in the Lower Jordan Basin* (Nir Becker).

Several members of the research group have furthermore conducted a number of research journeys to the case study areas in the Middle East and in Central Asia. As part of these stays, numerous regional experts from science, administration and non-governmental organisations were involved in discussions and field trips, allowing for in-depth insights into the specific regional contexts. Findings of these research journeys formed an important basis for the further analyses and eventually paved the way for this volume.

1.5 Acknowledgements

The editors of this book and the members of the research group would like to express their gratitude to the following experts from the case study areas, who made a significant contribution to the success of the project:

Jordan (October 2013): Daniel Busche (Deutsche Gesellschaft für Internationale Zusammenarbeit GIZ, Jordan), Niklas Gassen (Federal Institute for Geosciences and Natural Resources BGR, Hanover), His Excellency Saad Abu Hammour (Ministry of Water & Irrigation, Jordan Valley Authority, Amman, Jordan), Marwan Al Raggad (University of Jordan, Water and Environment Research and Study Centre, Amman, Jordan), Amer Salman (Faculty of Agriculture, University of Jordan, Amman, Jordan), Ali Subah (Ministry of Water and Irrigation, Amman, Jordan), Manuel Schiffler (KfW Development Bank, Berlin) and Matthias Toll (Federal Institute for Geosciences and Natural Resources BGR, Hanover).

Israel and Palestinian Territories (December 2013): Ruth Arnon (Israel Academy of Sciences and Humanities, IASH), Menahem Yaari (Israel Academy of Sciences and Humanities), Abdelrahman Alamarah (Palestinian Hydrology Group for Water & Environmental Resources Development, Ramallah), Nir Becker (Department of Economics, Tel-Hai College, Upper Galilee, Israel), Gidon Bromberg (Friends of the Earth Middle East), Rebhy A. El Sheikh (Palestinian Water Authority), Eran Feitelson (Department of Geography, Advanced School for Environmental Studies, The Hebrew University of Jerusalem, Israel), Oded Fixler (Israel Water Authority), Barak Herut (Israel Oceanographic & Limnological Research, Israel), Ahmad M. Hindi (Palestinian Water Authority, Ramallah), Bob Lapidot (Israel Academy of Sciences and Humanities, IASH), Clive Lipchin (Center for Transboundary Water Management, Arava Institute, Israel), Doron Markel (Monitoring and management Lake Kinneret and its watershed, Water Authority, Israel), Klaus Schelkes (Federal Institute for Geosciences and Natural Resources BGR, Hanover), Abraham Tenne (Desalination Division, Water Authority, Israel) and Menahem Yaari (Israel Academy of Sciences and Humanities, IASH).

Fergana Valley, Uzbekistan (May 2014): Iskandar Abduallaev (The Regional Environmental Centre for Central Asia, CAREC, Almaty, Kazakhstan), Akhmedov Djamaleddin (Water Users Association "Kadyrjon Azamjon", Fergana Valley, Uzbekistan), Victor A. Dukhovny (Scientific Information Center of the Interstate Coordination Water Commission of the Central Asia, SIC ICWC, Tashkent, Uzbekistan), Ahmad Hamidov (Humboldt Universität zu Berlin, Germany), Akmal Karimov (International Water Management Institute, IWMI, Tashkent, Uzbekistan), Shavkat Kenjabaev (Scientific Information Center of the Interstate Coordination Water Commission of the Central Asia, SIC ICWC, Tashkent, Uzbekistan), Alfiya Khaliullina (SIC ICWC), Markus Müller (Deutsche Bank AG, Frankfurt), Maraliev Maraim (Water Users Association "Kadyrjon Azamjon", Fergana Valley, Uzbekistan), Bolot Moldobekov (Central Asian Institute of Applied Geosciences, CAIAG, Bishkek, Kyrgyz Republic), Azim Nazarov (Nazar Business and Technology Limited Consulting NBT, Tashkent, Uzbekistan), Olivier Normand (Regional Rural Water Supply & Sanitation Project, Fergana City, Uzbekistan, Swiss Agency for Development and Cooperation), Nailja Rezjapova (Friedrich Ebert Foundation, Tashkent, Uzbekistan), Mukhtar Ruziev (SIC ICWC), Vitaliy Stepanov (SIC ICWC), Galina Stulina (Scientific Information Center of the Interstate Coordination Water Commission of the Central Asia, SIC ICWC, Tashkent, Uzbekistan), Elena Tsay (SIC ICWC) and Thierry Umbehr (Swiss Agency for Development and Cooperation, Swiss Cooperation Office Uzbekistan).

Additional advice was provided by the following external guests and experts: Volker Frobarth (Programme Director "Transboundary Water Management in Central Asia", GIZ), Abror Gafurov (German Research Centre for Geosciences GFZ, Potsdam), Wolfgang Kinzelbach (ETH Zurich), Stefano Parolai (German Research Centre for Geosciences GFZ, Potsdam), Daniela Scheetz (Climate and Environmental Foreign Policy, Sustainable Economy, Federal Foreign Office, Berlin), Joop de Schutter (UNESCO-IHE, Institute for Water Education, Delft), Jenniver Sehring (Deutsche Gesellschaft für Internationale Zusammenarbeit, GIZ), Hinrich Thölken (Climate and Environmental Foreign Policy, Sustainable Economy, Federal Foreign Office, Berlin), Katy Unger-Shayesteh (German Research Centre for Geosciences GFZ, Potsdam), Thomas Vetter (Climate and Environmental Foreign Policy, Sustainable Economy, Federal Foreign Office, Berlin) and Michael Weber (German Research Centre for Geosciences GFZ, Potsdam).

The scientific and administrative management of the Berlin-Brandenburg Academy of Sciences and Humanities gave its utmost support to the project. The research group particularly owes great thanks to Wolf-Hagen Krauth, Regina Reimann, Andreas Schmidt, Ute Tintemann and Janina Amendt.

For the final preparation of the manuscript of this book and for invaluable organisational and administrative support, special thank goes to Anna Kaiser, Yvonne Dinter and Edward Ott. We would also like to thank Dietmar Kraft, who shouldered the scientific coordination tasks in the early phase of the project.

For the thorough editing and proof-reading of the manuscript, the group thanks Ian Whalley and for the technical review of the contributions, Iskandar Abdullaev (CAREC, Regional Environmental Centre for Central Asia, Almaty), Ines Dombrowsky (DIE, German Development Institute, Bonn), Uwe Grünewald (Brandenburg University of Technology Cottbus-Senftenberg) and Yoav Kislev (Hebrew University of Jerusalem).

For the support in the organisation of the research trip to the Fergana Valley, we owe particular thanks to the team of the Scientific Information Center of Interstate Coordination Water Commission (SIC ICWC), especially to Victor Dukhovny, Galina Stulina and Shavkat M. Kenjabaev (SIC ICWC/SANIIRI).

Over and above the basic funding of the project by the Berlin-Brandenburg Academy of Sciences and Humanities, the activities of the working group were supported by various institutions. We are especially grateful to the German Academic Exchange Service (DAAD), the Hermann und Elise geborene Heckmann Wentzel Foundation and the National Academy of Science and Engineering (acatech), the Swiss Academy of Engineering Sciences (SATW), the Israel Academy of Sciences and Humanities (IASH) and the Austrian Academy of Sciences (ÖAW).

1.6   Members of the Interdisciplinary Research Group
Society – Water – Technology

Members of the research group were: Oliver Bens (German Research Centre for Geosciences GFZ, Potsdam), Petra Dobner (Martin-Luther-Universität Halle-Wittenberg), Rolf Emmermann (German Research Centre for Geosciences GFZ, Potsdam), Hans-Georg Frede (Justus-Liebig-Universität Gießen), Manuel Frondel (Rheinisch-Westfälisches Institut für Wirtschaftsforschung RWI, Essen), Carl Friedrich Gethmann (Universität Siegen), Gerhard Glatzel (Universität für Bodenkultur, BOKU, Vienna), Hermann H. Hahn (Heidelberg Academy of Sciences and Humanities, HAW), Bernd Hansjürgens (Helmholtz Centre for Environmental Research UFZ, Leipzig), Bernd Hillemeier (Technische Universität Berlin), Reinhard Hüttl (spokesperson of the IRG, German Research Centre for Geosciences GFZ, Potsdam), Hermann Kreutzmann (Freie Universität Berlin), Hans-Joachim Kümpel (Federal Institute for Geosciences and Natural Resources BGR, Hanover), Axel Meyer (Konstanz University), Timothy Moss (Leibniz Institute for Regional Development and Structural Planning IRS, Erkner), Helmar Schubert (Karlsruhe Institute of Technology KIT, Karlsruhe), Herbert Sukopp (Technische Universität Berlin), Klement Tockner (Leibniz Institute of Freshwater Ecology and Inland Fisheries IGB, Berlin), Menahem E. Yaari (Israel Academy of Sciences and Humanities) and Ugur Yaramanci (Leibniz Institute for Applied Geophysics LIAG, Hanover). The group was supported by the scientific and administrative coordination team consisting of Christine Bismuth, Yvonne Dinter, Sebastian Hoechstetter, Anna Kaiser and Dietmar Kraft.

Berlin, in April 2015
The editors of this volume and the speakers of the clusters Fergana Valley and
Lower Jordan Valley

Chapter 2
Water Ethics – Orientation for Water Conflicts as Part of Inter- and Transdisciplinary Deliberation

Armin Grunwald

Abstract The notion of a water ethics has only emerged over the past 10 years. It is mainly motivated by environmental concerns and the observation of water conflicts. This chapter focuses on the ethical aspects of human interventions into water systems. It describes cultural, moral and religious attitudes towards water and reviews the state of the art in this field. Its main objective is to conceptualise water ethics on the basis of the philosophical approach of discourse ethics and to draw conclusions for ethically responsible interventions into water systems and for dealing reasonably with water conflicts. Far from promising "miracles" from water ethics, the specific added value of ethical considerations lies in providing the orientation for ongoing debates on water challenges by not only applying substantial principles, but by offering suitable procedures as well.

Keywords Water ethics • Value of water • Sustainable development • Environmental justice • Equity • Responsibility • Western World • Islam • Christianity • Human right to water and sanitation

2.1 Objectives and Approach

The aim of water ethics is directed at responding to different challenges which have become obvious over the last decades: conflicts of interest in using scarce water resources, creeping river, sea and ocean pollution, devastating human interventions into sensitive water systems (drying-up of the Aral Sea and the Dead Sea as dramatic examples). These challenges have arisen out of human interventions into water cycles, to human changes of the chemical composition of waters by various forms of private and industrial use of water, and to human interventions into natural landscapes. The

A. Grunwald (✉)
Institute for Technology Assessment and Systems Analysis (ITAS),
P.O. Box 3640, 76021 Karlsruhe, Germany
e-mail: armin.grunwald@kit.edu

© The Author(s) 2016
R.F. Hüttl et al. (eds.), *Society - Water - Technology*, Water Resources Development and Management, DOI 10.1007/978-3-319-18971-0_2

"human factor" in water systems is at the core of ethical considerations. It focuses on the impact of human interventions and consequences of this impact, assesses these interventions with respect to ethical criteria and explores the options for developing this human factor into a more responsible direction.

The notion of a water ethics has emerged over the past 10 years only and is still relatively unknown in the field of applied ethics. The extent of its body of literature is rather small up to now. It consists of a heterogeneous set of approaches to the value of water in different respects (cultural, religious, ecological and economical) and associated with the debates on sustainable development and environmental justice. To present, the largest part of water ethics has been motivated by environmental concerns and related to the field of environmental ethics.

This chapter focuses on the ethical aspects of human interventions into water systems. And with the following objectives it intends to:

- Describe cultural, moral and religious attitudes on and perceptions of the field of water, in particular, value assigned to water
- Review the state of the art in the field of water ethics and provide an overview and orientation of not only basic diagnoses, argumentation patterns, proposed solutions, but also of shortcomings and criticisms
- Conceptualise water ethics on the basis of the philosophical approach of discourse ethics (Habermas 1991) for providing orientation on how to identify responsible strategies for dealing with water challenges
- Categorise the field of water conflicts with respect to ethically relevant criteria and provide pointers on how to deal with those conflicts reasonably and responsibly

The overview provided, the proposal for water ethics developed, and the conclusions given towards possible contributions to solving water conflicts constitute a theory-based approach to water ethics. It shows that society cannot expect "miracles" from water ethics. Rather, the specific added value of ethical considerations lies in providing the orientation for ongoing debates on water challenges by not only applying substantial principles, but by offering suitable procedures as well. In order to exploit these benefits, ethical inquiries in the field of water must be embedded into inter- and trans-disciplinary approaches bringing scientific disciplines together with stakeholders, decision-makers and citizens in the regions under consideration.

2.2 The Value of Water in Different Cultures

Water as a crucial precondition of life has gained prominent importance in almost all cultures and religions. High value is assigned to water, in particular in regions with high scarcity of water, is often accompanied by strong commandments on its usage. In describing briefly the most relevant cultural attitudes towards water I do not follow the usual classification where Christianity is mostly identified with the Western World. Instead, I will argue that the modern attitude to water in the

Western World developed its own and predominantly economic perspective while the water-related values of Christianity are closely related to those of Judaism and Islam. The case studies presented in this volume do not affect regions dominated by Eastern Asian religions I will focus in this section on the monotheistic religions and Western Modernity.

2.2.1 Judaism, Christianity, and Islam

The three monotheistic religions going back to Abraham and Moses, Judaism, Christianity and Islam, share common roots not only in spiritual respect. They all have their roots in areas with a harsh desert climate and water scarcity: Arabia, the Near East and Saharan North Africa. Not surprisingly, high value is assigned to water by all of them. The Bible in its Old and New Testament and the Quran are full of symbols related with water, pointing to water as source of life, or using analogies with water as metaphors for spiritual messages.

There are only a few indicators in the Bible on environment protection. Usually, the story of the Genesis is understood not only as permission, but even as a commandment to humans to dominate nature: "Be fruitful, and multiply, and replenish the earth, and subdue it; and have dominion over the fish of the sea, over the fowl of the air, and over everything living thing that moveth upon the earth" (Genesis 1:28). However, it has often been stated that simplified criticism misses the point (Armstrong and Armstrong 2005). Rejecting the assignment of an *eigenvalue* to nature and considering the value of nature only in relation to human needs is clearly an anthropocentric perspective, but does not allow or even request careless dominion of waters and nature in general (Bartholomew I Ecumenical Patriarch of Constantinople 2010). In contemporary Christianity caring for water resources is seen as part of the duty of stewardship of nature (Pradhan and Meinzen-Dick 2010, p. 48). In Judaism, the Talmud perspective on the environment states that while we may use the world for our own needs, we may never irresponsibly damage or destroy the environment.

In Islam, in particular, water is a pivotal issue (Faruqui et al 2001). Scarcity has always influenced the perception of water by Muslims, and it has, accordingly, shaped their behaviour and customs. The water of rain, rivers and fountains is seen, against this experience, as a symbol of Allah's benevolence: "He sends down saving rain for them when they have lost all hope and spreads abroad His mercy" (Quran 25:48). Water is considered a gift from Allah and should be freely available to all. Any Muslim who withholds unneeded water sins against God: "No one can refuse surplus water without sinning against God and against man." That means that "true Muslim believers cannot grab water in excess to their needs since they are obliged to allow free access to any amounts of water beyond these needs" (Al-Awar et al. 2010, p. 32). Every human being, not only Muslim, is given right to drink to assure survival. There is a right of irrigation but domestic use has precedence over agricultural or industrial use.

Regarding water radically as a common good makes it difficult to assign monetary value to it: following Islam, water should neither be bought nor sold (Al-Awar et al. 2010, p. 33). But how could then a water supply infrastructure be implemented at all? The solution to this problem is that "in spite of its original nature as common good, individuals have the right to use, sell, and recover value-added costs of infrastructure for water supply services" (Al-Awar et al. 2010, p. 33). This means that water belongs to the community and no one is allowed to own it unless labour was provided or an effort was taken to make it usable, or to distribute it. This does not create a right of ownership over water, but rather creates a property on the *value added* to water by labour, thus enabling pricing and trade (Gilli 2004).

2.2.2 Western Modernity

The capitalist economic system, which has developed over the past two centuries, has resulted in a materialistic culture. Value has been understood as *monetary* value to an increasing extent. Determining the monetary value of products and services is usually left to the rules of the marketplace: in the interplay between offer and demand, the adequate price will emerge. The "invisible hand" (Adam Smith) in the back of the participants at the marketplace will then take care that, if all participants will go for maximum profit for themselves, the collective will arrive at the optimal outcome. The reference of value determination is no longer an external authority such as nature, God or a holy book but economy. In the assignment of value to water this means: "Water has an economic value in all its competing uses and should be recognised as an economic good" (Young et al. 1994, p. 4).

Key to the economic view of water is demarcating it as a "resource". Classical economy understands nature as an ensemble of resources for human use: land, raw materials, minerals, biodiversity – and water. It was the U.S. American geographer and secretary of the U.S. Inland Waterways Commission, William McGee, who made this point in a famous paper (1909) expressing very clearly (and early) Western thinking about water as an economic resource which could and should be managed. In modern word this reads: "Managing water as an economic good is an important way of achieving efficient and equitable use, and of encouraging conservation and protection of water resources" (Young et al. 1994, p. 4).

But this is not the whole story. Already McGee and the community he worked with were also open to communitarian ideas and participation in the field of water (Schmidt and Shrubsole 2013), based on the conviction "that all the water belongs to the people" (McGee 1911, p. 822). Thus it becomes clear that Western Modernity is two-fold with an inherent tension: on the one hand, it reduces the value of water to that of an economic resource, but on the other it also acknowledged that water has a political and democratic side (which may be taken as a predecessor of the "human right on water" declared by the United Nations (UN) in 2010, see below).

This leads to the discussion on the common good-property of water in the Western World (Ostrom et al. 2003; Trawick 2010). For a long time, the supply of water was seen as task of public services. Large water supply and sewage infrastructures were built by municipalities and states. The value behind this is the public good property of water transformed into the state's duty to provide access to water and sanitation to everybody. However, neoliberal trends from the 1980s on have led to giving more emphasis to issues such as deregulation and liberalisation even in the field of water supply. In some countries, the borderline between state competencies and market affairs has been shifted in favour of the latter, expressing a shift in regarding the value of water towards an economic product. The installation of the human right on water by the UN in 2010 (see below) might be regarded as a counter-movement re-focusing on the common good character of water. A permanent balancing process is required between considering water either as a common or a private good.

2.2.3   The Need for Water Ethics Beyond Value Assignment

Thus, value assignments to water are an essential element of important religions and cultures. However, they are of limited use only when facing modern challenges to water systems. Given the arguments below, water ethics must go beyond traditional value assignments.

The first argument is that all the value assignments given in cultures and religions are, in spite of the fact that they may include wisdom and reflect cultural experience, in some way "out of the sky". Their reference often is a holy book such as the Bible, the Talmud or the Quran. The statements on water given there are binding only for members of the respective cultures, traditions and religions. Regarding water as a gift of Allah and following all the consequences for dealing with water given in the Quran and the Sharia is binding only for Muslims, and similar with all other religious wisdom and commandment. The limited range of validity of those value assignments also limits the applicability to solve real-world problems in the modern, pluralistic and secular world.

The second argument is closely related to this. Value assignments may be contested and controversial among different cultural groups (Priscoli Delli et al 2004). In particular, cultural and religious values assigned to water do not help in solving water challenges crossing cultural and religious borders. Cultural and religious values may provide inter-cultural and inter-religious conflicts with valuable insight but cannot be a self-evident basis for conflict resolution. For example, in case of a water conflict, for instance between Islamic tradition and the Western capitalist approach, neither Quran nor neoliberal rules of the market have priority *per se*. Both value systems depend on particular normative convictions. Dealing with inter-religious and inter-cultural water conflicts needs a third perspective – and the promise of philosophical ethics is to provide such a perspective by grounding its argumentation not in particular tradition but rather in an argumentation claiming a more general validity. In case of conflict, ethics must question the values at place and look beyond. It cannot take religiously motivated value assignments to water as granted but has to scrutinise the normative bases and resulting rules for human behaviour with respect to their argumentative validity.

A third argument focuses on strong limitations of value assignments for meeting major water challenges even within the same culture. For example, the Asian cultures express deep respect of water for life in general in highly abstract philosophical terms. While these may allow for a deeper understanding of water, life and nature, they will usually not be helpful in solving specific water conflicts, in orientating a more efficient use of water facing scarcity, and in deriving priorities in case of competing usage demands. Also a statement of the kind "Water offers a medium for creating a culture of peace" (Young et al. 1994, p. 4) is not helpful in spite of its truth: the opposite case could happen as well, and water could give rise to conflict instead to peace. The problem with high expectations on value assignments is that they are vague and not operable for specific contexts. They even do not prevent conflict in a morally homogeneous culture or religion. Conflicts of crude interest may occur between members of the same cultural or religious community. Then, obviously, the value assignments cannot help at all because all of the conflicting actors will share the same values. For example, some of the classical water conflicts in the Near East occurred between Islamic states (Iraq, Turkey, Syria, Jordan).

Summing up it becomes clear that the value dimension of water has to be considered carefully in water ethics because of its relevance to cultural attitudes to and perceptions of water. However, values assigned by religions, cultures and traditions are particular and not *per se* ethically legitimate, can often not prevent conflicts or orientate conflict resolution but have to be critically scrutinised with respect to their argumentative grounding. Situations in which existing values do not suffice to orientate action (e.g. in the case of conflicts between different value systems) shall be denoted as situations showing *normative uncertainty*. Ethics has developed exactly to facilitate coping by argumentation with such situations (Gethmann and Sander 1999).

2.3 Water Ethics

Based on a brief and critical review on existing literature on water ethics and a reflection on the very subject of any water ethics, the basic approach will be developed and structured into a substantial and a procedural branch. The section is completed by considerations on the precondition of this approach being applicable.

2.3.1 Review of the Literature on Water Ethics

The body of literature on water ethics is rather small and heterogeneous in terms of the ethics approach chosen. A first bundle of papers take the value of water in different cultural and religious respects into consideration (e.g. Priscoli Delli et al. 2004; UNESCO 2011). Most of the pieces on water ethics explain the demand for developing and implementing such an ethics. High expectations are raised such as:

In short, we need a water ethic – a guide to tight conduct on the face of complex decisions about natural systems we do not and cannot fully understand (Postel 2010, p. 222).

Poul Harremoës (2002) expects that a water ethics – in his understanding a more responsible and reflected behaviour of water users orientated to the precautionary principle – would make over-regulation by the state unnecessary. Most of the approaches, however, assume in broader sense that a water ethics shall help to protect and conserve water systems as the basis of present – and even more – future life on Earth. The call for a new water ethics expresses far-ranging concerns about increasing degradation of water systems – and is an indicator of severe dissatisfaction with protection measures implemented so far and also with our decision-making capabilities on water issues today.

The largest part of water ethics until the present has been related to the field of environmental ethics. An early and influential work by the U.S. American ecologist Aldo Leopold (1949) postulating a "land ethic" was the point of departure. His motivation was to develop a new type of ethic, where the boundaries of the communities shall be extended from humans to include soils, waters, plants and animals (Priscoli Delli et al. 2004, p. 9). Armstrong (2006; 2009) used this postulate as template to develop a water ethic and reasoned:

A thing is right if it preserves or enhances the ability of the water within the ecosystem to sustain life; and wrong if it decreases that ability (Armstrong 2006, p. 13).

However, also the reference to life as such or to "each organism" might not be very helpful because measures usually support some forms of life but would threaten others. *The* life as such cannot be sustained – only specific forms of life (ecosystems, organisms etc.). Therefore, additional criteria would be required to allow for the determination of priorities. Armstrong, however, refuses to start this debate. This is indeed the most acrimonious question of any water ethics, which has to deal with such typical conflicts, ambiguities or even dilemmas. There is no simple way to "sustain life" but we have to ask questions such as "sustain what life under which circumstances and at what price?"

Some authors look for an overall answer to this question in the direction of ranking the value of the natural principally higher than the value of the artificial. Human intervention, in this perspective, only can diminish the value of nature. Even the image of nature restoration is bad because it is human intervention. Statements of the kind

The ethical principle of stewardship teaches respect for creation and moral responsibility for that creation. However, it also calls for wise use of creation and complete unwillingness to modify nature (Priscoli Delli et al. 2004, p. 16).

express a romantic picture and ignore completely that "complete unwillingness to modify nature" is a strange postulate looking at the reality of human intervention into nature. Instead of a "complete unwillingness to modify nature" we need the willingness to take responsibility for our modifications of nature and to act accordingly. The valid question of water ethics is not how to sustain life (Armstrong above) and to leave nature untouched but is for the responsibility of human intervention into water systems.

Also other work in the context of water ethics supports this "anthropocentric turn" and the rejection of romantic eco-belief. In particular, considerations of water systems under the postulate of sustainable development (e.g. Parodi 2008; Lehn and Parodi 2009) have unavoidably an anthropocentric focus because they are dealing with human needs today and in future (WCED 1987; Grunwald and Kopfmüller 2012). In order to avoid misunderstanding: "anthropocentric" in this sense does not imply taking the position of human dominion and complete control over nature. The point here only is that the sustainability focus on human needs rejects fundamental bio-centric positions (such as taken by Leopold 1949) and instead asks how to sustain ecosystems and their services and functions in order to sustain or improve the possibilities of meeting human needs today and in the future.

Some sets of ethical principles have been proposed for water ethics (Groenfeldt 2013). The UNESCO (2011, pp. 18ff) unfolds the normative dimension of water ethics along a number of principles stemming partially from law and partially from ethics:

- Principle of human dignity and the right to water
- Principle of equity in availability and applicability of water
- Principle of eco-centric ethics
- Principle of vicinity
- Principle of frugality
- Principle of transaction
- Principle of multiple and beneficial use of water
- Principle of mandatory application of quantity and quality measures
- Principle of compensation and user pays
- Principle of polluter pays
- Principle of participation
- Principle of equitable and reasonable utilisation

This list seems to be comprehensive but might be perceived as confusing and overladen. It could be used as a checklist to identify ethically relevant issues of a water challenge under consideration and might thus have a heuristic function. However, it is not clear in which way this list could be made operable to be used in processes of deliberation and decision-making. What is missing in this respect is a system of criteria of weighing these principles in case of conflict, and a procedural proposal how this could be done.

2.3.2 Subjects of Water Ethics

One of the major lessons learned when considering the existing work on water ethics is its subject. As has become clear the subject of water ethics is not – as many people might expect – "water" as such but is *human intervention* into water systems and cycles. Water itself is subject to life, to the economy, to religious thought and to cultural attitudes. Ethics, however, always considers options how to act or to decide.

Human interventions into water systems and cycles (including the option not to intervene) are subject to water ethics (this picture puts water ethics in close neighbourhood to the ethics of technology (Parodi 2008; Potthast 2013).

Consequently, a differentiated typology of human interventions into water systems regarding criteria for ethical reflection would be required but is still desiderate. Rough characterisations would distinguish between intended and non-intended interventions. Intended interventions into water systems include direct measures such as building dams for different purposes, deviating rivers and the installation of huge irrigation systems, but also indirect measures regulating or de-regulating water trade and other elements of water governance. Non-intended interventions comprise of the side effects of human action resulting out of other actions. Examples are the creeping groundwater pollution by herbicides and antibiotics from medical and agricultural use, the heating of rivers by nuclear power plants or industry and the pollution of the oceans.

In a situation of normative uncertainty (see above) – for example: the planned regulation of a river leads to protest by ecologists; the planned construction of a hydroelectric power station is rejected by local people; water usage for irrigation in the upstream area of a river causes or increases scarcity of water in the downstream area – the task of ethical reflection is to analyse the normative foundations of the different and possibly conflicting options on the table to provide support in their assessment and comparison, and to contribute to deliberation and decision-making.

Water ethics thus involves a necessary anthropocentric perspective by looking at human intervention, intended or non-intended. Saying this may cause immediate protest by ecologists and sections of bio- and eco-ethics (e.g. Leopold 1949; Katz 1997; Armstrong 2006). Frequently, the anthropocentric perspective of former morals is blamed for negative developments in the natural environment we witness today: loss of biodiversity, climate change, pollution of waters, etc. Usually, other perspectives than an anthropocentric ethics are proposed to improve the situation, such as eco- or bio-centred ethics. However, the basis discourse model of philosophical ethics (Habermas 1991) cannot be other than anthropocentric because only humans can participate in discourse and deliberation. For example, animals or water systems cannot take the role of an active discourse partner. We humans have to take stewardship over them to bring their assumed "perspectives" into the ethical discourse. In this sense water ethics must also be anthropocentric – which, however, should not be confused with postulating for human dominion over nature. Issues of water protection in the interest of future generations, of maintaining biodiversity or of functioning eco-systems may be subject to this discourse as well as conflicts between upstream and downstream users. "Anthropocentric" means that it is up to us humans to take responsibility seriously – because, as far as we know, there is no other species on Earth able to do this. Thus, exploring and assessing options for taking over responsibility for water systems and water cycles by human interventions into water systems is the target of water ethics.

Because human intervention into water systems usually is done by implementing technologies (dams, river regulation, water distribution grids, irrigation technologies, geo-technologies, etc.) including their societal elements such as regulations, rules for action, acceptance patterns, the best place of water ethics in the system of applied ethics (Nida-Rümelin 2005) is, accordingly, the *ethics of technology* (Grunwald 2013) (with strong links with environmental ethics).

Beyond this water ethics is related also to philosophy of nature, with ethics in the field of the economy, with risk ethics and also with bio-ethics – and, of course, with the overall debate on sustainable development. Additionally, in an ethical discourse on water issues further groups of people must be involved: scientists (e.g. from geography, hydrology and engineering), social scientists (e.g. from cultural studies and economics) and, perhaps, humanities and cultural sciences bringing in knowledge about inter-cultural and inter-religious understanding. Furthermore, dealing with "real world" problems in an ethical discourse needs to involve groups and people affected, stakeholders and decision-makers etc. (see the principle of participatory water governance below).

2.3.3   Substantial and Procedural Aspects of Water Ethics

What we can learn from the existing papers on water ethics is that sets of principles might be used as checklists, e.g. to analyse water management options and to compare them. These principles express specific ideas, requirements and norms on how we currently imagine options of responsible interventions into water systems, or which types of interventions we would not consider responsible. They could be used to determine criteria of responsible action.

Those principles are "substantial" in a sense – they promise to provide guidance in cases of normative uncertainty. However, there are severe problems unanswered. First the question arises which system of principles we would like to adopt for what reasons. Second, those systems do not tell us how to proceed in cases of tension and trade-offs – and the experience shows that the field of sustainable development is full of them (Grunwald and Kopfmüller 2012). In the specific field of water it is easy to imagine conflicts, for example, between today's use of water and the assumed interests of future generations. Third, high uncertainties in the assessment of the consequences of today's actions for future water and ecosystem developments will prevent any fixed substantial system of principles from being helpful in the long run (Grunwald 2007). And fourth, any principles will remain abstract to the specific constellation of a given water conflict, while ideas for solutions to the conflicts always have to include contextual specificities and need participation of groups and persons affected (Schmidt and Shrubsole 2013).

Thus, the set of substantial principles has to be in a sense, flexible, must not be too strict and has to be complemented by added procedural elements. The combination of substantial and procedural elements allows the combination of providing orientation and being flexible and adaptive to specific situations.

Drawing on the literature on water ethics available (mainly UNESCO 2011; Groenfeldt 2013), the author would like to propose to use the following set of principles (closely related with the proposal made by Groenfeldt 2013) to be used in water ethics discourses as guiding heuristics:

2.3.3.1 Human Right to Water and Sanitation

The Human Right to Water and Sanitation was declared by the UN in 2010 (UNESCO 2011). It mirrors the fact that access to water, as well as to sanitation, are necessary preconditions of human life. In a sense the right to water is an implication of the postulate of human dignity not only because water is necessary for survival but also because water is essential for food production, for energy and also for cultural issues. In ethical respect, a clear imperative follows to ensure the fulfilment of this right today and in the future. This principle closely relates to the debate on water as a common good and also allows reference to the value dimension of water, e.g. in Islam.

2.3.3.2 Sustaining Ecosystem Functions

Ecosystem functions are essential for a functioning natural environment – in the way of providing ecosystem services according to human needs and for "keeping nature alive" (Groenfeldt 2013) as well. This principle overlaps strongly with the intergenerational dimension of the imperative of sustainable development (Grunwald and Kopfmüller 2012), which calls for long-term responsibility in maintaining the natural conditions of human life (Jonas 1979). But it also includes other ecosystem functions such as cultural or religious ones, or aesthetic arguments (Ott 2010). The precautionary principle (Harremoës et al. 2002) also has its place here in face of the huge uncertainties in long-term developments of water systems.

2.3.3.3 Responsible Use of Water

The Human Right to Water is a right to use water. This use, however, is restricted by the other principles: the actual use of water has to be arranged in a way that (1) the future fulfilment of the right to water is not endangered and (2) ecosystem functions will be sustained. This normative situation may limit the industrial and agricultural use of water in specific cases. Facing high uncertainties concerning long-term effects of human interventions it might be a postulate of responsibility to take care of the resilience of water systems. Responsibility also includes caring about the safety of water engineering, e.g. in the construction of dams.

2.3.3.4 Participatory Water Governance

The principles given above do not allow direct derivation or deduction of ethically justified advice. On the one hand, this will be prevented frequently by the virulence of conflicts between the principles (e.g. the well-known conflict between the economic use and the common good character of water). The judgement with respect to one principle might be positive but negative to another one. On the other, the principles are rather abstract and must be "contextualised", i.e. made operable for specific cases. This is an issue of hermeneutics, interpretation, deliberation, balancing and weighing – a political activity in its wider sense. Theories of democracies as well as ethics postulate that these activities should be performed in a participatory manner, involving people concerned and affected, as well as taking into account future generations and ecosystems by applying an advocatory approach.

These principles are, due to ethical theory, universal. They serve as "regulative ideas" and shall orientate ethical deliberation. But they cannot be used for simply deducing the ethically best or optimal solution. The process of identification needs the "real deliberation" with people concerned and affected as well as with stakeholders and decision-makers (Schwemmer 1987). Assessing optional water interventions with regard to ethical criteria, and this is the "procedural" message of discourse ethics, should be arranged in accordance with principles of fairness and equity (see Habermas 1991; Renn and Webler 1998) including the requirements:

- Participants in communicative exchange are using the same linguistic expressions in the same way.
- Participants agree to modify their own positions in case of better arguments given by others.
- No relevant argument is suppressed or excluded by the participants.
- Everyone entitled to participate, and everyone is equally entitled to introduce new topics or express attitudes needs or desires.
- No force except that of the better argument is exerted.
- No validity claim is exempt in principle from critical evaluation in argumentation (this in particular also holds for religious belief).

Processing the discourse in this way produces its result – and there is no abbreviation available, no algorithm, which could "calculate" the result without the "real deliberation" in a discourse. Habermas' basic idea is that the validity of ethical advice cannot be justified by an isolated individual thinking about the world, or by deriving it from abstract principles in an "ivory tower". Instead, the validity of a normative statement can only be justified in processes of inter-subjective argumentation in the real world. Water ethics shows itself as a specific kind of discourse ethics on water management intervening into water systems.

If the argumentation itself shall provide valid normative statements the focus is shifted to the conditions and presuppositions of the discourse. Criteria must be given to ensure that in a discourse taking place in the real world, the conditions are fulfilled to allow the "power of the argument" only to determine its course. Misuse of the discourse for ideological purposes, persuasive speech instead of argumentation, asymmetric access to important knowledge and other exertions of power in contrast to argumentative power have to be excluded.

In a discourse on water issues arguments shall be exchanged, challenged and defended following strict rules of fairness and equity, which shall ensure the justice of the procedure. All people involved or affected have the same right to participate and to bring in their arguments. This approach applies both to the discourses on water use and to decision-making in water governance.

Beyond considering these principles as regulative ideas to give orientation to ethical reflection, another function of these principles should be mentioned. They can also be used to educate and build capacity in the field of water as they include the main lines of argumentation in discussing any interventions into water systems. In this way the principles also could contribute to raising awareness about water problems, to avoid or overcome biased or limited views on those interventions, to enrich public debates and to shape a more responsible collective consciousness.

2.4   Cultural and Social Preconditions of the Ethical Discourse

The above-mentioned considerations raise the question how the preconditions of ethical discourses could be fulfilled – and what could be done in case they are not fulfilled. A sociological suspicion is that the idea of ethics is purely counterfactual in the sense that it might be an artefact without any empirical relevance: wishful thinking in an idealised world. However, things are more complex. The target of ethics – looking for an argument-based and peaceful way of conflict solving – is built on a strong normative foundation (Habermas 1991). It provides us with well-founded ideas of how conflicts should be solved – and these ideas serve as a kind of benchmark against which we can evaluate our status quo situation. This comparison then can give rise to change in the status quo – in this respect the power of normative thinking can even be revolutionary.

However, the question remains how the fulfilment of the necessary preconditions of ethical discourses could be supported. Counteracting aspects are, obviously:

- Religious fundamentalism – in an ethical discourse every normative position must be subject to criticism and possible change which is incompatible to fundamentalism.
- Prejudice due to race, gender, culture, etc. – ethical discourses need mutual respect among the discourse partners including the conflicting opponents independent from racial, cultural or gender issues.
- Fundamentalist commitments to own and partisan positions without any willingness to modify them in case of better arguments in favour of other positions.

Looking at the empirical situation it quickly becomes clear that the preconditions of ethical discourses will mostly be not fulfilled. However, that is no forceful argument against ethical discourses raises the question how the fulfilment of such conditions could be supported. The challenge is to convince people to participate and to accept the rules of an ethical discourse. At the core of this challenge is to convince people to accept the "discourse risk" which means: you never know at the beginning of the discourse what the result will be – and possibly the result will be contradictory with your initial position. Why should people accept this risk? From an ethics point of view good arguments should be provided instead of psychologically trying to simply persuade people. And there are several arguments, in particular in the field of water:

- People usually understand that they are not alone in the world – solidarity is often a concern.
- In particular, solidarity in the field of water supply can be built on deep-ranging intuitions and on cultural traditions (e.g. in Islam), which recommend or even request not to endanger supplying water to other human beings.
- To be more specific, the very nature of water being a necessary precondition of (human) life is a very strong argument that could hinder actors in egoistically exploiting water resources at the expense of threatening other actors with too little water for living.
- In some cases, there might also be a utilitarian argument for ethical discourses: considering for example, upstream and downstream competition for using water resources it might be a good idea to identify possible win-win situations. Cooperation might open up added value for all the downstream and upstream regions in comparison to a purely competitive situation.

Besides providing good arguments for trying to solve conflict by means of ethical reasoning and discourse, it is essential to establish trust among the actors involved (Habermas 2008). To this end, cultural and value resources (see above) should be exploited to maximum extent. For example, it has been shown that Islamic tradition provides a lot of chances to mobilise and motivate people to think about the responsible use of water resources (Gilli 2004). Analogously, cultural resources and traditions should also be considered as a means of constructing adequate preconditions for ethical discourses.

The basic precondition of an ethical discourse is to respect other human beings as morally autonomous persons (Immanuel Kant). Thus, taking care about the preconditions of an ethical discourse on the usage of water is an issue to create insight by argumentative reasoning rather than trying to persuade people. Obviously, the expectability of creating insight by arguments depends on cultural issues and the status of education. Thus, at the end, it is an issue of education and capacity building that contribute in fulfilling the preconditions of an ethical discourse – at least from a mid and long-term perspective.

2.5 Water Conflicts

If all the principles identified above would be met simultaneously by specific water management strategies or other human interventions into water systems, there wouldn't be any problem in ethical respect. But usually there are counteracting influences, trade-offs and inherent conflicts. Meeting one or two of the principles frequently will endanger meeting the others. Normative uncertainties, in particular conflicts on the required priority-setting between competing claims and their justification, arise and are subject to ethical analysis and discourse. In making the ethical water discourse operable, a lot of concretisations and contextualisation must be applied, according to scales in space and time, but also with respect to different types of normative uncertainties to be tackled. A typology of water conflicts, of argumentation patterns used in specific constellations and of typical normative uncertainties involved would be desired. In this section, I would like to discuss briefly some typical conflicts arising in water systems interventions and ask for the possible contribution of ethics to deal constructively with them. Before doing this, the scope of possible conflict types shall be sketched.

Discourses around water have some specific properties related to the medium under consideration. Water and water systems are different from soils and other compartments of the environment. Water is almost always flowing: it passes landscapes and private property, soils and regions (e.g. see above for the Islamic view on this). It cannot be consumed chemically – in using water it remains the same compound of hydrogen and oxygen. However, human use of water adds other materials and chemicals and so usually leads to degradation of its quality, or it changes landscapes and impacts human settlements, ecosystems and the biodiversity.

Water conflicts have different origins and can be characterised by different parameters. An ethical characterisation would consider possible or intended interventions into water systems with a focus on questions such as:

- What is at stake? Are the aims of the intervention justified? Which non-intended effects could arise?
- Who or what would be the winners and losers (not only with respect to humans but also regarding elements of the biosphere and the natural environment in general)?
- Could rights (e.g. the Human Right on Water but also cultural or religious rights) be violated as unintentional side effects?
- What can be said about the distribution of gains and losses, of costs and benefits, of chances and risks among people alive today or between contemporary and future generations?
- Which uncertainties are involved, which characteristics do they show, and should the precautionary principle (Harremoës et al. 2002) or another type of precautious approach be applied?
- Which timescales are involved? There are huge differences in the timescales of the impact of human interventions in water systems and the recovery of the water systems. This is important in particular in taking over long-term responsibility
- Are there alternatives available to the intervention proposed meeting the same targets at lower "moral costs"?

Basically, the "moral constellation" of the water conflict under consideration must be clarified and then can be used for categorising different types of conflict. This constellation in particular will determine who and which positions should be represented in the ethical discourse. It also must be clarified whether advocates of non-human interests or of future generations should be present.

This "moral constellation" strongly depends on the range of the conflict under consideration and its possible implications in space and time. Extension in space and time influences heavily the range of moral positions and of actors to be involved. A first and rough differentiation could be:

- Conflicts on the use of water in a strongly limited region without impacts beyond the borders of the region and beyond present time as well. In this case the actors from the corresponding region directly affected are entitled to solve the conflict according to ethical rules.
- Conflicts in a strongly limited area between the economic use of water and long-term ecological requests of sustaining ecosystems and biodiversity – in this case advocates of those desires must be included.
- Conflicts on the use of water today with assumed interests of future generations, e.g. in case of degrading water resources by usage or of degrading landscapes. In this case advocates of future generations must be included in the ethical discourse.
- Conflicts on large-scale interventions into water systems (geo-engineering Potthast 2013), which per se will have a global, an inter-cultural and an inter-temporal dimension.

This approach can only demarcate the task in front of water ethics. In the literature available (see above), the task of classifying water conflicts in ethical respect has not even been touched. However, it is necessary to do so because it is exactly this type of differentiation that allows water ethics to become specific and to genuinely contribute to dealing rationally and responsibly with conflicts. In particular, this type of differentiation allows establishing relations between reflective ethics and the governance in the respective field. The ethical discourse as a "real deliberation" (see above) forms the link between theoretical ethical analysis and reflection on the one side, and empirical deliberation and decision-making on the other. This is also the place to locate debates on the assignment and distribution of responsibilities.

Classifying water conflicts is, in this sense, necessary to enable actors to determine adequate strategies in responding to challenges and involving stakeholders. However, the classification and subsequent assignments of an individual conflict to a classification scheme will be an act in itself, which is not value-neutral and might itself be an issue of controversy. Thus those determinations are part of the ethical discourse. This also might include reaching a consensus that the assignment usually will not fully meet all of the situational aspects but rather some of them, which have been identified crucial for problem-solving.

2.6   Water Ethics and Specific Water Conflicts

Water ethics is an emerging field of ethical reasoning. Its motivation is fuelled by high concerns about the stability of water cycles, about quality issues of water, about increasing water scarcity in many world regions, about an increasing pressure on water systems by agriculture and industry but also about growing population – shortly speaking: concerns about the sustainability of water systems and about possible damage or breakdown of ecosystems as consequences of water problems (Lehn and Parodi 2009).

Increasing pressure on water systems leads to competing claims for water usage and to corresponding conflicts. Insofar as these water conflicts refer to moral conflicts or normative uncertainties, ethical reasoning might help in better understanding and contribute to solving the conflict. Human values are present in these conflicts nonetheless, in the form of cultural or religious values based on traditions, or in the form of economic values in the Western approach of regarding water as a resource.

Human values are also present in water technologies or other intervention measures into water systems because technology is not value-free but inherently normative as can be learned by philosophy of technology (Grunwald 2013). The complex interplay between our normative views on water, water usage and water conflicts on the one hand, and normative aspects of water technology, on the other, need to be uncovered transparently in order to allow for an open and enlightened debate on our responsibility in this field.

Water conflicts arise in a different respect: as conflicts between today's generation and future generations, between users of water facing scarcity, in particular between upstream and downstream populations, between different types of usage (irrigation, industrial use, household use, etc.) and so forth. Ethics cannot identify "best solutions" in these conflicts and then force the conflict partners to accept. Ethics only can help to better understand the normative structure of water conflicts and support deliberation about responsible solutions. Ethics may contribute to a bottom-up process of conflict solving by conducting an ethical discourse with all groups involved – and which advocates for those who cannot participate directly such as future generations or ecosystems.

Ethically legitimate principles cover the most relevant concerns on water issues and transform them into guidance for such interventions into water systems, which are compatible to the imperative of sustainable development or even support this imperative. However, there is no direct way from these (or other) principles to specific recommendations in the case of a challenge at hand. Often, the principles must be weighed, balanced and prioritised – a complex challenge, which can be met only by carefully considering the individual case under consideration. Responsible solutions can be identified only in a genuine process of ethical deliberation involving persons, groups, representatives of moral positions and advocates of non-human (such as eco-systems) of future generations. The ethical discourse as a process of real deliberation and following legitimate rules is the place where ethical theory, water systems analysis, environmental sciences and engineering inter-disciplinarily meet with context-specific requirements and trans-disciplinarily with specific actors of the field.

The research project out of which this volume emerged focused on two specific water systems and the involved water conflicts: the Fergana Valley in Uzbekistan and the Lower Jordan Valley with Israel, Jordan and Palestine as involved regions. The results of the inter-disciplinary research and analysis are presented in the following chapters. They may be considered as first and knowledge-providing steps towards a "real deliberation" with people concerned and affected as well as with stakeholders and decision-makers.

References

Al-Awar F, Abdulrazzak M, Al-Weshah R (2010) Waters ethics perspectives in the Arab Region. In: Brown PG, Schmidt J (eds) Water ethics. Foundational readings for students and professionals. Island Press, Washington, DC, pp 29–38

Armstrong A (2006) Ethical issues in water use and sustainability. Area 38:9–15. doi:10.1111/j.1475-4762.2006.00657.x

Armstrong AC (2009) Viewpoint – further ideas towards a water ethic. Water Altern 2:138–147

Armstrong AC, Armstrong MB (2005) A Christian perspective on water and water rights. In: Tvedt T, Oestigaard T (eds) A history of water: the world of water, 3rd edn. Tauris, London/New York, pp 367–384

Bartholomew I Ecumenical Patriarch of Constantinople (2010) Byzantine heritage. In: Brown P, Schmidt J (eds) Water ethics. Foundational readings for students and professionals. Island Press, Washington, DC, pp 25–28

Faruqui N, Biswas A, Bino M (2001) Water management in Islam. United Nations University Press, Tokyo

Gethmann CF, Sander T (1999) Rechtfertigungsdiskurse. In: Grunwald A, Saupe S (eds) Ethik in der Technikgestaltung. Praktische Relevanz und Legitimation. Springer, Berlin, pp 117–151

Gilli F (2004) Islam, water conservation and public awareness campaigns. Paper presented at the 2nd Israeli-Palestinian-international academic conference on water for life, Antalya, 10–14 October 2004

Groenfeldt D (2013) Towards a new water ethic. In: Bhaduri A, Flinkerbusch E, Holtermann T et al (eds) The Bonn declaration on global water security. Global Water System Project, Bonn, pp 14–15

Grunwald A (2007) Working towards sustainable development in the face of uncertainty and incomplete knowledge. J Environ Policy Plan 9:245–262. doi:10.1080/15239080701622774

Grunwald A (2013) Handbuch Technikethik. Metzler, Stuttgart

Grunwald A, Kopfmüller J (2012) Nachhaltigkeit, 2nd edn. Campus, Frankfurt am Main

Habermas J (1991) Erläuterungen zur Diskursethik. Suhrkamp, Frankfurt am Main

Habermas J (2008) Hat die Demokratie noch eine epistemische Dimension? Empirische Forschung und normative Theorie. In: Habermas J (ed) Ach, Europa. Suhrkamp, Frankfurt am Main, pp 138–191

Harremoës P (2002) Water ethics – a substitute for over-regulation of a scarce resource. Water Sci Technol 45:113–124

Harremoës P, Gee D, MacGarvin M (2002) The precautionary principle in the 20th century. Late lessons from early warnings. Sage, London

Jonas H (1979) Das Prinzip Verantwortung: Versuch einer Ethik für die technologisierte Zivilisation. Suhrkamp, Frankfurt am Main

Katz E (1997) Nature as subject: human obligation and natural community. Rowman & Littlefield, Lanham

Lehn H, Parodi O (2009) Wasser – elementare und strategische Ressource des 21. Jahrhunderts. Umweltwissenschaften und Schadstoffforschung 21:272–281. doi:10.1007/s12302-009-0052-6

Leopold A (1949) A sand county almanac, and sketches here and there. Oxford University Press, Oxford

McGee WJ (1909) Water as a resource. Ann Am Acad Polit Sci 33:37–50

McGee WJ (1911) Principles of water-power development. Science 34:813–825. doi:10.1126/science.34.885.813

Nida-Rümelin J (2005) Angewandte Ethik. Die Bereichsethik und ihre theoretische Fundierung. Metzler, Stuttgart

Ostrom E, Stern PC, Dietz T (2003) Water rights and the commons. In: Brown P, Schmidt J (eds) Water ethics. Foundational readings for students and professionals. Island Press, Washington, DC, pp 147–154

Ott K (2010) Umweltethik zur Einführung. Junius, Hamburg

Parodi O (2008) Technik am Fluss. Philosophische und kulturwissenschaftliche Betrachtungen. Oekom verlag, München

Postel S (2010) The missing piece: a water ethic. In: Brown P, Schmidt J (eds) Water ethics. Foundational readings for students and professionals. Island Press, Washington, DC, pp 221–226

Potthast T (2013) Geo- und Hydrotechnik sowie Bergbau. In: Grunwald A (ed) Handbuch Technikethik. Metzler, Stuttgart, pp 284–289

Pradhan R, Meinzen-Dick RS (2010) Which rights are right? Water rights, culture, and underlying values. In: Brown P, Schmidt J (eds) Water ethics. Foundational readings for students and professionals. Island Press, Washington, DC, pp 39–58

Priscoli Delli J, Dooge J, LLamas R (2004) Water and ethics: overview. http://www.international-waterlaw.org/bibliography/articles/Ethics/Overview.pdf. Accessed 28 Jan 2015

Renn O, Webler T (1998) Der kooperative Diskurs – Theoretische Grundlagen, Anforderungen, Möglichkeiten. In: Wilhelm U (ed) Abfallpolitik im kooperativen Diskurs. Bürgerbeteiligung bei der Standortsuche für eine Deponie im Kanton Aargau. ETH Zürich, Zürich, pp 3–103

Schmidt JJ, Shrubsole D (2013) Modern water ethics: implications for shared governance. Environ Values 22:329–379. doi:10.3197/096327113X13648087563746

Schwemmer O (1987) Handlung und Struktur. Suhrkamp, Frankfurt am Main

Trawick P (2010) Encounters with the moral economy of water: general principles for successfully managing the commons. In: Brown P, Schmidt J (eds) Water ethics. Foundational readings for students and professionals. Island Press, Washington, DC, pp 155–165

UNESCO (2011) Water ethics and water resource management. http://unesdoc.unesco.org/images/0019/001922/192256e.pdf. Accessed 28 Jan 2015

WCED (1987) Report of the World Commission on Environment and Development: our common future. http://www.un-documents.net/our-common-future.pdf. Accessed 28 Jan 2015

Young GJ, Dooge JC, Rodda JC (1994) Global water resources issues. Cambridge University Press, Cambridge

Part II
Major Water Engineering Projects – Challenges, Problems, Opportunities

Chapter 3
Major Water Engineering Projects: Definitions, Framework Conditions, Systemic Effects

Sebastian Hoechstetter, Christine Bismuth, and Hans-Georg Frede

Abstract This chapter aims at providing an overview of major water engineering projects as they are perceived within the scope of this volume. Furthermore, general principles of water technologies and their role in water conflicts and for solving water-related problems are outlined. An evaluation framework of technological measures for water management is meant to serve as a guiding concept for the in-depth analysis of the two study regions dealt with in this volume, i.e. the Fergana Valley and the Lower Jordan Valley.

Keywords Water technology • Definition of MWEPs • Trends in water engineering • Path dependency • Evaluation framework • Water storage • Water distribution • Water use • Integrated water resource management • Water engineering

S. Hoechstetter (✉)
Helmholtz Centre Potsdam - GFZ German Research Centre for Geosciences,
Telegrafenberg, 14473 Potsdam, Germany
e-mail: sebastian.hoechstetter@gmx.de

C. Bismuth
Interdisciplinary Research Group Society - Water - Technology,
Berlin-Brandenburg Academy of Sciences and Humanities,
Jägerstraße 22/23, 10117 Berlin, Germany

Helmholtz Centre Potsdam - GFZ German Research Centre for Geosciences,
Telegrafenberg, 14473 Potsdam, Germany

H.-G. Frede
Institute of Landscape Ecology and Resource Management (ILR),
Justus-Liebig-Universität Gießen, Heinrich-Buff-Ring 26-32, 35392 Gießen, Germany

R.F. Hüttl et al. (eds.), *Society - Water - Technology*, Water Resources
Development and Management, DOI 10.1007/978-3-319-18971-0_3

3.1 Definition of Major Water Engineering Projects – A Proposal

Major water engineering projects (MWEPs) have been the subject of a large number of studies and analyses in a variety of scientific disciplines. However, a precise, standardised and generally accepted definition is missing so far.

Thinking of MWEPs, one of the first associations coming to the mind of many people might be large dams and reservoirs. A rather simple definition of large dams has been provided by ICOLD (International Commission on Large Dams). ICOLD has defined them as dams with a height over 15 m. Other institutions (such as the World Commission on Dams) additionally designate dams as "large" when they have a height between 5 and 15 m and a reservoir capacity of more than 3×10^6 m^3. Similar specifications, however, are not available for other MWEPs.

Large-scale technological approaches, however, are not limited to dams and reservoirs, but also comprise irrigation management, industrial usage and many other purposes. Therefore, an attempt to a general definition of MWEPs is made in this chapter. We argue that the planning, implementation and evaluation of such projects should not be reduced to merely technical aspects. Instead, a more generic approach is pursued here, and the implications of MWEPs for economy, society and the environment are considered as well.

"Water engineering" has been defined by Lotti (1980) as the subject that includes (among others) "everything connected with the use of water resources, namely: the study of the element 'water' (hydrology and hydraulics); the definition of its uses and the structures needed for these; flood control; the maintenance of water quality standards; socioeconomic aspects; and the relationship between water resources and the environment."

Accordingly, within the scope of this volume, major water engineering projects are regarded as technical operations and construction schemes related to these fields with a spatial, temporal and economic extent that is likely to result in relevant and far-reaching effects on society and the environment. Their specific characteristics affect many sectors of public life. The most prominent feature of MWEPs, however, is their significant effect on and their (usually permanent) intervention into natural hydrological regimes.

The most important purposes of MWEPs relate to the following types usage (according to Lotti 1980):

- Water supply
- Irrigation
- Industrial use
- Energy production, conversion and storage
- Inland navigation
- Fishing
- Recreation

As a result, major water engineering projects feature both potentially beneficial and adverse outcomes and can be described by the following common characteristics:

- MWEPs are of high national or even transnational relevance.
- They exhibit a high degree of complexity, resulting in a high demand in financial and human resources. As a consequence, they may involve major financial obligations or even risks.
- Expert knowledge and highly skilled personnel are needed for the establishment and the operation of MWEPs.
- Various stakeholders and social interest groups are affected by MWEPs. Thus, such projects imply the necessity of a reconciliation of different interests of individuals, groups and the society as a whole, national and international.
- The consequences of these projects are in many cases irreversible, i.e. the initial state of the system (ecological, constructional, social) cannot be restored.
- MWEPs tie up a large amount of resources and, therefore, limit the choice of options for action both for the presence and for the future.

3.2 Water Technologies – Uses and Functions

Within the scope of this volume, three main fields of application of water technologies are of particular relevance and will be discussed in the following, comprising:

- Water storage
- Water distribution
- Water use

The main emphasis of the considerations on the effects of water technologies in this context is placed upon the potentially critical and adverse aspects, as these issues tend to be underestimated in many cases and need to be included in comprehensive assessments of MWEPs.

3.2.1 Water Storage

One of the major functions of dams, i.e. prominent examples of MWEPs, is to have water in store for purposes at a later point in time. Reservoirs provide water for consumption and irrigation and for the production of energy. Furthermore, dams serve as a means of flood protection. While in the history of European industrialisation, the construction of reservoirs in mountainous regions at the source of larger river basins was an accepted and wide-spread procedure with a potential to stimulate the economic development, such projects are being seen more critical from today's viewpoint – particularly concerning the downstream impacts on ecology and hydrology. Nevertheless, large reservoirs are capable of contributing to the necessary and continuous provision with water and energy. Reservoirs and dams were a means of securing the increasing demands of these resources in many industrialised areas of the world. In this regard, they may open up opportunities in terms of the economic and infrastructural development of urban and rural regions.

On the other hand, reservoirs comprise a number of economic and ecological drawbacks. For example, by their nature they tend to be situated in remote areas far away from the economic centres, resulting in considerable transport expenditures for infrastructure and pumping. They are likely to exhibit negative effects on the global climate: Methane emissions – a driver of climate change – from large reservoirs can be significant (Lima et al. 2007; Barros et al. 2011). These emissions are due to the rotting of organic matter (e.g. from the vegetation and the soils flooded during the filling phase of the reservoirs) in anaerobic conditions. Furthermore, on the global scale, the water losses from reservoirs via evaporation are even larger than the amounts extracted for human consumption (UNEP 2008) (Fig. 3.1).

Reservoirs are known to have substantial impacts on aquatic biodiversity, mainly by constituting a movement barrier to various organisms, but also because flow regimes and habitat characteristics are changed in a distinct way (Vörösmarty et al. 2010). While upstream the increasing sedimentation caused by dams gradually diminishes the storage capacity of the reservoirs, further downstream the lack of transported sediments leads to the deterioration of the river bed. Consequently, groundwater levels are decreasing and lower sedimentation rates may result in a retreat of the coastal lines at the estuaries.

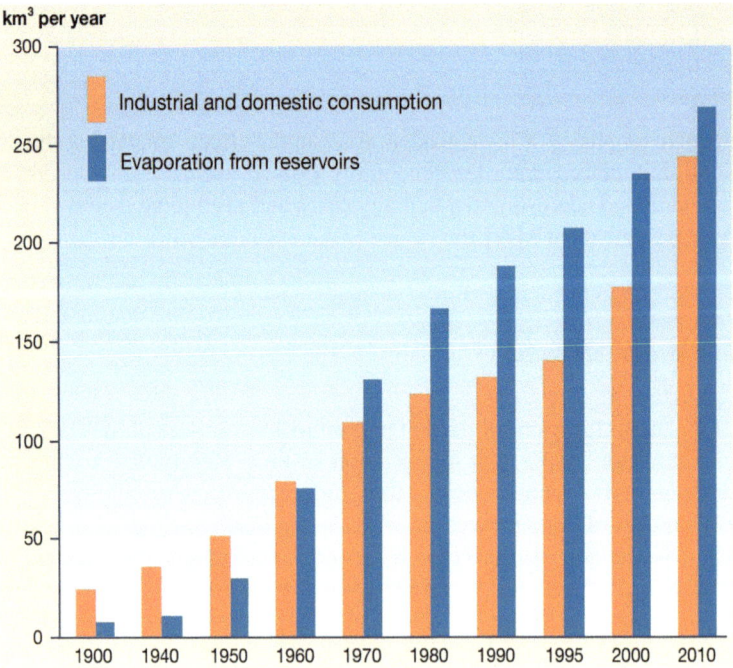

Fig. 3.1 Evaporation from reservoirs compared to global consumption rates of selected sectors. Projections indicate that both global water use and evaporation will continue to increase (Source: Igor A. Shiklomanow, State Hydrological Institute (SHI, St. Petersburg) and United Nations Educational, Scientific and Cultural Organisation (UNESCO, Paris), 1999. Design: Philippe Rekacewicz, http://www.grida.no/graphicslib/)

Apart from that, sediments serve as natural fertiliser for occasionally flooded arable lands along the rivers. This missing nutrient input has to be compensated by artificial fertilisers. As important humic matter is missing as well, deterioration of the soil quality and its water storage capacity is frequently the consequence. The changes in the groundwater levels around the reservoirs are likely to act as a trigger of soil salinisation.

As can be seen from the description of these potentially adverse effects, large-scale water storage facilities also impact on the secondary ways of usage. Along rivers, technical measures and the water withdrawal of upstream resources have implications on the users situated downstream.

Another major criticism against the establishment of dams refers to the fact that during the construction of large reservoirs, considerable amounts of fertile lands are flooded and inhabitants are displaced. Particularly the rights of indigenous populations have been violated in this way, with the compensation measures frequently remaining unsatisfactory (Robinson 2003).

In addition, the proliferation of water-borne diseases like malaria and bilharzia is a problem linked to the construction of large dams and reservoirs, which may pose substantial risks to human health (World Commission on Dams 2000).

3.2.2 Water Distribution

The stage that usually follows the storage and collection of water is its distribution to the end users via channels and water pipes. Such transport and distribution systems contribute to improving negative water balances. This is the case not only in dry regions but also in areas exhibiting a humid climate, e.g. when the required water quality cannot be achieved for the amounts needed in densely populated areas. An example is Lake Constance, which supplies a large quantity of the drinking water demand for the federal state of Baden-Wuerttemberg in South Germany. The geographical disconnection of water abstraction and water use over long distances may lead to a detachment of the responsibility for the protection of water resources from the users of the source. The compensation of those regions which provide the service is in many cases not sufficient and not based on objective figures and can thus lead to tensions between the provisioning (rural) areas and the consuming (urban) centres. For dams and reservoirs, this circumstance has been examined by the World Commission on Dams, stating that those who receive the (economic and social) benefits of water distribution infrastructure (such as urban dwellers, farmers and industries) are typically not the same groups that bear the social costs (World Commission on Dams 2000).

Transport of water always implies a certain loss due to leakages and evaporation, especially in open channels. This problem can be studied in a very pronounced way when assessing the water supply of megacities in developing and emerging countries, where about one third of the delivered water volume can be lost via leaky water pipes (Niemczynowicz 1996). Unless natural slopes can be used, the transport over large distances may require considerable amounts of energy – a crucial factor for the determination of the provision costs. This is particularly relevant for the effectiveness of large irrigation systems (this aspect will be dealt with in detail in Part III of this volume).

Another problematic aspect of large water distribution networks is that the cross-connection of formerly separate water systems can lead to undesired ecological effects and the introduction of non-endemic species into an ecosystem. Man-made waterways (resulting in the interconnection of watersheds) have been mentioned to be among the most important dispersal vectors of invasive aquatic species (Leuven et al 2009).

3.2.3 Water Use

In general, societies and individuals use water resources in four major ways. We distinguish between water for human consumption, water for food production purposes, water for industrial uses and water for the generation of (electrical) energy.

On a global scale, 70 % of the freshwater resources are used for agricultural purposes, and only 19 % for industrial and 11 % for domestic purposes (including drinking, washing, food preparation and sanitation) (FAO 2010).

Consequently, water needs to be treated according to the respective intended type of usage. While the water for human consumption purposes requires elaborate treatment technologies, irrigation water in most cases remains untreated and is applied directly to the arable land. Water for industrial processes is treated according to the requirements of the intended usage.

A large proportion of the freshwater resources used by humans are fed into agricultural irrigation systems. These systems can be classified according to their efficiency. A simple definition of the term irrigation efficiency refers to the ratio of the irrigation water consumed by the crops of an irrigation farm or project to the water diverted from a river or other natural water source into the irrigation canal (Jensen 2007). Efficiency varies considerably among different technologies. For example, subsurface micro- or trickle-irrigation in combination with mulch is among the most efficient schemes at present. Nevertheless, the actual efficiency levels highly depend upon the crop water consumption patterns, soil types, cultivation methods and the automation grade (the most technically sophisticated systems use automatised soil moisture measurements in combination with computer-controlled water and fertiliser applications and remote sensing systems).

Large scale irrigation schemes are in many cases installed for the cultivation of so-called "cash crops" (crops for export), while crops for the local needs are in many cases cultivated under rain-fed conditions (an example for this circumstance, the Fergana Valley in Central Asia, is dealt with in detail in Part III of this volume). Cash crops contribute to the "virtual water" balance of nations and regions by binding water in exported crops and thus to increasing the (negative) water balance of the exporting country. "Virtual water trade" is a term used to include the latent/hidden flows of water into water balances, e.g. if food or other products are transferred from one place to another (e.g. Hoekstra and Mekonnen 2012). For example, the consumption of water for livestock farming and the export of meat can be substantial.

Furthermore, the use of water for environmental needs and environmental services is gaining rising importance (Petts 2009). While formerly the environmental needs of natural ecosystems have not been acknowledged as a factor to be taken into account, today e.g. the protection of wetlands and the ecological restoration of river ecosystems are increasingly gaining attention, particularly in well-developed countries. In order to establish this perspective as an eco-hydrological concept, the term "environmental flows" (or E-flows) has been introduced. An environmental flow is defined as the water regime provided within a river, wetland or coastal zone to maintain ecosystems and their benefits where there are competing water uses and where flows are regulated. It has been argued that water resources need to be managed to provide environmental flows, either by means of modifications of infrastructure or changes in water allocation policies and entitlements (Dyson et al 2003).

3.3   Principles, Trends and Framework Conditions of Major Water Engineering Projects

Water technologies have been examined from various perspectives and from different disciplinary angles already. This sub-chapter tries to summarise and condense the findings and experiences made within the scope of major water engineering projects in different parts of the world. Based upon these general observations, we try to formulate general "principles" of such projects. Overarching trends in technology-related water management are defined in order to illustrate the relevance of the topic and the need for interdisciplinary research and assessment approaches. It is important to note that the focus in this regard is on the aspects that tend to be underestimated in many current evaluations and impact assessments of MWEPs. Their undisputedly numerous positive and beneficial effects for economic and social development are usually not as underrepresented as are the uncertainties and risks associated with their implementation.

The statements presented here can be perceived as the conceptual backbone of the practical parts of this volume and as presuppositions and general observations of the previous work of the authors. Furthermore, these principles and trends serve as the basis for an evaluation framework, considering the main compartments of the relevant social, economic and ecological sub-systems that are affected by technological measures of water management. Aspects of this framework will be covered throughout the following chapters of this volume.

- *Long-term path dependencies caused by MWEPs may affect the future options for sustainable development of regions and nations.*

The provision of water for human needs requires (large-scale) infrastructures, particularly in countries exhibiting a dry and arid climate. In many cases, the type of infrastructure chosen for a specific region determines the availability of further technological options and the remaining radius of action – now and in the future. This

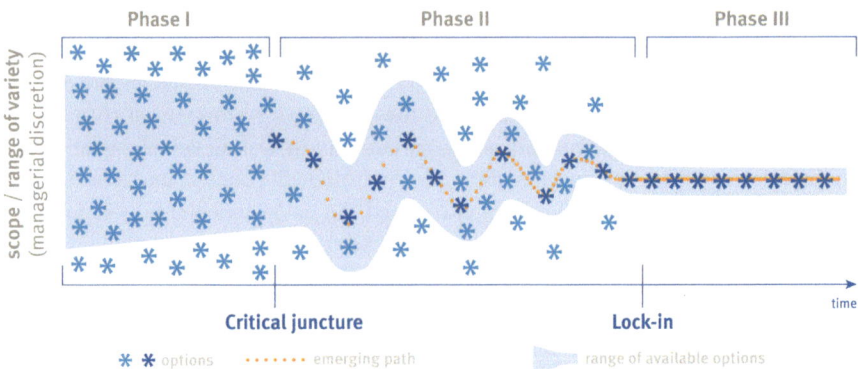

Fig. 3.2 Schematic representation of the emergence of development paths: decisions for particular options limit the future scope of action. The illustration refers to the fact that decisions taken at "critical junctures" reduce the amount of available options for actions of future decisions. After a so-called "lock-in", the present situation can be regarded as irreversible (Source: Altered according to Sydow and Schreyögg (2010, p. 8), previously published in German by Hoechstetter et al. (2013))

circumstance is referred to as "path dependency" (see Fig. 3.2; for details about this concept, refer to Chapter 8 (Moss and Dobner 2015, in this volume, pp. 101–111). For instance, existing waste water collection infrastructures may limit the options for establishing more decentralised waste water treatment installations and minimise the chances to use faecal-free, low-polluted waste water for domestic or industrial purposes after appropriate processing and treatment. This is mainly due to the high investment costs of infrastructure projects and their long period of amortisation. This circumstance is studied by the example of the Red Sea – Dead Sea Conveyance Project in the Middle East (see Part IV of this volume).

- *MWEPs have economic consequences for investors and the society; long-term financial commitments connected to MWEPs may constitute a potential burden.*

Most water management technologies in MWEPs require specific organisational structures for planning, operation, control and supervision. The costs for setting up and maintaining such structures are underestimated in many cases or simply disregarded as "external costs". That is, costs that have to be carried by third parties and that are not charged via markets. The costs for an effective operation of MWEPs can be considerable. They may have a direct effect on the provision costs of water to the end users. Moreover, they bind financial resources for a long time and thus limit the options for action in the future – thus constituting another source of path dependencies. Studies have suggested that under certain circumstances, waiving of large-scale technical measures (e.g. channels, barriers, dams) in favour of smaller-scale decentralised solutions (e.g. rehabilitation measures) can lead to positive net economic benefits when all external costs are appropriately assessed (Becker et al 2014). According estimations can only be based on an analysis of every individual case.

- *MWEPs potentially have effects on both water quality and water quantity.*

Criteria of water quality and quantity may be directly linked to each other, especially when we look at the usage of surface water: While, on the one hand, quantitative aspects can be solved by an improved management of quality, intelligent management of water amounts can, on the other hand, have positive impacts on water quality. For instance, the use of fertilisers in (irrigation) agriculture or the establishment of closed water cycles in industrial processes may have direct impacts on the water quality. The higher the water quality at the outlet of a process, the larger the amounts of water that are available for the end users. However, the relationship between water quality and quantity is a very complex one, its nature depending strongly on the individual catchment, the type of the aquatic ecosystem under consideration and the socio-economic framework conditions (Australian Government 2013).

- *MWEPs are embedded into local and regional framework conditions.*

There is a wide range of technical solutions for different water-related problems, but they are not always adapted to the specific regional needs or capacities. Biochemical treatments of waste water, for instance, have to be adjusted to the specific climatic conditions in order to assure satisfying treatment results. The complexity of technical solutions has an impact on their potential field of application. Local knowledge and capacities as well as issues of long-term maintenance have to be taken into consideration when technological decisions are made. Integrated views and approaches can open up synergies in this regard. For instance, water provision and disposal of waste water may be regarded as interlinked sectors, based on the actual local framework conditions. Attempts to reuse domestic wastewater for diverse purposes, such as for agricultural irrigation, have proven to be successful in many parts of the world already (Oron et al 1999).

- *MWEPs are frequently embedded in an international/interstate context; complex legal issues pose a challenge to international water law.*

Flowing water does not stop at political borders. Therefore, international aspects are a crucial variable for the implementation of MWEPs. Water law is evolving more and more into a part of environmental law under the leading principle of sustainability (Farrajota 2009). This is true for both national and international law and reflects that rational water law has to be compatible on all levels of regulation. Integrated water resource management (IWRM) incorporates basic elements of modern (European) water law, such as the focus on river basins as management objectives or public participation during the planning and establishment of water management framework plans. In general, there are three largely undisputed rules of customary international water law concerning non-navigational uses of international water resources. These are the rule of equitable and reasonable utilisation, the no-harm rule and the duty to cooperate. Apart from that, the content of framework conventions and regional water treaties forms a heterogeneous legal foundation, on the basis of which interstate conflicts need to be dealt with.

- *Technological solutions for water-related problems may both impede or foster more far-reaching systemic changes of the water sector.*

In the short and medium term, technical solutions can provide the desired results for a certain water-related problem. In the long run, however, the focus on a merely technical approach for solving water-related problems can hamper more extensive and sustainable systemic changes, such as modifications in water consumer patterns, changes in agricultural cultivation practices towards a greater climatic adaptability or the introduction of economic incentive instruments. This may be due to the above-mentioned path dependencies. For instance, Libecap (2011) has illustrated how past arrangements to meet conditions of the time have constrained contemporary economic opportunities, using measures implemented in the American West in the late nineteenth and early twentieth centuries and their effect on today's water markets as an example.

- *MWEPs can result in external effects that need to be considered.*

Water technologies imply external effects, i.e. concern other individuals, groups or societies are affected without being compensated. Examples include a positive or negative carbon footprint or potentially negative and positive effects on the environment. These effects and their financial, social, economic and ethical consequences need to be included in an overall assessment of MWEPs. Most MWEPs tend to be supply-oriented and neglect external effects. However, state intervention regulation or demand-oriented measures like the introduction of water saving devices or pricing policies may turn out to be more sensible when they are compared to the establishment of MWEPs (such as the installation of sea water desalination plants in the case of Israel, Becker 2010).

- *MWEPs affect public and private stakeholders on all societal levels.*

For a successful implementation of technical solutions, a broad participation and an involvement of the end users in the planning process and beneficiaries throughout all phases of the planning and operation process is likely to have a positive effect. Participation can help to build trust and enhances the cooperation between the users and planners. With concern to an overall project perspective, the early participation of users can help to reduce the costs for implementation and operation. For example, Newig et al. (2005) have shown that the participation of interested parties and the broader public can be an important instrument to both deal with and, where possible, reduce uncertainty associated with the implementation of the European Water Framework Directive (WFD), which involves technological and systemic measures on different societal levels.

3.4   An Evaluation Framework of MWEPs

Taking into account the definition and the general principles mentioned above, the complex and far-reaching impact (both desired and undesired) of major water engineering projects becomes obvious. Therefore, transdisciplinary perspectives and integrated assessments are needed in order to comprehensively capture the impact of such projects on different societal and ecological sectors and levels. An according "evaluation framework" that covers the main systems and processes involved and that assesses the outcomes of implemented projects is presented here and will be reflected by the course of action in the case studies (Parts III and IV of this volume).

The concept has been inspired by the analytical framework designed for social-ecological systems by Ostrom and Cox (2010). Based upon this fundamental work, primary classes of entities have been defined, which are embedded in a social, economic and political setting. The primary elements of this framework (see Table 3.1) are the resource systems and the physical framework conditions, the governance systems, social systems/stakeholder concerns and issues of the operation, management and maintenance of MWEPs. In addition, the "outcomes" or the measurable impact of MWEPs can be assessed along the categories of social, ecological and economic performance. In general, an assessment of all the sectors and compartments mentioned appears to be necessary to secure a successful

Table 3.1 Evaluation framework major water engineering projects

Systems and processes			
Resource systems and physical framework conditions	Governance systems	Social systems and stakeholder concerns	Technology systems
Geographical attributes and boundaries	Rules and norms	Group attributes	Technological attributes
Existing and complementary infrastructure	Property right regimes	Leadership	Achievement of objectives and goals
Economic value of relevant ecosystem services	Governance mode	Rules of participation	Monitoring (environmental, social, economic)
Productivity and effectiveness	Monitoring and sanctioning processes	Capacity and social capital	Management and operation
Resource properties	Participatory processes	Resource dependence	Human resources
Resource challenges (e.g. climate and demographic change)	Network structure (centrality, connectivity, number of levels)	Acceptance of and familiarity with technology used	
Outcomes			
Social performance *(e.g. equity, accountability, sustainability)*	Ecological performance *(e.g. exploitation, ecosystem resilience and resistance, biodiversity, sustainability)*	Economic performance *(e.g. efficiency, cost-recovery, cost-benefit ratio, external costs)*	

Source: In its outline based upon Ostrom and Cox (2010) and Ostrom (2009)

implementation and operation of water engineering projects and may serve as a general evaluation guideline.

Considering and investigating every aspect of this analytical framework will in many cases go beyond the scope of standard evaluation and monitoring approaches, depending on the size, scope and context-dependency of the project of concern. Therefore, according analyses will have to concentrate on the main relevant subsystems and outcomes applying in specific situations. Nevertheless, we are convinced that general recommendations can be derived if this interpretative and comprehensive framework is applied. Important issues with regard to the volume at hand are the institutional frame settings and the challenges that MWEPs pose to the management of natural resources. Particular emphasis will be on the evaluation and weighing up of the benefits and costs and their distribution among the different stakeholders – e.g. when comparing the implementation of the Red Sea – Dead Sea Conveyance Project in the Middle East and its potential alternatives (see Part IV of this volume).

References

Australian Government (2013) Characterising the relationship between water quality and water quantity. Department of Sustainability Environment Water Population and Communities, Canberra

Barros N, Cole JJ, Tranvik LJ et al (2011) Carbon emission from hydroelectric reservoirs linked to reservoir age and latitude. Nat Geosci 4:593–596. doi:10.1038/ngeo1211

Becker N (2010) Desalination and alternative water-shortage mitigation options in Israel: a comparative cost analysis. J Water Res Prot 02:1042–1056. doi:10.4236/jwarp.2010.212124

Becker N, Helgeson J, Katz DL (2014) Once there was a river: a benefit-cost analysis of rehabilitation of the Jordan River. Reg Environ Chang 14:1303–1314. doi:10.1007/s10113-013-0578-4

Dyson M, Bergkamp G, Scanlon J (eds) (2003) Flow – the essentials of environmental flows, 2nd edn. IUCN, Gland

FAO (2010) Aquastat – the FAO's global water information system. Food and Agriculture Organization of the United Nations (FAO). http://www.fao.org/nr/water/aquastat/main/index. stm. Accessed 10 Feb 2015

Farrajota MM (2009) International cooperation on water resources. In: Dellapenna JW, Gupta J (eds) The evolution of the law and politics of water. Springer, Dordrecht, pp 337–352

Hoechstetter S, Bens O, Bismuth C (2013) Konflikte um die Georessource Wasser in Zentralasien. System Erde 3:50–55. doi:10.2312/GFZ.syserde.03.02.8

Hoekstra AY, Mekonnen MM (2012) The water footprint of humanity. Proc Natl Acad Sci 109:3232–3237. doi:10.1073/pnas.1109936109

Jensen ME (2007) Beyond irrigation efficiency. Irrig Sci 25:233–245. doi:10.1007/s00271-007-0060-5

Leuven RSEW, Velde G, Baijens I et al (2009) The River Rhine: a global highway for dispersal of aquatic invasive species. Biol Invasions 11:1989–2008. doi:10.1007/s10530-009-9491-7

Libecap GD (2011) Institutional path dependence in climate adaptation: Coman's "Some unsettled problems of irrigation.". Am Econ Rev 101:64–80. doi:10.1257/aer.101.1.64

Lima IBT, Ramos FM, Bambace LAW (2007) Methane emissions from large dams as renewable energy resources: a developing nation perspective. Mitig Adapt Strateg Glob Chang 13:193–206. doi:10.1007/s11027-007-9086-5

Lotti C (1980) Water engineering. Water Int 5:18–27. doi:10.1080/02508068008685882

Moss T, Dobner P (2015) Between multiple transformations and systemic path dependencies. In: Huettl RF, Bens O, Bismuth C, Hoechstetter S (eds) Society – water – technology: a critical appraisal of major water engineering projects. Springer, Dordrecht, pp 101–111

Newig J, Pahl-Wostl C, Sigel K (2005) The role of public participation in managing uncertainty in the implementation of the Water Framework Directive. Eur Environ 15:333–343. doi:10.1002/eet.398

Niemczynowicz J (1996) Megacities from a water perspective. Water Int 21:198–205. doi:10.1080/02508069608686515

Oron G, Campos C, Gillerman L et al (1999) Wastewater treatment, renovation and reuse for agricultural irrigation in small communities. Agric Water Manag 38:223–234. doi:10.1016/S0378-3774(98)00066-3

Ostrom E (2009) A general framework for analyzing sustainability of social-ecological systems. Science 325:419–422. doi:10.1126/science.1172133

Ostrom E, Cox M (2010) Moving beyond panaceas: a multi-tiered diagnostic approach for social-ecological analysis. Environ Conserv 37:451–463. doi:10.1017/S0376892910000834

Petts GE (2009) Instream flow science for sustainable river management. J Am Water Resour Assoc 45:1071–1086. doi:10.1111/j.1752-1688.2009.00360.x

Robinson WC (2003) Risks and rights: the causes, consequences, and challenges of development-induced displacement. The Brookings Institution – SAIS project on internal displacement, Washington, DC

Sydow J, Schreyögg G (2010) Understanding institutional and organizational path dependencies. In: Schreyögg G, Sydow J (eds) The hidden dynamics of path dependence. Institutions and organizations. Palgrave Macmillan, Houndmills/New York, pp 3–12

UNEP (2008) Vital water graphics – an overview of the state of the world's fresh and marine waters. In: Global water policy and strategy, , 2nd edn. http://www.unep.org/dewa/vitalwater/article32.html. Accessed 10 Feb 2015

Vörösmarty CJ, McIntyre PB, Gessner MO et al (2010) Global threats to human water security and river biodiversity. Nature 467:555–561

World Commission on Dams (2000) Dams and development: a new framework for decision-making. Earthscan Publications, London/Sterling

Chapter 4
A Global View on Future Major Water Engineering Projects

Klement Tockner, Emily S. Bernhardt, Anna Koska, and Christiane Zarfl

Abstract Human activities have altered how the world functions. During the past decades, we have globally, fundamentally, in the long-term, and in most cases irreversibly modified all spheres of earth. This new epoch, often referred to as the Anthropocene, is just in its early stages. Indeed, there is general agreement that the transformation of our globe takes speed, with consequences that we can hardly imagine but that may threaten our own survival. This goes along with the general idea that major infrastructure projects are a sign of technological progress and believed to stimulate economic development and to improve living conditions for humans. In the present essay, a representative inventory of future major engineering projects, either planned or under construction in aquatic systems worldwide, shows that the rapid transformations of the Anthropocene are particularly evident in the freshwater domain. Worldwide examples of very large dams, major interbasin water-transfer and navigation projects, as well as large-scale restoration schemes, underline the dimensions of and the challenges associated with future megaprojects that will change our freshwater environment. Opportunities to mitigate the consequences of megaprojects based on the lessons learnt from projects in other infrastructure sectors range from ecological engineering to smart water investments that are adjusted to the respective national, social, economic, and environmental conditions.

K. Tockner (✉)
Leibniz-Institute of Freshwater Ecology and Inland Fisheries (IGB),
Müggelseedamm 310, 12587 Berlin, Germany

Department of Biology, Chemistry and Pharmacy, Freie Universität Berlin,
Altensteinstraße 6, 14195 Berlin, Germany
e-mail: tockner@ibg-berlin.de

E.S. Bernhardt
Department of Biology, Duke University, Durham, NC, USA

A. Koska
Leibniz-Institute of Freshwater Ecology and Inland Fisheries (IGB),
Müggelseedamm 310, 12587 Berlin, Germany

C. Zarfl
Center for Applied Geoscience, Eberhard Karls Universität Tübingen,
Hölderlinstraße 12, 72074 Tübingen, Germany

© The Author(s) 2016
R.F. Hüttl et al. (eds.), *Society - Water - Technology*, Water Resources
Development and Management, DOI 10.1007/978-3-319-18971-0_4

47

Keywords Hydropower dams • Interbasin transfer projects • Navigation • Restoration • Biodiversity • Global change • Anthropocene • Terraforming

4.1 Introduction

"Welcome to the Anthropocene", entitled *The Economist* on 21 May 2011. The Anthropocene indicates a geological epoch that follows the Holocene. Since the beginning of the industrial revolution, human activities have altered how the world functions. We have globally, fundamentally, in the long-term, and in most cases irreversibly modified the geosphere, hydrosphere, atmosphere and particularly the biosphere (Steffen et al. 2011; Rockström et al. 2014). And we are just at the beginning of this new epoch. There is general agreement that the transformation of our globe takes speed, with consequences that may threaten our own survival. Indeed, we are probably not able to imagine, or at least it sounds like science fiction (e.g. Pendell 2010), which alterations we will face in the coming decades to centuries. This includes, for example, the widespread creation of synthetic organisms, the exploitation of the ocean floor, the large-scale loss of coastal areas, the collapse of deltas, and a 4–8 °C warmer globe. We urgently need to understand the future dimension of this epoch and the consequences for the environment and the humans alike. And we need to consider fundamentally new strategies on how to cope with the immense challenges we are facing.

"Terraforming", i.e. the remaking of the earth surface, is not science fiction. Indeed, it is taking gear too. In China, for example, whole mountain ranges are levelled off to create space for new cities (Li et al. 2009). Mining activities reshape increasingly the earth surface because of an increasing demand for minerals and the concurrent depletion of resources. We dry-up entire river basins, truncate the global fluvial sediment transport, alter the global biogeochemical cycles and transform forests, steppes and deserts for crop and biomass production (e.g. Hooke and Martín-Duque 2012, Table 1). Terraforming not only requires land but also consumes huge amounts of energy and water.

At the same time, our planet is facing a major water crisis. Population growth and economic development are strongly increasing the global freshwater demand, while climate change further exacerbates the uneven distribution of water. The water crisis is spreading through all sectors, from sanitation, drinking water supply, agriculture and energy. Therefore, the signals of the Anthropocene are particularly evident in the freshwater domain (Table 4.1). Nutrient enrichment, exploitation of fossil groundwater reservoirs, fragmentation of river networks, alteration of the flow, sediment and thermal regimes, shrinking deltas and the accelerating erosion of freshwater biodiversity are clear signs of the rapid transformation of aquatic systems.

The water sector represents an immense market and the projected global expenditure on water and waste-water services is steadily increasing: from USD 576

Table 4.1 Signs of the Anthropocene in freshwater systems

		Pressures	Status and projection	References
Geosphere	Delta regions	Reduced sediment input, increasing subsidence rate, salinisation, sea water rise, demographic pressure (e.g., actually 500 million people live in deltas)	Surface area vulnerable to flooding may increase by 50 % in twenty-first century. Several deltas (e.g. Indus, Ganges) are facing complete collapse in near future	Syvitski et al. (2009) and Giosan et al. (2014)
	Sediment balance	Retention behind dams, land-use change, soil erosion	100 billion metric tons of sediments and 1–3 billion metric tons of C sequestered in reservoirs (status 2005); storage may double within the coming decades due to a boom in dam construction	Syvitski et al. (2005) and Zarfl et al. (2015)
	Coastal regions	Nutrient input, climate change	Exponential increase of dead zones (oxygen-depleted regions) since the 1960s; today about 400 coastal dead zones cover more than 245,000 km^2	Diaz and Rosenberg (2008)
Hydrosphere	River flow	Climate change, water abstraction, land-use change, damming, interbasin transfer	Increase in extreme events, rapid expansion of temporary streams, widespread change of flow regimes	Nilsson et al. (2005), Vörösmarty et al. (2010), and Acuña et al. (2014)
	Groundwater	Exploitation for irrigation, demographic development, climate change, pollution	From 1960 until today, groundwater depletion increased from 126 ± 32 to 283 ± 40 km^3 a^{-1} (India, USA, Pakistan, China and India jointly share 71 % of global groundwater withdrawal)	Giordano (2009) and Wada et al. (2010)
	Water quality	Urbanisation, agriculture intensification, industry, synthetic substances	Worldwide contamination of freshwater systems with thousands of industrial and natural chemical compounds, eutrophication	Camargo and Alonso (2006), Schwarzenbach et al. (2006), and Vörösmarty et al. (2010)
Biosphere	Biodiversity	Habitat degradation, flow regulation, pollution, climate change, invasion	10–20,000 freshwater species are estimated extinct or imperiled; Freshwater vertebrate populations declined by 73 % between 1970 and today; 63 % of all freshwater megafauna species listed as threatened	Strayer (2006) and WWF (2014)
	Ecological novelty	Species invasion, novel stressors	Faunal homogenisation, emerging pathogens, GMOs and synthetic organisms	Jeschke et al. (2013)
Atmosphere	Global carbon cycle	Pollution, habitat degradation, damming	Increased greenhouse gas emissions from near-natural, altered, and artificial freshwaters	Wehrli (2011), Raymond et al. (2013) and Zarfl et al. (2015)

billion in 2006, to USD 772 billion in 2015, to USD 1,038 billion in 2025 (Ashley and Cashman 2006). And the projected expenditure on water infrastructure as percentage of GDP will increase too, from 0.75 % in 2015 to more than 1 % in 2025. These values do not include the major water engineering projects that are either planned or under construction. For example, the construction of 3,700 future hydropower dams may require an investment of about USD 2 trillion, excluding operation costs as well as the costs caused by social and environmental damages (Zarfl et al. 2015). A primary challenge in designing and operating major water infrastructure projects will be to balance the economic benefits while preventing social costs and the loss of natural ecosystem services.

In this chapter, we provide a comprehensive albeit in no case complete inventory of future major engineering projects, so-called megaprojects, that are either planned or under construction in freshwater systems worldwide. We focus on very large dams, major interbasin water-transfer and navigation projects, as well as on large-scale restoration schemes. The main goal is to raise awareness about the dimension of and the challenges associated with future megaprojects. We discuss opportunities to mitigate the consequences of megaprojects based on the lessons learnt from projects in other infrastructure sectors.

4.2 Major Engineering Projects in the Water Sector

Major engineering projects are large-scale and complex projects that typically cost much more than USD 1 billion, require years to decades to be developed and constructed, affect large areas – very often across political and geographical boundaries, involve many public and private stakeholders, induce transformational processes, and may impact millions of people (Flyvbjerg 2014). In the water sector, such megaprojects encompass interbasin water-transfer projects, large-scale wetland drainage and irrigation schemes, navigation canals, drinking water facilities and sewage treatment plants for large cities, large dams, flood control and coastal protection measures, and major restoration schemes (Table 4.2). Furthermore, many small engineering projects may have cumulative effects that are similar to the effects caused by individual megaprojects.

The monetary scale of the investment is often inversely correlated with the potential for future adaptation and modification. Indeed, the lifespan of major water infrastructure projects is a century, and more, therefore new ideas and creativity now get "fixed". It means that the decisions we make now will heavily constrain the options we will have later.

4.2.1 Interbasin Water-Transfer Projects

Interbasin transfer projects (IBTs) are considered as an approved engineering solution meeting the accelerating demands for water to secure food production, support economic development and reduce poverty. To compensate for the increasingly

Table 4.2 Selected major water engineering projects globally (name, type, expected construction costs, planned construction time, short description and potential consequences)

Project	Type	Brief description	Costs [billion USD]	Timeline	(Environmental) Impacts	References
Nicaragua Canal *Nicaragua*	Navigation	286 km long navigation canal (90 km through Lake Nicaragua), 27.6 m deep, 520 m wide	40	2015-	Relocation of indigenous people (hundreds of villages); no environmental feasibility study released so far, but impacts expected on Lake Nicaragua (salt intrusion, sedimentation, invasive species, pollution) and destruction of around 400,000 ha of pristine rainforests and wetlands	Meyer and Huete-Pérez (2014)
Emergency Water Transfer Project (Tarim River Restoration Project) *China*	Restoration/ transfer	Artificial canal to divert water (annual average: 320×10^6 m^3) from Lake Bostan and the Daxihaizi Reservoir to the Tarim River	1.3	2000–2006	Transfers depend on the hydrological condition of the Kaidn-Konqui River system (Bostan Lake); ecosystem integrity strengthened	Li et al. (2009), Zhang et al. (2010), and Sun et al. (2011)

(continued)

Table 4.2 (continued)

Project	Type	Brief description	Costs [billion USD]	Timeline	(Environmental) Impacts	References
Grand Melen Project *Turkey*	Transfer	water transfer from Grand Melen Stream to Istanbul through a 180 km long transmission line; annually 268×10^6 m^3 at Stage I, 1.18×10^9 m^3 total at Stage IV	2.15	Stage I finalised in 2011	One town and 16 villages are expected to be covered by water with the project	WWF Germany (2008)
Lesotho Highlands Water Project *Lesotho/South Africa*	Transfer	five dams and about 200 km of tunnels; transfer of water from Orange/Senqu River (Lesotho) to Vaal River (South Africa) $\sim 2,000 \times 10^6$ m^3 per year	8	1986–2020	displacement of 17 villages, loss of agricultural land for 71 villages, and degradation of water quality; began without an environmental impact assessment for the overall project	WWF Global Freshwater Programme (2007)
The Bay Delta Conservation Plan *USA*	Transfer/ restoration	two tunnels, each 40 ft high and 35 miles long, under the Sacramento-San Joaquin River Delta	25	Construction start in 2017	could decrease fresh water flows to San Francisco Bay, could harm endangered fish in a different part of the estuary, will permanently transform the Delta	California Department of Water Resources (2014) and Safe the Bay (2014)

Project	Type	Description		Restoration	Details	Reference
Emscher River Master Plan *Germany*	Restoration	A 30-year regional regeneration program involves water quality improvements and physical rehabilitation of the river network	4.4	Restoration start in 1991	Once an open sewer in one of the most populated areas in Europe, the catchment of the Emscher River is one of the most ambitious restoration projects actually carried out in Germany	Schwarze-Rodrian and Bauer (2005)
Restoration and remediation of coal mining areas *Germany*	Restoration	Avoidance and source treatment, natural attenuation of acidity loads, remediation in constructed wetlands and underground treatment, in-lake and outflow treatments	9 (during the past 20 years)	Restoration of coal mining areas	The restoration of open-cast mining areas in Germany, as well as in Poland and other countries, is a long-lasting task because the large areas affected, delay effects, and the challenge to combine geo-engineering with ecological engineering approaches. Formation of novel ecosystems and communities.	Geller et al. (2013)

uneven distribution of water, IBTs are regaining popularity (WWF Global Freshwater Programme 2007). However, many of the IBTs either planned or under construction are too big and complex to imagine their consequences, and they may distort the economic, social and environmental conditions of entire countries and regions. Donor and recipient systems will be affected alike.

The South-North Project in China, the Indian Rivers Linking Project, the Transaqua Project in Africa, the Sibaral Project in Central Asia and many more projects are at various stages of development and implementation. Projects that are in a very preliminary stage of planning may rapidly emerge and gain wide support when political and social circumstances change, or when major disasters occur (Table 4.2).

With the objective to provide water for more than 500 million people, the *South-North IBT* in China is one of the largest water engineering projects already under construction (Berkoff 2003). By 2050, about 45×10^9 m³ water per year will be diverted through three branches from the Yangtze basin to northern and western China. The estimated costs are about USD 60 billion. The 1,264 km long Central Route was opened in 2014. 330,000 people were resettled. Moreover, plans exist for a much larger transfer project, namely to divert up to 200×10^9 m³ water from the major rivers in SW China, including the upper sections of the Mekong, Brahmaputra and Salween Rivers, to the water-thirsty regions in northern and eastern China. This project is actually on halt because of the transboundary nature of the affected river basins.

The *Indian Rivers Linking Project (ILR)* might become the largest water infrastructure project ever undertaken globally (Shah et al. 2008; Bagla 2014). It is planned to build 30 links and about 300 reservoirs and to connect 37 Himalayan and Peninsular rivers to form a gigantic water grid system on the Indian subcontinent. The canals are 50–100 m wide and 6 m deep to allow navigation. In total, 178×10^9 m³ water per year will be redistributed. For comparison, the annual discharge of the Rhine River at its mouth is 75×10^9 m³. The total length of the planned canal network is 15,000 km, 30 million ha of newly irrigated area are expected to be created, and 35 GW hydropower should be produced (although a significant proportion of the energy will be required for the transfer of the water). The costs at this stage can only be estimated to amount three times the total costs of China's South-North water-transfer scheme.

Transaqua, the largest water infrastructure project planned in Africa, is intended to divert 100×10^9 m³ (in average 3,200 m³/s) from the Congo Basin, through a 2,400 km navigable canal, to the Chari River and finally to Lake Chad. It is intended to stabilise the lake area at about 7,500 km² and create large irrigation areas north of the lake. Planning dates back to the late 1970s. Actually, the project re-emerges to the surface. The estimated costs are USD 23 billion. An alternative and much smaller option, the Obangi Water-Transfer Project, is expected to transfer 320 m³/s to Lake Tschad (Freeman and DeToy 2014). Up to now, no feasibility study has been carried out. However, there exists hope that the Chinese, within the frame of their Silk Road Fund, may invest into this project.

The almost complete drying of the Aral Sea is one of the largest global environmental disasters. Ideas to divert water from Siberia to Central Asia already emerged during the Tzarist period in the late nineteenth century. The concrete planning of the

Sibaral Project (from Siberia to Aral Sea), together with plans to nourish the Volga River through a transfer from western Siberian rivers, started during the Soviet era, but had been abruptly stopped in 1986 by Michael Gorbachev. More recently, it is enjoying again favour among various actors in Central Asia and in Russia as well. The archived construction plans for the 2,540 km long Sibaral canal are actually unearthed from the various institutes previously involved in the planning (Micklin 1977; Pearce 2009; Singh 2012). Sibaral is an example where the consequences of poor catchment management are expected to be solved through an immense engineering megaproject, which again is associated with probably very high economic, social, and environmental risks and costs, although a calculation of the costs and risks must remain a very rough estimate at this stage.

For North America, at least 15 separate projects have been proposed but not (yet) realised during the past century to reshape the continental water courses (Forest and Forest 2012). The most popular and ambitious proposal has been the *North American Water and Power Alliance (NAWAPA)*, which would have reconfigured the water courses through dams, canals, pipelines, etc. It is very hard to imagine the dimensions of this project (e.g. Barr 1975; Micklin 1977). It has been proposed to divert 20 % of the flow from the northern rivers, mainly from the Peace and Yukon rivers in Alaska and British Columbia, southward to a huge, 800 km long excavation called the Rocky Mountain Trench. From there, water would be diverted to the Great Lakes, to SW USA and finally to Mexico. The annual volume of water provided could be up to 300×10^9 m^3 per year, the estimated costs are between USD 420 billion and USD 1.4 trillion, and three million jobs are projected to be created. The concept had been called "grand and imaginative" (Abelson 1965), while Luton (1965) replied: "[…] let us wait until we know our doom is at hand, and when our last realisable ambition is to amaze future archeologists".

For a long time, plans have existed in *Australia* to move water from the water-rich northern areas to the southern parts of the continent (e.g., Kimberley to Perth Scheme, Bradfield Scheme, South-East Queensland water grid). However, local solutions such as improved use efficiency, recycling of water, desalinisation and reduced consumption prove to be economically, socially and environmentally much more sustainable than the long-distance transfer of water (Australian Government 2010).

Sudan and Egypt jointly began the construction of the *Jonglei Canal* in the 1970s (Salman 2011). The canal was meant to increase the downstream flow of the Nile waters by diverting water away from the vast wetlands where a high proportion of water is lost by evapotranspiration. The project, which was funded to a large extent by the World Bank, stopped in 1983 at about 100 km short of completion when the civil war between North and South Sudan started.

There are several other large-scale projects in the planning and construction phase throughout the African continent, including projects in Botswana, Namibia Lesotho, Morocco and other areas. Similar projects exist for southern Europe, Greece and Spain in particular, and for Turkey, but also for South America, in particular Brazil.

4.2.2   Navigable Waterways

The global navigable waterways encompass a network of 700,000 km that connects river basins across geographic regions. In the European Union, more than 50,000 km navigable rivers and canals create a dense web of waterways. Today, you may travel from southern France to western Siberia without entering the sea. Waterways facilitate the spreading of exotic and invasive organisms, which may lead to a homogenisation of freshwater biodiversity. Plans exist to upgrade and enlarge existing waterways, such as the Danube-Main-Rhine Waterway, and to create new waterways. At the same time, most of the planned interbasin transfer projects (Table 4.2) will support inland navigation too.

In South America, the 3,440 km Parana-Paraguay waterway, called *Hidrovia*, will connect Cáceres (Brazil) with Nueva Palmira (Uruguay). The main aim is to facilitate the export of soybeans, minerals, timber and other commodities from the interior. Extensive wetlands, in particular the Pantanal, will be affected by this project (Huszar et al. 1999; Gottgens et al. 2001). About 7.3×10^6 m^3 of sediment are dredged to enforce and straighten the rivers for navigation and to build ports. The calculated investment costs for a total of 88 individual projects are estimated at USD 4 billion (www.iirsa.org).

4.2.3   Hydropower Mega Dams

Currently, at least 161 very large hydropower dams, with a capacity of more than 1,000 MW each, are globally either under construction or planned. For illustration, a dam of 1,000 MW could, if to 100 % efficiently working, provide the annual electric consumption for 1.2 million people, assuming the electric consumption per capita in Germany in 2011 (World Bank 2014). This corresponds to sufficient energy to fully power a city almost as large and wealthy as Munich (2011: 1.38 million). The total capacity of the future mega dams amounts to 440 GW; most of the dams are planned in developing countries and emerging economies (Fig. 4.1), where consumption rates per capita are much lower than in Germany.

Not surprisingly, the global distribution of these future dams is not equal (Fig. 4.1). The largest number of large dams will be constructed in the Asian countries of China, Pakistan and Myanmar, in the South American countries of Brazil and Peru as well as the African countries of the Democratic Republic of Congo, Ethiopia and Nigeria.

What does this mean for the ecology? The Yangtze basin will be fragmented by the highest number of hydropower dams in Asia (33 % of the continental total). In South America, the Amazon basin will receive 61 % of the future hydropower dams of the continent providing 69 % of the planned capacity. And in Africa, mainly the Zambesi basin will be confronted with new dams (35 % of continental total) while the Grand Inga hydropower dam alone and planned in the Congo basin will provide 55 % of the capacity planned in Africa. The Yangtze and the Zambesi basins have already been classified as heavily fragmented, whereas the Amazon and the Congo

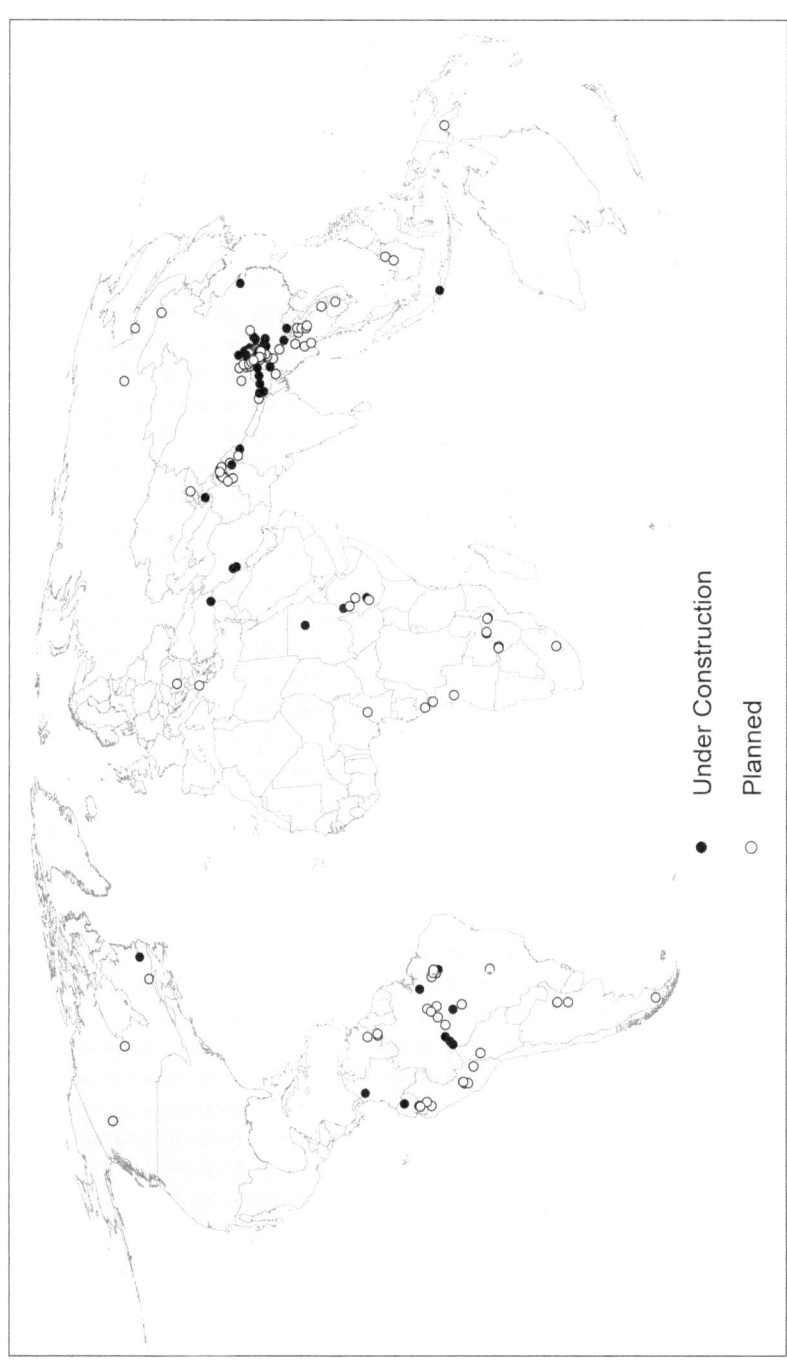

Fig. 4.1 Global distribution of future hydropower dams >1,000 MW that are under construction or planned, status 2014 (Data: Zarfl et al. (2015))

basins are among the moderately fragmented river systems (Nilsson et al. 2005). This means that these additional huge hydropower dams might change the fragmentation degree. There are also basins that have been classified as not fragmented so far that will face an increase in fragmentation, e.g. the Salween basin in Asia with 11 hydropower dams planned. Fragmentation leads to habitat destruction and prevents migration of aquatic organisms which might result in a decrease of biodiversity. In addition, hydropower dam operation will change discharge patterns, transport of sediments and water temperature.

For some of the "smaller" very large dams (1,000–2,000 MW) data on the size of the planned reservoir is known, which ranges up to 1,230 km^2 for the Mupata George in the Zambesi basin. This exceeds the size of Lake Yssel in the Netherlands and just indicates the area that might be required for huge hydropower dam constructions. One of the largest reservoirs by surface area, Lake Volta in Ghana, is flooding more than 8,500 km^2. This does not only affect the environment like flooding of terrestrial habitats, but also has social impacts due to the required relocation of the residential human population.

4.2.4 Large-Scale Restoration Projects

While major investments in new water redistribution and hydropower projects are underway across the emerging economies, governments throughout North America, Europe and Australia are devoting increasing efforts into removing or heavily modifying the major water infrastructure projects of the last century. Increasing recognition of the environmental harm to fisheries and biodiversity caused by hard infrastructure has spurred growing interest in dam removal and channel reconfiguration efforts to restore historic river structures and ecological processes.

As an inevitable consequence of increased environmental degradation and anticipated future environmental change, the demand for ecosystem restoration is rapidly increasing. "Develop now, clean-up later" has been the motto of many rapid emerging economies. The clean-up phase will be expensive and will take decades to centuries in most cases. Indeed, we probably are not able to imagine the future costs of ecosystem restoration, the damage in respect to loss of ecosystem services and the need to install extremely expensive engineering measures to compensate for the impacts of the past management decisions.

One of the most ambitious projects is the restoration of the *Mississippi River Delta*, following hurricane Katrina. It is planned to run for 50 years and estimated to cost between USD 500 million and USD 1.5 billion per year. And it will only stop future land loss, and not lead to the full recovery of the vast amount of wetlands already gone (Giosan et al. 2014). A recent study by Batker et al. (2010) shows that the investment to sustainably restore the Mississippi River Delta would return several time more of values in respect to provided ecosystem services. Indeed, the Mississippi River Delta Ecosystems provide economically valuable services, including hurricane storm protection, water supply, climate stability, food, furs, waste treatment, wildlife habitat, recreation and other benefits. These services are valued at USD 12–47 billion per year.

The *Comprehensive Everglades Restoration Plan* consists of over 68 civil works projects that will be implemented over a time period of 30 years. It will cost USD 8.3 billion, with the stated goals of improving water quality and reduced flooding of urban and agricultural areas. The main goal, however, is to restore the hydrology of the Everglades, including the Kissimmee River and Okeechobee Lake, in order to maintain these unique ecosystems (Sklar et al. 2005; LoSchiavo et al. 2013). The implementation plan is adaptive, benefiting from the increasing knowledge and the lessons learnt so far. A strong scientific framework allows for a clear setting of goals, an understanding of how the system works and an identification of the uncertainties and risks associated with the project (http://www.evergladesplan.org).

The *Four Major Rivers Restoration Project* in South Korea is a major water infrastructure project that addresses the environmental challenges faced by the Han, Nakdong, Geum and Yeongsam Rivers (Cha et al. 2011). The objectives include improved water storage and flood control, enhanced water quality and ecosystem health, provision of recreational space for local residents and improved cultural values of the rivers. It can primarily be considered an engineering project, rather than a restoration project, where natural processes will be mimicked and combined with constructed measures such as the building of dams and reservoirs. This project has provoked major opposition from both scientists and environmentalists because of the primary focus on hard engineering solutions and the distortion of scientific data for political purposes (Normile 2010). At the same time, this project is indented to form the nucleus for a national management programme to restore more than 10,000 km of local streams and rivers. At the same time, it is expected to stimulate the economy and create directly and indirectly up to 340,000 new jobs. The costs for this restoration project are estimated at USD 18 billion.

Other major restoration projects include the restoration of the Iraqi's marshes, once one of the largest and most valuable wetlands worldwide and identified as the historic garden Eden (Zahra Douabul 2012), restoring the balance of the Murray-Darling basin in Australia, or the manifold restoration work carried out across Europe, North America and Japan (Bernhardt et al. 2005; Nakamura et al. 2006). The European Water Framework Directive (WFD), for example, triggers major activities in the water sector in order to achieve a good ecological status or potential for its rivers, lakes and coastal zones. Indeed, a high proportion of the European river network must be restored in order to meet the ambitious goals set by the WFD. The restoration of the Rhine River, which already started much earlier than the implementation of the WFD, is a very good example for what may be achievable if the political will and the public pressure are high enough. After the Sandoz Accident in 1986, which released about 20 tons of pesticides and insecticides into the Rhine, immense financial resources have been invested to transform the Rhine from an open sewer to a river with a highly improved water quality and ecological state. However, the Rhine Programme also emphasis the limitations of restoration because the biological communities today are dominated by non-native species and thus very different from the communities that occurred before major pollution.

4.3 Discussion

We are facing a "pandemic array" (see www.gwsp.org) of transformations in the global water cycle, including fundamental changes in physical characteristics and biogeochemical and biological processes. Demographic and economic development increases the pressure on freshwaters and, together with an increasing frequency of floods and droughts, causes a dramatic increase in water stress and insecurity of water availability. At the same time, water security – the availability of freshwater in the eligible quantity and quality and at the right time – is a prerequisite for human wellbeing and ecosystem integrity.

Consequently, water security is one of the greatest challenges we are facing globally. Water shortage will require significant shifts in the way this precious resource will be managed. This is particularly the case if we want to manage freshwaters as a hybrid system, as both a medium for life and a resource for humanity. It is an interdisciplinary and cross-sectorial challenge linking economic, social, cultural and ecological systems. One of the proposed solutions is an engineering approach, although alternative ways to manage water as a resource and as a medium need to be established. However, in areas and during periods of water shortage, environmental damage and social consequences may receive less attention.

Megaprojects, such as major water engineering projects, are a sign of "high modernisms" (Scott 1998), an ideology that builds on self-confidence about technological progress. These projects are considered as a scaling-up of earlier successful projects and as a continuation of century-old practices (Forest and Forest 2012). Major infrastructure projects are believed to stimulate and guide economic development and to improve the living conditions for humans. At the same time, re-shaping the landscape is considered as a sign of progress as well as of regional and global power. In the Soviet Union, nuclear explosions were used to support major engineering projects, and their use had also been proposed for the NAWAPA project in North America. In this respect, Forest and Forest (2012) analysed the powerful role of visual rhetoric of water-transfer maps. Maps generalise and simplify, may ignore political boundaries, represent technology as an unproblematic approach and present water as a virtual entity. Therefore, we need to be very careful and responsible in using maps and visualisation tools in the decision making of major water engineering projects.

A fundamental problem with the planning and construction of large infrastructure projects is pervasive misinformation about costs, benefits and risks involved. A comprehensive analysis of large infrastructure projects in the UK and US (with emphasise on transportation infrastructure projects) demonstrated that in most cases the "unfittest proposals", which underestimated costs and overestimated benefits, were approved (Flyvbjerg 2007). The causes of misinformation and risk are mainly cognitive and political biases such as optimism bias and strategic misrepresentation. This is also most likely true for water-related megaprojects, such as for large dam and IBT projects. Therefore, Lovallo and Kahnemann (2003) and Flyvbjerg (2007) recommend the use of a "reference-based forecasting" approach, taking an outside-view derived from information from a class of similar projects to reduce inaccuracy and bias. Even more important is accountability, both public-sector accountability through transparency and public control and private-sector accountability via competition and market control (Flyvbjerg et al. 2003). Indeed, there is hope that widespread mismanagement of large-infrastructure projects is decreasing because democratic governance is improving around the world.

However, many of the future mega-projects in the water sector are transboundary projects that are planned or under construction in less democratic, economically fragile and politically often unstable countries, governed by weak national and international water management organisations. In these countries international disputes over water issues are more likely, as clearly shown for the Central Asian water conflict, particularly between Kyrgyzstan and Uzbekistan over the Syrdarya water resources (e.g. Bernauer and Siegfried 2012). Solutions for megaprojects under these conditions may include the establishment of an outside international body for management, coordinated water resource management across political and sectoral borders, the application of a "reference-based forecasting" approach (see above), involvement of all stakeholders through a participatory approach, accountability and critical questioning. According to Flyvbjerg (2007), a key principal should be that the costs of making wrong forecasts should fall on those making the forecast.

Several of the listed megaprojects are so-called "zombie projects" (Gleick et al. 2014) because they were once proposed, killed for one reason or another and brought back to life, even if they are socially, politically, economically and environmentally unjustified. Indeed, they may resurrect when so-called "windows-of-opportunity" open, mainly as a consequence of major disasters, such as nuclear disasters, severe droughts, long-lasting water shortages, famines, etc. Many of these "zombie projects" are technically feasible; however, the associated costs are immense and may cause dramatic social, economic and environmental damages. In most cases, only the planning and construction costs are considered, and these costs are already systematically underestimated. The follow-up costs need to be covered mostly by the general public and may therefore harm the economy of entire countries.

A potential alternative concept is ecological engineering, which encompasses a variety of approaches for working with nature. This approach is often cheaper and more effective than hard engineering solutions at accomplishing specific goals and may include the restoration of floodplains, coastal zones and upland areas. The Mississippi Delta Repair Program, following hurricane Katarina, has been recently approved by the U.S. federal government. It applies a restoration rather than a hard engineering approach, and the benefits may be immense on the long-run considering the lower maintenance costs as well as the multiple services provided by the restored delta ecosystems. Similar approaches have been successfully applied in the lower Rhine River and the delta system in the Netherlands (e.g. providing more room for rivers instead of always higher dikes), as well as in Australia (see above). Many of the major future water infrastructure projects are planned in developing countries and emerging economies, where low-tech efforts rather than expensive engineering projects are required to meet the major challenges in the water sectors. Smart water investments in both developing and developed countries may include

water-efficiency technologies, better wastewater treatment plants capable of producing high quality waters, improved piping and distribution systems, lower energy
desalination systems, improved monitoring tools, low-water-using crop types and
much more (Gleick et al. 2014).

References

Abelson PH (1965) Water for North America. Science 147:113
Acuña V, Datry T, Marshall J et al (2014) Why should we care about temporary waterways. Science
343:1080–1081
Ashley R, Cashman A (2006) The impacts of change on the long-term future demand for water
sector infrastructure. In: OECD (ed) Infrastructure to 2030: telecom, land transport, water and
electricity. OECD Publishing, Paris, pp 241–349
Australian Government (2010) Moving water long distances: grand schemes or pipe dreams? Department
of Sustainability, Environment, Water, Population and Communities. Canberra Act 2601
Bagla P (2014) India plans the grandest of canal networks. Science 345:128
Barr L (1975) NAWAPA: a continental water development scheme for North America? Geography
60:111–119
Batker D, De la Torre I, Costanza R, et al (2010) Gaining ground: wetlands, hurricanes, and the
economy. The value of restoring the Mississippi Delta. Earth Economics Project Report
Berkoff J (2003) China: the South – North water transfer project – is it justified? Water Policy
5:1–28
Bernauer T, Siegfried T (2012) Climate change and international water conflict in Central Asia.
J Peace Res 49:227–239
Bernhardt ES, Palmer MA, Allan JD et al (2005) Synthesizing U.S. river restoration efforts.
Science 308:636–637
California Department of Water Resources (2014) Bay delta conservation plan. http://baydeltacon-
servationplan.com/Home.aspx. Accessed 11 Mar 2015
Camargo JA, Alonso Á (2006) Ecological and toxicological effects of inorganic nitrogen pollution
in aquatic ecosystems: a global assessment. Environ Int 32:831–849
Cha YJ, Shim M-P, Kim SK (2011) The Four Major Rivers Restoration Project. Paper presented at
UN-Water international conference, Zaragoza, 3–5 October 2011
Diaz RJ, Rosenberg R (2008) Spreading dead zones and consequences for marine ecosystems.
Science 321:926–929
Flyvbjerg B (2007) Policy and planning for large-infrastructure projects: problems, causes, cures.
Environ Plann B Plann Des 34:578–597
Flyvbjerg B (2014) What you should know about megaprojects and why: an overview. Proj Manag
J 45:6–19
Flyvbjerg B, Bruzelius N, Rothengatter W (2003) Megaprojects and risk: making decisions in an
uncertain world. Cambridge University Press, Cambridge
Forest B, Forest P (2012) Engineering the North American waterscape: the high modernist map-
ping of continental water transfer projects. Political Geogr 31:167–183
Freeman L, DeToy D (2014) EIR brings Transaqua Plan, BRICS to Lake Chad Event. Exec Intell
Rev 41:28–36
Geller W, Schultze M, Kleinmann R, Wolkersdorfer C (2013) Acidic pit lakes. The legacy of coal
and metal surface mines. Springer, Berlin/Heidelberg

Giordano M (2009) Global groundwater? Issues and solutions. Annu Rev Environ Resour 34:153–178

Giosan L, Syvitski JPM, Constantinescu S et al (2014) Climate change: protect the world's deltas. Nature 516:31–33

Gleick P, Heberger M, Donnelly K (2014) Zombie water projects. In: Gleick P (ed) The world's water volume 8: the biennial report on freshwater resources. Island Press/Center for Resource Economics, Washington/Covelo/London, pp 123–146

Gottgens JF, Perry JE, Fortney RH et al (2001) The Paraguay-Paraná Hidrovía: protecting the Pantanal with lessons from the past. Bioscience 51:301–308

Hooke RL, Martín-Duque JF (2012) Land transformation by humans: a review. GSA Today 22:4–10

Huszar P, Petermann P, Leite A et al (1999) Fact or fiction: a review of the Hydrovia Paraguay-Paraná official studies. World Wildlife Fund (WWF), Toronto

Jeschke J, Keesing F, Ostfeld R (2013) Novel organisms: comparing invasive species, GMOs, and emerging pathogens. Ambio 42:541–548

Li Y, Chen Y, Zhang Y et al (2009) Rehabilitating China's largest inland river. Conserv Biol 23:531–536

LoSchiavo AJ, Best RG, Burns RE et al (2013) Lessons learned from the first decade of adaptive management in comprehensive Everglades restoration. Ecol Soc 18:70

Lovallo D, Kahnemann D (2003) Delusions of success. How optimism undermines executives' decisions. Harv Bus Rev 7:56–63

Luton DB (1965) NAWAPA. Science 149:163

Meyer A, Huete-Pérez JA (2014) Conservation: Nicaragua Canal could wreak environmental ruin. Nature 506:287–289

Micklin PP (1977) NAWAPA and two Siberian water-diversion proposals: a geographical comparison and appraisal. Sov Geogr 18:81–99

Nakamura K, Tockner K, Amano K (2006) River and wetland restoration: lessons from Japan. Bioscience 56:419–429

Nilsson C, Reidy CA, Dynesius M et al (2005) Fragmentation and flow regulation of the world's large river systems. Science 308:405–408

Normile D (2010) Restoration or devastation. Science 372:1568–1570

Pearce F (2009) Russia reviving massive river diversion plan. New Scientist, 9 February.www. newscientist.com/article/dn4637. Accessed 9 Apr 2015

Pendell D (2010) The Great Bay. Chronicles of the collapse. North Atlantic Books, Berkeley

Raymond PA, Hartmann J, Lauerwald R et al (2013) Global carbon dioxide emissions from inland waters. Nature 503:355–359

Rockström J, Falkenmark M, Allan T et al (2014) The unfolding water drama in the Anthropocene: towards a resilience-based perspective on water for global sustainability. Ecohydrology 7:1249–1261

Salman SMA (2011) The new state of South Sudan and the hydro-politics of the Nile Basin. Water Int 36:154–166

Schwarzenbach R, Escher B, Fenner K et al (2006) The challenge of micropolluants in aquatic systems. Science 313:1072–1077

Schwarze-Rodrian M, Bauer I (2005) Masterplan Emscher Landschaftspark 2010. Klartext-Verlag, Essen

Scott JC (1998) Seeing like a state: how certain schemes to improve the human condition have failed. Yale University Press, New Haven/London

Shah T, Amarasinghe UA, McCornick PG (2008) India's river linking project: the state of debate. In: Amarasinghe UA, Sharma BR (eds) Strategic analyses of the National River Linking Project (NRLP) of India, series 2. Proceedings of the workshop on analyses of hydrological, social and ecological issues of the NRLP. International Water Management Institute (IWMI), Colombo, pp 1–21

Singh KK (2012) Re-designing geography through inter-linking of rivers: a feasibility study. Int J Sci Environ 1:358–362

Sklar FH, Chimney MJ, Newman S et al (2005) The ecological–societal underpinnings of Everglades restoration. Front Ecol Environ 3:161–169

Steffen W, Persson Å, Deutsch L et al (2011) The Anthropocene: from global change to planetary stewardship. Ambio 40:739–761

Strayer DL (2006) Challenges for freshwater invertebrate conservation. J N Am Benthol Soc 25:271–287

Sun Z, Chang N-B, Opp C, et al (2011) Evaluation of ecological restoration through vegetation patterns in the lower Tarim River, China with MODIS NDVI data. Ecolo Inform 6:156–163

Syvitski JPM, Vörösmarty CJ, Kettner AJ et al (2005) Impact of humans on the flux of terrestrial sediment to the global coastal ocean. Science (New York, NY) 308:376–380

Syvitski JPM, Kettner AJ, Overeem I et al (2009) Sinking deltas due to human activities. Nat Geosci 2:681–686

Vörösmarty CJ, McIntyre PB, Gessner MO et al (2010) Global threats to human water security and river biodiversity. Nature 467:555–561

Wada Y, van Beek LPH, van Kempen CM et al (2010) Global depletion of groundwater resources. Geophys Res Lett 37:1–5

Wehrli B (2011) Climate science: renewable but not carbon-free. Nat Geosci 4:585–586

World Bank (2014) Electric power consumption (kWh per capita). In: World development indicator. http://data.worldbank.org/indicator/EG.USE.ELEC.KH.PC. Accessed 11 Mar 2015

WWF (2014) Living planet report 2014. Species and spaces, people and places. World Wildlife Fund (WWF), Gland

WWF Germany (2008) Drought in the Mediterranean – recent developments. World Wildlife Fund Germany, Frankfurt am Main

WWF Global Freshwater Programme (2007) Pipedreams? Interbasin water transfers and water shortages. WWF Global Freshwater Programme, Zeist

Zahra Douabul AA (2012) Restoration versus re-flooding: Mesopotamia Marshlands. J Waste Water Treat Anal 03:3–8

Zarfl C, Lumsdon AE, Berlekamp J et al (2015) A global boom in hydropower dam construction. Aquat Sci 77:161–170

Zhang J, Wu G, Wang Q, Li XY (2010) Restoring environmental flows and improving riparian ecosystem of Tarim River. J Arid Land 2:43–50

Chapter 5
Neglected Values of Major Water Engineering Projects: Ecosystem Services, Social Impacts, and Economic Valuation

Bernd Hansjürgens, Nils Droste, and Klement Tockner

Abstract Major water infrastructure projects like dams can provide substantial benefits such as food and drinking water security, hydropower generation, and flood control. But these benefits may come at a (too) high cost of large scale ecological alterations or adverse social impacts such as involuntary resettlements. If these costs are neglected, an investment decision will hardly be efficient. In this chapter, we will stress the necessity to make these "neglected values" visible and demonstrate how this can be achieved through economic valuation.

Keywords Major water engineering projects • Dams • Ecosystem services • Social impacts • Economic valuation • Ecological trade-offs • Distributional impacts • World Commission on Dams • Benefits • Costs

B. Hansjürgens (✉)
Department of Economics, Helmholtz Centre for Environmental
Research – UFZ, Permoserstraße 13, 04318 Leipzig, Germany
e-mail: bernd.hansjuergens@ufz.de

N. Droste
Helmholtz Centre for Environmental Research – UFZ,
Permoserstraße 13, 04318 Leipzig, Germany

K. Tockner
Leibniz-Institute of Freshwater Ecology and Inland Fisheries (IGB),
Müggelseedamm 310, 12587 Berlin, Germany

Department of Biology, Chemistry and Pharmacy, Freie Universität Berlin,
Altensteinstraße 6, 14195 Berlin, Germany
e-mail: tockner@igb-berlin.de

© The Author(s) 2016   65
R.F. Hüttl et al. (eds.), *Society - Water - Technology*, Water Resources
Development and Management, DOI 10.1007/978-3-319-18971-0_5

5.1 Controversial Discussions About Benefits and Costs of Major Water Engineering Projects

> The main challenge for water and energy resource developers in the 21st century will be to improve options assessment and the performance of existing assets. This will require open, accountable and comprehensive planning and decision-making procedures for assessing and selecting from the available options (World Commission on Dams 2000, p. 166).

Major water engineering projects (MWEPs) such as dams are subject to a controversial debate about their role for development. Especially where water is scarce and poverty widespread, water infrastructure can improve the livelihood of people. Large dams can facilitate access to water in water-scarce regions by improved ground water levels and increased flows in downstream areas during water scarce periods; thereby increasing food security (Shah and Kumar 2008). In addition to irrigation, water supply and flood control, electricity generation by hydropower has been a major driver for many large-scale water infrastructure projects (Biswas and Tortajada 2001). These large-scale projects are often co-financed by international organisations such as the World Bank and increasingly by private investors (Moore et al. 2010; Zarfl et al. 2015).

To assess the performance of such investments, the World Bank jointly with the International Union for Conservation of Nature (IUCN) initiated a multi-stakeholder dialogue in 1997. The final report of the World Commission on Dams (WCD) concluded that "dams have made an important and significant contribution to human development, and the benefits derived from them have been considerable. In too many cases an unacceptable and often unnecessary price has been paid to secure those benefits, especially in social and environmental terms, by people displaced, by communities downstream, by taxpayers and by the natural environment" (World Commission on Dams 2000, p. xxviii). While the core principles and strategic priorities of the report were broadly accepted, the guidelines for practical implementation resulted in dissent (Moore et al. 2010). Especially professional associations such as the International Commission on Large Dams (ICOLD) criticised the report for omitting benefits of large-scale dam constructions and calling for procedures that would deter large dam investments (Varma 2001).

The ongoing debate on large-scale water infrastructure remains controversial. Shah and Kumar (2008) analyse large dams in 145 countries and find that the main arguments for large dams are food and drinking water security, hydropower generation, and flood control. The criticism against large dams generally focuses on environmental, financial, economic, and human rights issues (ibid.). A very recent study finds (1) on average large cost and schedule overruns in the construction of large dams and (2) that risks are not sufficiently taken into account; therefore, smaller scale projects with less associated risks are strongly recommended (Ansar et al. 2014).

Whether a dam (or any similar large water infrastructure project) yields societal welfare gains or losses mainly depends on its specific characteristics. But as a general principle any assessment aiming towards a comprehensive analysis has to include *all related costs and benefits* of different infrastructure options.

Along this basic principle, we will give an overview of large dams that have emerged worldwide during recent years. We will then focus on the costs and benefits of MWEPs with special attention to the often neglected values of ecosystem services and distributional effects for society.

5.2   The Emergence of Major Water Engineering Projects Worldwide: Large Dams on the Advance

Hydropower production is a very well-established technique of the electricity system. Worldwide, out of 37,600 dams higher than 15 m more than 8,600 dams for hydropower generation are in operation, contributing about 20 % to the global electricity production. A period of intense dam building has been observed from the 1930s to the 1970s in North America, Japan and Western Europe. Major hydropower dam building in industrialised countries has now slowed, partly because the best sites have already been exploited within these countries, but also due to a greater understanding of the often unexpected social, economic, and environmental costs (Poff and Hart 2002; Lehner et al. 2011; ICOLD 2014).

Large dams mean high risks. Indeed, on average, the construction of large dams took much longer than planned and the expenses were twice as high as calculated. Furthermore, social and ecological costs were often not included because they are difficult to be calculated. Therefore, large dams, as well as other large infrastructure projects, are considered economically ineffective and should not be favoured (Ansar et al. 2014).

Despite the expected risks, we actually face an unprecedented boom in hydropower dam construction worldwide, primarily in developing countries and emerging economies (Zarfl et al. 2015, Fig. 5.1). While the expected construction of more than 3700 major dams may almost double the global electricity production from hydropower, it also may reduce the number of the last remaining large free-flowing rivers by about 20 %, in particular in South America. Many of the future dams are planned in areas with an exceptional high freshwater biodiversity. The Mekong, Amazon, and Congo basins are biodiversity hot spots that together contain about 1/5 of the global freshwater fish diversity. In particular, these basins will be heavily impacted by future hydropower development. Similarly, the Balkan area and Turkey face a major boom in dam construction; both regions are major centres of freshwater biodiversity.

Dam construction is becoming more and more a global business. A recent analysis on involved investors demonstrates that an increasing number of dam projects are financed by internationally operating companies (Zarfl et al. 2015). The construction costs of the 3700 dams planned or under construction may amount up to USD 2 trillion within the coming 10–20 years, excluding running and maintenance costs. Considering the fact that the costs and the construction timelines are systematically underestimated, we may expect costs that are at least twice as high. This is in line with Ansar et al. (2014) who find a mean cost overrun of 96 % in the construction of dams. At the same time, there are doubts that the projected increase in hydropower production will close the so-called electricity gap, i.e. providing access

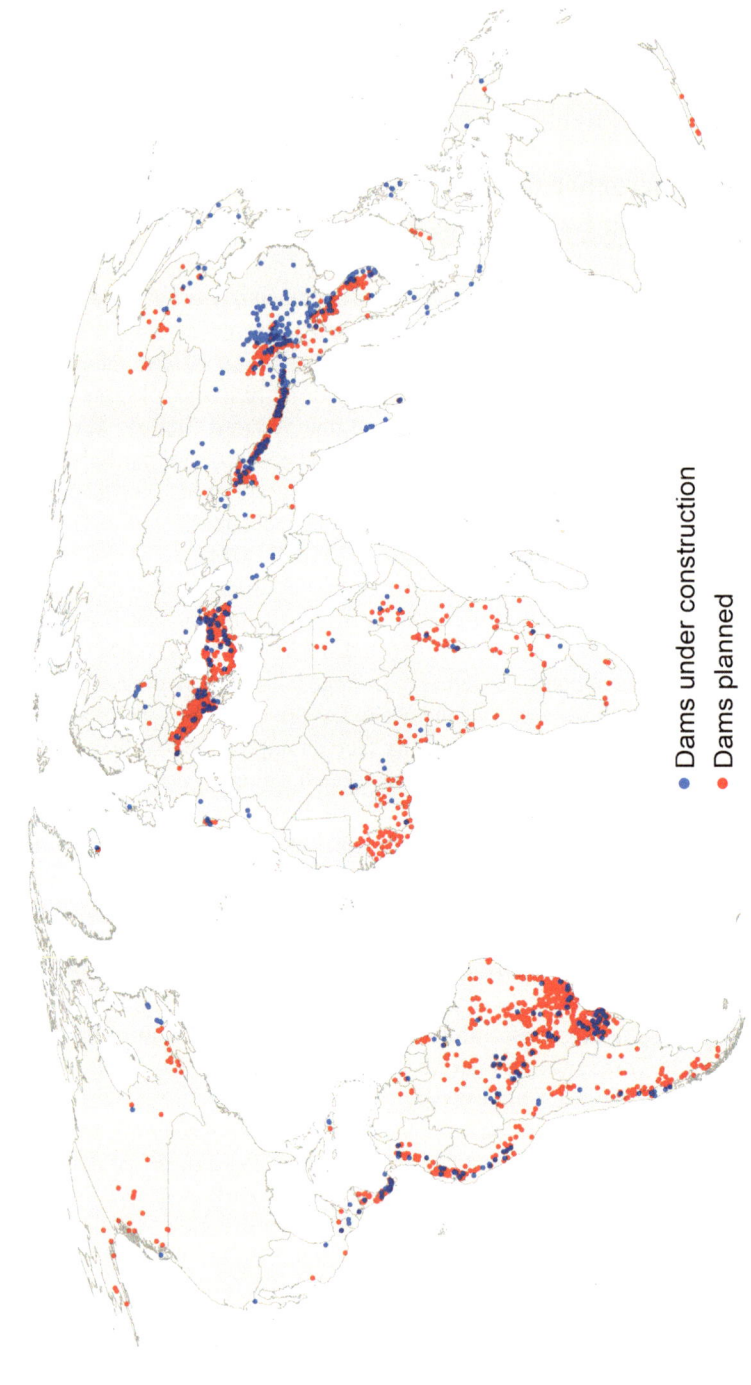

Fig. 5.1 Large dams planned and under construction (Zarfl et al. 2015)

● Dams under construction

● Dams planned

to 1.4 billion people who still remain disconnected from electricity supply. Large dams will primarily provide energy for industry and the mining sector, while the establishment of a extensive grid network would be required to serve rural human communities. Therefore, small, decentralised systems might probably be more effective to close the electricity gap.

Water management has historically emphasised services that depend on infrastructure such as navigation, irrigation, and hydropower (Auerbach et al. 2014). However, infrastructure projects create trade-offs and affect the services provided by natural ecosystems such as fishery yield, floodplain agriculture, cultural aspects, and the intrinsic values of biodiversity (ibid.). Indeed, one needs to be very careful in considering the trade-offs of maximising only a particular set of services. Today, navigation or hydropower production are considered as ecosystem services, although they impose major trade-offs with the service provided by a healthy ecosystem, e.g. a free-flowing river (Auerbach et al. 2014). Therefore, evaluating the benefits and trade-offs of large water infrastructure projects needs to consider both the ecosystem services provided by free-flowing rivers as well as the services provided by technical and engineered structures.

Mitigation of the impacts of large dams and other infrastructure projects may be very expensive. Because most future large dams will be constructed in developing countries and emerging economies, it will be particularly difficult to cover the associated costs. The question is not, however, if we should build new dams or not. The questions are where to build dams, how to construct them, and how to operate them. Therefore, there is an urgent need to further develop existing standards such as the Hydropower Sustainability Assessment Protocol (International Hydropower Association 2010), which must consider the economic, social, and ecological consequences of future dams. Furthermore, present dams are primarily evaluated individually, while ignoring the cumulative effects of multiple dams.

5.3 Making Ecosystem Services and Distributional Concerns Visible and Incorporating Them into Decision Making

5.3.1 The Concept of Total Economic Value

Although water resources are vital for the functioning of any economy, they continue to be depleted and degraded at an unsustainable rate (Birol et al. 2006, p. 106).

With a rising world population, the demand for water, food, and energy increases. The so-called "water-food-energy nexus" is a fundamental and increasing challenge for society (Russi et al. 2013). Water is a source of life for humans and nature. Without drinking water humans cannot survive. Without water any production of food or biomass becomes impossible. It is a resource that can be used for multiple purposes such as health and sanitation, agriculture and aquaculture, renewable energy generation and storage, among others.

To some extent, the different options in water uses are mutually exclusive and constitute a potential for conflicting interests. For example, water consumed for drinking is no longer available for other uses or users. It is economically rational to minimise trade-offs and maximise net gains, particularly through the application of multifunctional management approaches.[1] Management decisions on water resources therefore have to consider these trade-offs (Falkenmark and Rockström 2006). Assessments that provide reliable data can help to inform decisions on efficient resource use (Poff et al. 2003). In this respect, it is decisive to include *all* water-related services into consideration. To account for all associated costs and benefits of different uses, an overarching value framework is required.

Economics can provide such an overarching framework.[2] The concept of total economic value (TEV) represents a general framework to assess all values water delivers to humans. It is an anthropocentric view[3] where values go far beyond direct use values. Economic values do not consider just the (market) values of directly used goods and services (e.g. drinking water, water for irrigation), but also those subject to an indirect use, such as ground or surface water regulating services for agricultural production or flood control (e.g. high water tables, providing wetlands for a protection against flooding). In addition to these use-values, the TEV covers values assigned to a non-use of the resource such as bequest, altruistic, or existence values. Bequest values point to the fact that people have a benefit if certain water services are important for future generations (their children and grandchildren). Altruistic values are benefits people obtain from the fact that other people have water resources sufficiently at hand. And existence values point to the fact that people have a benefit just from the existence of water resources, e.g. certain species, irrespective of whether they see and enjoy this species. Between the use values and non-use values, there is the "option value" that points to the fact that keeping an option might be beneficial for humans (e.g. the option of future water benefits obtained from water tables where the benefits are not yet known). Figure 5.2 provides an overview of the TEV framework.

---

[1] Dams often serve a single goal. According to ICOLD Data (2014) over 70 % of the world's large dams are single purpose dams and half of them are constructed for irrigation.

[2] We see such an economic approach, which is based on a comprehensive understanding of advantages and disadvantages of a MWEP, in line with the framework developed in Sect. 3.1 of this volume. The economic valuation approach, as we see it, can serve as a comprehensive method taking the assessment principles derived in Sect. 3.1 into account.

[3] As will be shown below, we are fully aware that there exist also holistic approaches of water values where intrinsic values and additional ethical issues are considered, too (Young 2005a, b). These approaches include the aspect that water is not primarily considered as a resource for humans ("water as a means"), but also as a living entity with a non-economic value ("water as an end"). Freshwaters are among the most diverse, complex and dynamic ecosystems globally, at the same time they are more threatened than many other systems (e.g., Living Planet Index 2014). Therefore, many argue that there is an urgent need to balance the needs for humans (anthropocentric view) *and* nature (non-anthropocentric view) (e.g. Pahl-Wostl et al. 2013).

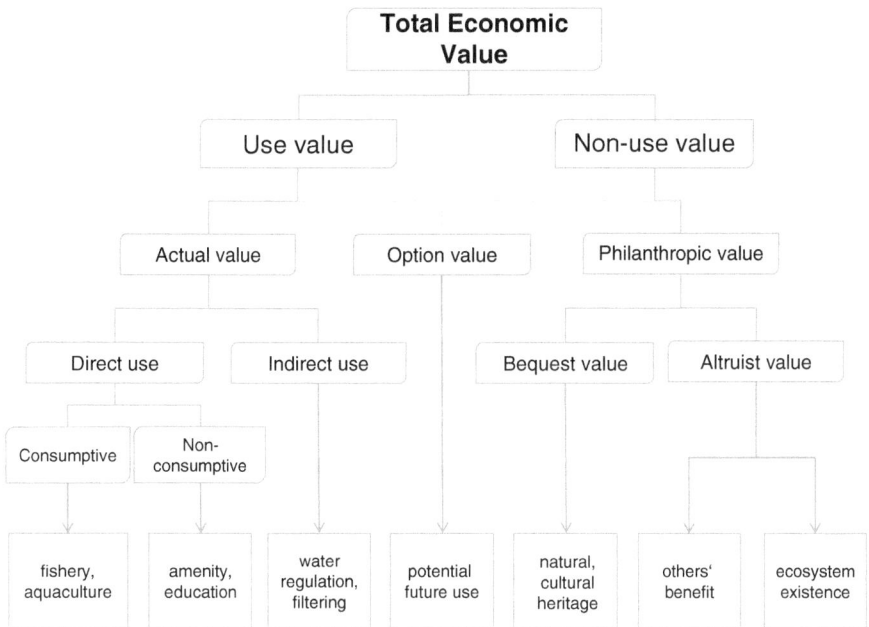

Fig. 5.2 Value types within the TEV approach (Adapted from TEEB 2010, p. 195)

Two remarks are worth mentioning here:

1. The economic approach of values of large water infrastructure projects is an anthropocentric one; benefits are derived from *people's preferences ("water as a means")*. This means that "intrinsic values" of water are not covered. If for example water infrastructure projects are rejected because people think that water has a value in itself ("water as an end"), this cannot be captured by economic valuation. Also values that a society as a whole might consider important (i.e. values that go beyond the aggregation of individuals' preferences) are not included.[4]

2. Nevertheless the economic valuation approach captures a rather broad set of values. The decisive point here is that *all* affected humans and *all* services of major infrastructure projects are to be included. Thus, ecological services (benefits) are integral part of this concept. This allows in particular to conceptually address ecological trade-offs, which are induced by major water engineering projects, and the identification of people who are negatively impacted by large projects through re-settlements or other *forms of disadvantages (costs)*.

---

[4] Sometimes it is difficult to distinguish between "self-interested" economic preferences and "common-interested" societal values. The distinction by Vatn (2009) might be helpful here, who speaks of "I-preferences" (people value, act, and decide as self-interested entities) and "We-preferences" (where people value, act, and decide as members of society, setting aside their self-interest and focusing more on the joint production of common goods).

5.3.2   Ecological Trade-Offs

> Natural biogeochemical processes and diverse communities of aquatic biota regulate fresh-
> water quantity and quality in ways that are not sufficiently acknowledged nor appreciated
> by the water resources management community (Arthington and Naiman 2010, p. 1).

In a majority of MWEPs, environmental impacts and consequences for ecosystem services and human well-being are underestimated (World Commission on
Dams 2000; Russi et al. 2013). The Three Gorges Project (Gleick 2009; Fu et al.
2010) or the Aral Sea basin (Cai et al. 2003) are examples of large scale environmental impacts that have been underestimated at early planning stages. Ecological
costs are difficult to estimate since they partly become apparent only in the longer
run. Nevertheless, it has become obvious (especially in the Aral Sea basin case) that
these ecological impacts have destroyed an entire landscape, with severe consequences not only for environmental but also for human health.

The conceptual framework of ecosystem services accounts for four categories of
services that people derive from ecosystem functions (Millennium Ecosystem
Assessment 2005a, b; Russi et al. 2013).

- *Provisioning services* are water for consumptive uses (drinking water, irrigation), non-
consumptive uses (hydropower, navigation), and food, medicine, and genetic resources.
- *Regulating services* are maintenance by water (filtration), climate (carbon
sequestration), or natural hazard regulation (flood and erosion).
- *Cultural services* are nonmaterial benefits such as recreation, learning, cultural
heritage, tourism, or existence values.
- *Supporting services* are those services required to maintain the overall functioning of ecosystems such as primary production and ecosystem resilience that
often have long-term effects on people's benefits (Russi et al. 2013).

While in the past provisioning services have regularly been captured by economic analysis, *it is the long-term impacts on ecosystems that have long been
neglected in economic assessment methods for costs and benefits*. Nevertheless,
these ecosystem services are important for people's welfare. The Millennium
Ecosystem Assessment (2005a) clearly showed how much people's well-being
depends on functioning ecosystems and their services, and that many ecosystems
are not managed sustainably. Management decisions that omit effects on the functioning of ecosystems and losses in their (provisioning, regulating, cultural, and
supporting) services are therefore incomplete and likely have deteriorating consequences for ecosystem service provision and human well-being.

One illustrative example might be the greenhouse gas emissions from creating
large reservoirs. While hydropower generally is considered a renewable source of
energy, newly created reservoirs emit substantial amounts of greenhouse gases
due to organic matter decomposition (World Commission on Dams 2000;
Fearnside 2002; Mäkinen and Khan 2010). Emission rates are highly site specific
and depend, for example, on reservoir age and latitude (Barros et al. 2011).
Indeed, emissions from reservoirs may even outnumber the emissions by a com-

parable fossil-fuel station by a factor of up to four (Giles 2006). Energy policies that do not take such emissions into account may cause external effects to the global climate and affected communities.

Another example is the loss of natural floodplains. In Germany, around 70 % of originally natural floodplains have been disconnected from rivers or lost by built infrastructure (Brunotte et al. 2009). This has led to severe – also economic – damages. Especially in rural areas, the ecosystem services of nutrient retention and carbon sequestration in restored natural floodplains often exceed the value of built infrastructure (Scholz et al. 2012). MWEPs that do not take into account these ecosystem services and only look at costs of man-made flood protection vs. costs of ecological restoration are hardly efficient in cost-benefit-considerations.

These are only selected examples. Costs and benefits are always project-specific and site-dependent. But in general, it can be concluded that any comprehensive assessment has to account for all ecosystem services (including ecological impacts) and the costs of their deterioration or the benefits that they may provide for humans. Any policy that is based on an incomplete assessment of costs and benefits (because it focuses only on tangible costs and benefits or on costs and benefits that merely affect the business sector) is likely to be incomplete. If MEWP shall be sustainable in the long run (meaning that the decision for the MEWP is not considered as a fault after a few years), comprehensive economic valuation is not only required, but essential for choosing the right decisions.

5.3.3 Social Conflicts

> Our language reflects [...] ancient roots: 'rivalry' comes from the Latin rivalis, or 'one using the same river as another.' Riparians – countries or provinces bordering the same river – are often rivals for the water they share (Wolf et al. 2005, p. 80).

Social conflicts are often closely related to the distribution of benefits and costs (Bernauer et al. 2012). In the real world, most MWEPs that generate benefits only to some groups in society and losses to others are subject to a potential conflict. These benefits and losses are very often associated with power relations. Ohlsson (2000) argues that the source of MWEPs conflict is not water scarcity itself but the required institutional change following large scale engineering to govern the scarce resource. The Human Development Report (UNDP 2006, p. 2) also states that the "scarcity at the heart of the global water crisis is rooted in power, poverty and inequality, not in physical availability".

Here, one important step to avoid or reduce conflicts is to explicitly address distributive impacts in decision support schemes. Who are the individuals or societal groups affected by the MWEPs? Are indigenous people or vulnerable groups particularly affected? What are alternatives? How can people be compensated? Thus it is decisive to include distributive aspects into (economic) analysis of major water infrastructure projects. Such distributive effects can be assessed quantitatively and qualitatively to build up a knowledge base for potential solutions (TEEB 2012, p. 31). In case distributional effects are not sufficiently taken into account potential conflicts may incur substantial economic (welfare) losses.

Involuntary or uncompensated re-settlements and losses of heritage sites may provide an example. Creation of large reservoirs for hydropower often requires that farmers and settlements are to be relocated. This entails not just the loss of nutrient rich high-yield marsh soils for farming, but often historically important settlement areas, cultural heritage sites, and aesthetic and recreational services. Some of these values may never be recovered – such as the loss of cultural heritage. Other values may be recovered such as the value of agricultural production. It is basically an ethical and often a political question of whether these values can be offset, compensated for or outweighed by benefits.

Another example may be provided by trans-boundary water conflicts caused by different goals and needs in different countries along the same river. Water is rival in its use – meaning that water used for a specific purpose is often not available for other uses. These trade-offs occur over space and time when hydropower is required upstream in winter, but downstream agricultural irrigation is required in summer like in the Syrdarya basin (see Chap. 7).

Economically speaking these "social costs" (impacts on ecosystem services, vulnerable groups or indigenous people that are not covered in market relations) have to be accounted for – otherwise an infrastructure investment decision like a dam might constitute a decision that turns out to be inefficient in terms of welfare. Whether this is practically feasible is not just a question of a proper assessment but also political will, power, and (democratic) institutions.

5.3.4   Potential and Limits of Economic Analysis

An informed decision on whether to build or not to build e.g. a large dam is not just a question of "yes" or "no" but also about different options to realise a certain goal, let it be flood control, food security, or electricity generation. Such a decision involves inevitable trade-offs. A well-designed analysis of different options and their associated costs and benefits can show which alternative actually yields the largest positive net benefit.

Figure 5.3 exemplifies hypothetical use scenarios for fresh water bodies. Optimising a single service or benefit dimension such as hydropower or ecosystem services does not necessarily yield an overall optimum. Finding balanced solutions that take into account multiple value dimensions may create the largest net benefits.

A comprehensive analysis that compares the advantages (benefits) and disadvantages (costs) of different scenarios for the use of water and wetland resources among multiple value dimensions can identify the best choice option. One policy goal could be secure water access and availability for a certain region. Another policy goal might be renewable energy production by hydropower. A third policy goal might be

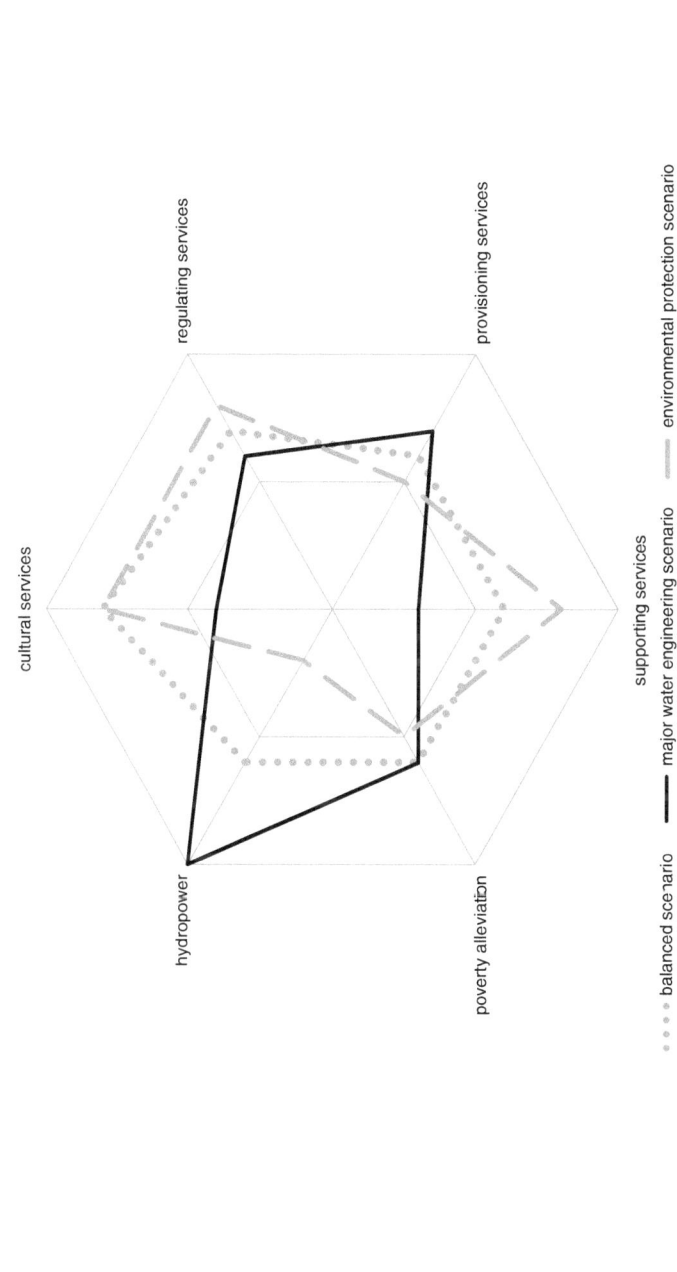

Fig. 5.3 Hypothetical scenarios of trade-offs between services and social impacts (Adapted from Millennium Ecosystem Assessment 2005b, p. 219)

sustaining a healthy environment. The decisive point here is that all impacts should be considered – and not only a single one. We see this as a major fault in past decisions on dam building where in many cases the focus was merely on energy production.

The actual management options to realise a single or multiple of these goals can look quite different and can incur different costs and benefits. It could be nature-based or technical, decentralised or centralised, small- or large-scale. After an identification of potential approaches, different options could be compared by cost-benefit-analysis (Hansjürgens 2004), multi-criteria-analysis (Hansjürgens 2011), or other methods (TEEB 2010, 2012).

If properly designed an economic analysis can thus be an important source for decision-making and can ideally ensure that all relevant costs and benefits are taken into account.

5.4   Concluding Remarks

Major water engineering projects (MWEPs) can have both substantial benefits and costs. Which one prevails is not just a question of the project but also of the assessment. If substantial trade-offs such as irreversible losses in ecosystem services or social costs are neglected, it will be perceived as beneficial, while the extent of losses remains hidden. Hence, based on such an incomplete informational basis, no large-scale investment decisions should be made. By already taking the neglected values of water into account during the planning phase of MWEPs, the faults of the past could be avoided.

We have presented an economic framework, namely the Total Economic Value (TEV), to assess the often neglected costs and benefits; and there is a range of potential valuation methods that can be used to assess the related costs and benefits of various options (TEEB 2010).

However, a definition of what values, stakeholders, and effects are relevant can never be an exclusively scientific task. The realisation of projects fundamentally depends on political and societal actors and embedding institutional structures. It therefore also requires approaches and engagement beyond science – one could call them transdisciplinary. To comprehensively assess the outcomes of water-related investment decisions, our key recommendation is to integrate science, stakeholders, and political decision makers within an integrative, participatory, and process-based assessment approach – especially if large-scale outcomes are to be expected.

References

Ansar A, Flyvbjerg B, Budzier A et al (2014) Should we build more large dams? The actual costs of hydropower megaproject development. Energy Policy 69:43–56

Arthington A, Naiman R (2010) Preserving the biodiversity and ecological services of rivers: new challenges and research opportunities. Freshw Biol 55(1):1–16

Auerbach DA, Deisenroth DB, McShane RR et al (2014) Beyond the concrete: accounting for ecosystem services from free-flowing rivers. Ecosystem Services 10:1–5

Barros N, Cole JJ, Tranvik LJ et al (2011) Carbon emission from hydroelectric reservoirs linked to reservoir age and latitude. Nat Geosci 4:593–596

Bernauer T, Böhmelt T, Koubi V (2012) Environmental changes and violent conflict. Environ Res Lett 7:015601

Birol E, Karousakis K, Koundouri P (2006) Using economic valuation techniques to inform water resources management: a survey and critical appraisal of available techniques and an application. Sci Total Environ 365:105–122

Biswas A, Tortajada C (2001) Development and large dams: a global perspective. Water Resour Dev 17:9–21

Brunotte E, Dister E, Günther-Diringer D et al (2009) Flussauen in Deutschland. Erfassung und Bewertung des Auenzustandes. Bundesamt für Naturschutz (BfN), Bonn – Bad Godesberg

Cai X, McKinney DC, Rosegrant MW (2003) Sustainability analysis for irrigation water management in the Aral Sea region. Agric Syst 76:1043–1066

Falkenmark M, Rockström J (2006) The new blue and green water paradigm: breaking new ground for water resources planning and management. J Water Resour Plan Manag 132(3):129–132

Fearnside P (2002) Avanca Brasil: environmental and social consequences of Brazil's planned infrastructure in Amazonia. Environ Manag 30:735–747

Fu B, Wu B, Lü Y et al (2010) Three Gorges project: efforts and challenges for the environment. Prog Phys Geogr 34:1–14

Giles J (2006) Methane quashes green credentials of hydropower. Nature 444:524–525

Gleick P (2009) Three Gorges dam project, Yangtze River, China. In: Gleick PH, Palaniappan M, Morikawa M et al (eds) The World's water 2008–2009. Island Press, Washington, DC, pp 139–150

Hansjürgens B (2004) Economic valuation through cost-benefit analysis – possibilities and limitations. Toxicology 205:241–252

Hansjürgens B (2011) Bewertung von Wasser in Landschaften – Konzepte, Ansätze und Empfehlungen. acatech – Deutsche Akademie der Technikwissenschaften, München

ICOLD (2014) Register of dams. http://www.icold-cigb.org/GB/World_register/world_register.asp. Accessed 14 Jan 2015

International Hydropower Association (2010) Hydropower sustainability assessment protocol. International Hydropower Association, London

Lehner B, Liermann CR, Revenga C et al (2011) High-resolution mapping of the world's reservoirs and dams for sustainable river-flow management. Front Ecol Environ 9:494–502

Mäkinen K, Khan S (2010) Policy considerations for greenhouse gas emissions from freshwater reservoirs. Water Alternat 3:91–105

Millennium Ecosystem Assessment (2005a) Ecosystems and human well-being: synthesis. Island Press, Washington, DC

Millennium Ecosystem Assessment (2005b) Ecosystems and human well-being: policy responses. Island Press, Washington, DC

Moore D, Dore J, Gyawali D (2010) The world commission on dams + 10: revisiting the large dam controversy. Water Alternat 3:3–13

Ohlsson L (2000) Water conflicts and social resource scarcity. Phys Chem Earth (B) 25:213–220

Pahl-Wostl C, Arthington A, Bogardi J et al (2013) Environmental flows and water governance: managing sustainable water uses. Curr Opin Environ Sustain 5:341–351

Poff N, Hart D (2002) How dams vary and why it matters for the emerging science of dam removal. Bioscience 52:659–668

Poff N, Allan J, Palmer M (2003) River flows and water wars: emerging science for environmental decision making. Front Ecol Environ 1:289–306

Russi D, ten Brink P, Farmer A et al (2013) The economics of ecosystems and biodiversity for water and wetlands. Gland, London/Brussels

Scholz M, Mehl D, Schulz-Zunkel C et al (2012) Ökosystemfunktionen von Flussauen – Analyse und Bewertung von Hochwasserretention, Nährstoffrückhalt, Kohlenstoffvorrat, Treibhausgasemissionen und Habitatfunktion. Bad Godesberg, Bonn

Shah Z, Kumar M (2008) In the midst of the large dam controversy: objectives, criteria for assessing large water storages in the developing world. Water Resour Manag 22:1799–1824

TEEB (2010) The economics of ecosystems and biodiversity: the ecological and economic foundations. Earthscan, London/Washington

TEEB (2012) The economics of ecosystems and biodiversity for local and regional policy makers. Earthscan, London/Washington

UNDP (2006) Human development report 2006 beyond scarcity: power, poverty and the global water crisis. UNDP Human Development Reports

Varma CVJ (2001) ICOLD's final position on the WCD report. http://www.unep.org/dams/documents/default.asp?documentid=453. Accessed 14 Jan 2015

Vatn A (2009) An institutional analysis of methods for environmental appraisal. Ecol Econ 68:2207–2215

Wolf AT, Kramer A, Carius A et al (2005) Managing water conflict and cooperation. In: Assadourian E, Brown L, Carius A et al (eds) The state of the world. Redefining global security. Worldwatch Institute, Washington DC, pp 80–95

World Commission on Dams (2000) Dams and development a new framework for decision-making. Earthscan, London

Young R (2005a) Determining the economic value of water: concepts and methods. Resources for the Future, Washington, DC

Young R (2005b) Water as an economic good. In: Brouwer R, Pearce D (eds) Cost-benefit analysis and water resource management. Edward Elgar, Cheltenham, pp 13–45

Zarfl C, Lumsdon AE, Berlekamp J et al (2015) A global boom in hydropower dam construction. Aquat Sci 77:161–170

Chapter 6
Water Governance: A Systemic Approach

Petra Dobner and Hans-Georg Frede

Abstract The article investigates the contributions of system theory and governance literature to implementing and managing MEWPs. Both approaches share the belief that problems should be addressed holistically and thus challenge the division of labor in the water sector. After sketching some major characteristics of system theory and governance, the findings are applied to MEWPs in Fergana and Jordan.

Keywords Governance • Millennium development goals • System theory • Major water engineering projects • Institutions • Public participation • Complexity • Soviet Union • Legacy • Fergana Valley • Red Sea • Dead Sea Project

6.1 Introduction

All commonplaces about water share the conviction that water is universal and essential. Water is life, indeed, and the availability of and access to water is of utmost concern for every organism and organization. Water issues are ubiquitous and all embracing. It is quite plausible to conclude that the technical and political management of water should be similarly comprehensive.

Quite contrary, though, is everyday practice. Typically, water issues are addressed from numerous partial standpoints. Technicians and engineers should solve infrastructural problems, natural scientists deliver information about the quality and availability of water, and political and social scientists are in charge of regulatory or social matters. This division of labour is neither peculiar nor surprising. Fragmented responsibilities seem a natural consequence of differentiation, segmentation, and specialization, the foremost features of modernity. The water sector is but one example of many for this dominant trend, but in this case the price is especially

P. Dobner (✉)
Institute of Political Science and Japanese Studies, Martin-Luther-Universität
Halle-Wittenberg, Emil-Aberhalden-Str. 7, 06099 Halle/Saale, Germany
e-mail: petra.dobner@politik.uni-halle.de

H.-G. Frede
Institute of Landscape Ecology and Resource Management (ILR),
Justus-Liebig-Universität Gießen, Heinrich-Buff-Ring 26-32, 35392 Gießen, Germany

© The Author(s) 2016
R.F. Hüttl et al. (eds.), *Society - Water - Technology*, Water Resources
Development and Management, DOI 10.1007/978-3-319-18971-0_6

high. Despite some progress, many parts of the world are constantly lagging behind the water-related Millennium Development Goals (MDGs) (United Nations 2006, 2007, 2008, 2009, 2012) not to mention the minor advances made in the overarching goal of the UN Charta to guarantee access to drinking water and safe sanitation for everybody as a human right or, even more ambitious, establish water security for all individual, agronomic, and economic purposes. With good cause, there are doubts that various water sector goals can ever be achieved as long as both scientific knowledge about and practical management of water are dominated by specialization and distinction.

Major water engineering projects (MWEPs) can act as examples for studying the problems the water sector faces in general. MWEPs are conceived of "complex socio-technical, social-ecological and political-economic systems or configurations" (Moss and Dobner 2015, in this volume, pp. 101–111). Hence their scientific assessment and management has to address a number of different, but interdependent issues simultaneously. As evident as this basic insight into the challenges of sustainably managing MWEPs is, it remains difficult to meet these demands in theory and practice. That MWEPs' knowledge is created in separate realms put up barriers to ambitions to holistically bridge unique scientific languages, knowledge bases, and approaches. Practitioners usually are well aware of the complexity and interrelatedness of problems associated with the installation and management of MWEPs, but fragmented responsibilities for technical, ecological, or cultural and social issues frequently prevent coordinated action. For example, while a MWEP may be technically feasible and economically desirable, social, cultural, or political resistance can throw up insurmountable obstacles for a project's implementation and/or its maintenance.

Understanding and managing the complexity of MWEPs properly remains a scientific and practical aspiration. In the following section, we address this problem by first looking at two different approaches: System Theory (6.2) and Governance (6.3), which from different perspectives both promise to link the separate fields of science and practice. Secondly, by applying our findings to MWEPs, and finally by drawing some conclusions on further research (6.4).

6.2 System Theory Revisited

In the middle of the twentieth century, a number of scientists from heterogeneous disciplines promoted a holistic approach under the label of (General) System Theory (see von Bertalanffy 1950). Taking on ideas from cybernetics, i.e. a scientific stream dealing with regulation and control of machines, living organisms and social organizations (Wiener 1948), system theory centres around the notion that any "complex of interacting elements" (von Bertalanffy 1950, p. 143) can be defined as a "system." A system is, therefore, any social, biological, technical, political, economic, psychological, or other kind of entity that consists of a number of connected parts. It is important to note that system theory also operates with the similarly abstract notion of "environment" to describe the fact that systems have borders and (usually) interact with and are influenced by their surroundings. Obvious differences between mice and men, organisms and organizations are deliberately neglected for the benefit of discovering general laws, abstract similarities, and analytical analogies.

System theory thus considers a country as a system as well as a large dam, or any other kind of MWEP. They are embedded in a specific environment, e.g. neighboring countries or the political and geographical circumstances they are placed in. By applying this general analytic framework on all kinds of objects, the theory claims that irrespective of their nature, systems are characterised by common features: foremost, the systemic view is fuelled by the conviction that "the operation of no one part can be fully understood without reference to the way in which the whole itself operates" (Easton 1957; see Easton 1965) and that, vice versa, the system in general depends on the interaction of all its parts. This view of a system and its elements is substantiated by further ideas about the ways the elements and the system are connected with each other and how they are embedded in their environment. Attention is directed to the *complexity*, i.e. the number of relations between the elements of a system as well as to its relations to its environment (Hill et al 1994, p. 22f.). The *dynamic* of a system is expressed as the intensity of changes the system undergoes in a certain amount of time. By stressing the complexity and dynamics of both the system and its environment, system theory also rejects simplified ideas about *causes and effects* and enhances the idea of multiple interdependencies instead. Advocates of system theory also point out the fact that *isomorphic laws* – such as the law of exponential growth – operate in as differing fields as e.g. biology and sociology. System theory should therefore enable methodological advancements throughout different scientific fields as "an important means of controlling and instigating the transfer of principles from one field to another" (von Bertalanffy 1950, p. 142).

With its core concern of understanding the wholeness of something (i.e. any "system") in terms of the interaction, interrelatedness, and interdependence of all its components within a changing environment, system theory seems to be well capable of addressing generally acknowledged growing complexities; something which holds true in the field of water in general, and MWEPs in particular. Nonetheless, system theory findings cannot be easily applied to them: firstly, system theory itself provides a major obstacle by employing relatively idiosyncratic terminology and alienating concepts such as "autopoiesis" (Varela et al. 1974), "recursivity," or "emergence." While system theory has spread some marks throughout scientific terminology, e.g. we refer to ecological or political "systems," this often does not go beyond the mere usage of the term "system" without further acknowledging the ambitious theoretical programme associated with it. Secondly, a systemic approach is inevitably complex and, therefore, naturally adverse to specialization and expertise in a distinct field. Thirdly, system theory's critics complain about the vagueness of the concept; the uncertainty, or arbitrariness in defining the line between system and environment, and the inherent conservatism of a theory dealing with structures and patterns, thus not leaving much room for action, active revolution, or at least change.

For all the above reasons, system theory has never succeeded as a widely accepted theoretical framework for the analysis of complex realities. Regardless of the correctness of these denouncements, there are two good reasons for a renewed interest in systemic thinking with respect to water issues: their complexity is undebatable and despite great efforts utilizing single disciplines, political and technical approaches, water related problems are persistent and will probably grow into one of the most challenging problems for the third millennium without coming close to any substantial and sustainable solution. Both reasons render it worthwhile to test the application of system theory's findings on the subject matter.

6.3   The Governance Approach

Governance has advanced to be one of the most popular concepts in the last two decades and thus the utilization of the governance discourse has also ousted the concept of water management (e.g. GWP 2003; Lebel et al. 2007; Bakker 2010; OECD 2011; Anne 2012; Perkins 2013). On the one hand, the popularity of the concept can well be explained by the widely shared expectation that governance enhances civil society's participation in making decisions and is more efficient at solving problems than governmental approaches. On the other hand, "governance" remains a highly amorphous notion and its promises still await empirical proof. Despite its widespread use, there is neither a shared definition of governance nor solid evaluations of governance practices.[1] There is some agreement though that governance involves all sorts of regulations irrespective of whether they have been created intentionally or not, of their source, and of their success. The governance discourse thus shifts the perspective from deliberate, hierarchical, and goal-oriented exercise of political control by a government to the far more ambiguous processes of regulatory and evolutionary (self-) governing. Leaving aside the conceptual and normative deficits of the concept,[2] governance may serve as a rough pattern for a new assessment of water issues, pointing "on the one hand [...] to the complex setting of water management in wider governance structures that have to be accounted when assessing water usage. On the other hand, it points to the necessity of good governance – rule of law, stakeholder participation, transparency, accountability, etc. – in the water sector" (Sehring 2009, p. 17f.) Summed up, water governance encompasses all the actions, regulations, subjects, objects, and goals, which are involved in order to guarantee sustainable access to water for everyone and for all purposes (see Huitema et al. 2009). It also points out the impact and relevance of *institutions*, i.e. explicit and implicit norms and rules, which shape patterns of behavior and often severely influence how apparently just technical or ecological realities are dealt with by humans.

---

[1] As examples of positive expectations see Hauff (1987), Deutscher Bundestag (2002), for critique see e.g. Dingwerth (2003), Brand et al. (2000), Brand and Scherrer (2003).

[2] See out of innumerous: Benz (2004), Dobner (2010), Mayntz (2005), Folke-Schuppert (2006).

There are four key guidelines of special interest in the literature on water governance, namely "polycentric governance, public participation, experimentation, and a bioregional approach" (Huitema et al. 2009) *Polycentric governance* relates to systems in which "political authority is dispersed to separately constituted bodies with overlapping jurisdictions that do not stand in hierarchical relationship to each other" (Skelcher 2005, p. 85, also Ostrom (1999), Dietz et al. (2003). *Public participation* stands for the cooperation between non-governmental stakeholders and governmental bodies. *Experimentation* is the demand to invent new practices and institutions and evaluate their potential for transfer to other problems, regions, or contexts. The *bioregional approach* finds its most prominent interpretation in the water sector in the call for river basin management. While all of these findings can be considered theoretically sound, unfortunately there is little known about their effectiveness so far: polycentricity and the bioregional approach have scarcely been evaluated because of "the lack of monitoring data and the attribution problem," participation "can contribute to the quality and legitimacy of decisions, but the connection to the formal decision process needs to be clearly specified" and about experimentation there is "not much known, but in other policy domains, experiments are often watered down to 'pilots'" (Huitema et al. 2009).

6.4 Dealing with Complexity

System theory and the governance discourse both raise awareness that complex systems such as MWEPs must rely on a multifaceted analysis of interacting elements. By emphasizing the interrelatedness of different issues, they both underline the necessity to overcome fragmented scientific perspectives on and isolated responsibilities for MWEPs. Their potential benefits direct research and practice in two different, but complementary ways: system theory provides potential cornerstones for an *analytic framework* to address the complexity of MWEPs, and the governance discourse stresses the need to adhere to *practical management* issues.

From a systemic perspective, MWEPs are but one example in the generalized concept of systems, and thus opens the way for transferring general insights about systems to this specific field. Taking up the selected list mentioned above: the *complexity* of MWEPs can be interpreted as the need to take the mutual interdependence of historic, technical, political, social, and so forth aspects of an MWEP into account in the estimation of its feasibility, durance, robustness, and sustainability. Going one step further, MWEPs can be considered as examples of *disorganized complexity*, i.e. a system "in which the number of variables is very large, and one in which each of the many variables has a behaviour which is individually erratic, or perhaps totally unknown" (Weaver 1948). Therefore, a MWEP's management depends on the knowledge and consideration of *all* its influencing factors, rendering any hierarchy counterproductive in the sense of which elements are more or less important.

This aspect challenges especially technocratic approaches: MWEPs may be technically well thought out and can be perfectly infrastructural or materially organized, but they still can fail because of social or political rejection. All existing and future infrastructures and institutions have to be established against or before the background of former decisions and institutions. The role of history therefore cannot be overestimated in its effect of how things work out. The strong political, technical, and financial support, e.g. for the irrigation systems; the leading technicians and organizations; and the dominant role cotton played in the Fergana Valley during Soviet times were all demonstrations of economic strength. Economic strength then was a currency for inner and outer legitimacy of both communism in general, and of the political Soviet Union system in particular. For good or bad, the leftovers of the former water technocracy are still highly influential on the performance and stability of the water sector nowadays. On the one hand, the remaining organizational structures and personal connections are a guaranteeing factor, possibly the only one, for the survival of the technical functionality of the complex irrigation system. On the other hand, the remaining structures are a severe obstacle for institutional reforms such as establishing water pricing, land ownership, or individual agricultural decisions. The soviet legacy is therefore an important factor in the persistence of viable institutions as well as being a severe hindrance to institutional change. Historical institutions must also be seen as a background for newer political decisions: the agricultural output of the Fergana Valley counts for 20 % of the state income of sovereign Uzbekistan, thereby relying on a system of water distribution which was stable only under the conditions of the former political union with Kyrgyzstan (which by now has other things in mind with the water desperately needed downstream for agriculture) and the agricultural system of a planned economy. Sustainable agricultural production was for ecological reasons – foremost the mismatch between water availability in the valley itself and the water greed of cotton – never achievable, and is unlikely in the future. Political decisions, of course, are intertwined with financial and economic aspects. As far as the new post-soviet economy is built on the ruins of an unsustainable agronomic revolution, it will be in the way of any capitalistic reform. But the intervention of capitalism, on the other hand, is one of leading demands of international donors such as the World Bank. Given the fact that former agricultural arrangements and the leftovers of soviet technology are the backbone of domestic economic income, any international intervention is doomed to be nothing other than a mock capitalist economy, severely hindered by soviet organizational and staffing structures and ideologies of the former republic.

Equally complex is the multifaceted context of the Red Sea – Dead Sea Project (RSDS). While it is certain that the shrinkage of the Dead Sea is severely impacting the surrounding landscape and seashore, the groundwater with salinization and depletion, and the economy and ecology of all riparian states; so far there are no agreements on the ecological and political outcomes of the project. Some promoters believe the technical feasibilities, others the goals – increasing and stabilizing water and power supply – are not realistic, nor are the potential damages to the environment foreseeable. Since the intervention is to take place in a highly variable climatological, hydrological, and geological surrounding, many factors should be taken into account, but there can be no certainty that the scientific models will provide an entirely complete picture of the consequences. Moreover, the potential benefits of the project would be unevenly distributed among the riparian states, thus increasing the danger of political turmoil, already one the most severe problems of the region. In the past lack of political agreement has already led to an underdevelopment of regulatory means: existing treaties are often ambiguous and weak when it comes to compliance control, adaption, and dispute settlement, and there is a little hope that the shortcomings of todays' regulations can be overcome within the framework of the ambitious new MWEP. It is obvious that to successfully achieve the set goals, if this is possible at all, very much depends on intervention that takes political, eco-logical, and social factors as much into account as the technical aspects of the endeavor.

This idea that, in general, systems operate within an environment draws our attention to another important factor of MWEPs: given that 263 of all river basins and more than 300 aquifers worldwide are transnational (UN-Water 2008), the oper-ation of MWEPs is often placed in contexts where all variables are not under the control of their management/operational teams. In case of the Fergana Valley, it is very clear that the Kyrgyz government and the needs of the Kyrgyz economy are not under control of the Uzbek bodies running the irrigation system in the valley, yet their influence on time and amount of water flow has major influence on the system. In the RSDS case, one its leading characteristics is political hostility fuelled by scar-city of available space and ecological resources for growing populations in the ripar-ian states which have little in common but a violent history and mutual distrust.

With their emphasis on polycentricity, public participation, a bioregional approach, and experimentation, governance literature and system theory both point to relevant factors beyond technical management. Dispersing institutional responsi-bility, ensuring participation, cooperating with neighboring states, and benefiting from experiments are hardly compatible with autocratic political environments. Indeed, all these findings from water governance literature call for a far more ambi-tious programme than just simply managing the technology of MWEPs; their call is for MWEPs to be embedded into democratic political environments.

Taking both approaches into account, the evaluation of MWEPs raises the following questions:

- Which relevant political, social, technical, economic, ecological, and interna-tional factors need to be taken into account in order to evaluate the implementa-tion and maintenance of a MWEP?
- What is the constitutional setting into which the MWEP will be implemented?
- Who takes part in decision making about the implementation and management of a MWEP? Are all shareholders involved and do they have an equal share in the decision-making progress?
- What are the political relations with neighboring states like in general? Are there legal means to resolve conflicts? Which established institutions are there to man-age transboundary issues associated with the project?
- Are there means to evaluate different approaches and identify good practices? Can they be transferred to, and how can they be adapted to another environment? Are local specifics taken into account in the transfer?

Technical solutions for water issues rely on political and institutional designs, which encourage individuals and personal networks to work in favor of them. No great technology – not even one which is well engineered – can fulfill its purposes satisfactorily if it is not supported and sustained by people and institutions. A *systemic view* helps to understand the interrelatedness of ecological needs, patterns of behavior, political and regulatory institutions, historical legacies, and economic and technological features. The *governance* perspective points to the necessity to create regulatory means which can address the complexity of motives both of individuals and collectives. *Institutional* thinking underlines the fact that immaterial rules and norms eventually *materialize*.

References

Anne I (2012) Water governance – challenges in Africa. Hydro-optimism or hydro-pessimism? Peter Lang, Bern/Berlin/Bruxelles

Bakker K (2010) Privatizing water. Governance failure and world's urban water crisis. Cornell University Press, Ithaca

Benz A (ed) (2004) Governance – Regieren in komplexen Regelsystemen. Eine Einführung. VS Verlag für Sozialwissenschaften, Wiesbaden

Brand U, Scherrer C (2003) Contested global governance. Konkurrierende Formen und Inhalte globaler Regulierung. Kurswechsel, Zeitschrift für gesellschafts-, wirtschafts-, und umweltpolitische Alternativen 1:90–103

Brand U, Brunnengräber A, Schrader L et al (2000) Global governance. Alternative zur neoliberalen Globalisierung. Westfälisches Dampfboot, Münster

Bundestag D (ed) (2002) Schlussbericht der Enquete-Kommission. Globalisierung der Weltwirtschaft – Herausforderungen und Antworten. Leske + Budrich, Opladen

Dietz T, Ostrom E, Stern PC (2003) The struggle to govern the commons. Science 302:1907–1912. doi:10.1126/science.1091015

Dingwerth K (2003) Globale Politiknetzwerke und ihre demokratische Legitimation: Analyse der World Commission on Dams. Global Governance Working Paper No. 6. www.glogov.org. Accessed 10 Apr 2015

Dobner P (2010) Wasserpolitik: Zur politischen Theorie, Praxis und Kritik globaler Governance. Suhrkamp, Frankfurt am Main

Easton D (1957) An approach to the analysis of political systems. World Polit 9:383–400. doi:10.2307/2008920

Easton D (1965) A systems analysis of political life. Wiley, New York

Folke-Schuppert G (2006) Governance-Forschung. Vergewisserung über Stand und Entwickungslinien, 2nd edn. Nomos, Baden-Baden

GWP (2003) Effective water governance: learning from the dialogues. Global Water Partnership, Stockholm

Hauff V (1987) Unsere gemeinsame Zukunft: Der Brundtland-Bericht der Weltkommission für Umwelt und Entwicklung. Eggenkamp Verlag, Greven

Hill W, Fehlbaum R, Ulrich P (1994) Organisationslehre 1: Ziele, Instrumente und Bedingungen der Organisation sozialer Systeme, 5th edn. UTP, Stuttgart

Huitema D, Mostert E, Egas W et al (2009) Adaptive water governance: assessing the institutional prescriptions of adaptive (co-)management from a governance perspective and defining a research agenda. Ecol Soc 14:26

Lebel L, Dore J, Daniel R, Koma YS (eds) (2007) Democratizing water governance in the Mekong region. Mekong Press, Chiang Mai

Mayntz R (2005) Governance Theory als fortentwickelte Steuerungstheorie? In: Folke-Schuppert G (ed) Governance-Forschung. Vergewisserung über Stand und Entwicklungslinien. Nomos, Baden-Baden, pp 11–20

Moss T, Dobner P (2015) Between multiple transformations and systemic path dependencies. In: Huettl RF, Bens O, Bismuth C, Hoechstetter S (eds) Society – water – technology: a critical appraisal of major water engineering projects. Springer, Dordrecht, pp 101–111

OECD (2011) Water governance in OECD countries: a multi-level approach. OECD Publishing, Paris

Ostrom E (1999) Die Verfassung der Allmende. Jenseits von Staat und Markt. Mohr Siebeck, Tübingen

Perkins P (ed) (2013) Water and climate change in Africa. Challenges and community initiatives in Durban, Maputo and Nairobi. Routledge, Oxon/New York

Sehring J (2009) The Politics of water institutional reform in neo-patrimonial states: a comparative analysis of Kyrgyzstan and Tajikistan. VS Verlag für Sozialwissenschaften, Wiesbaden

Skelcher C (2005) Jurisdictional integrity, polycentrism, and the design of democratic governance. Governance 18:89–110. doi:10.1111/j.1468-0491.2004.00267.x

United Nations (2006) The millennium development goals report 2006. United Nations, New York

United Nations (2007) The millennium development goals report 2007. United Nations, New York

United Nations (2008) Millenniums-Entwicklungsziele. Bericht 2008. United Nations, New York

United Nations (2009) The millennium development goals report 2009. United Nations, New York

United Nations (2012) The millennium development goals report 2012. United Nations, New York

UN-Water (2008) Transboundary waters: sharing benefits, sharing responsibilities. UN-Water, Geneva

Varela FG, Maturana RH, Uribe R (1974) Autopoiesis: the organization of living systems, its characterization and a model. Biosystems 5:187–196. doi: 10.1016/0303-2647(74)90031-8

von Bertalanffy L (1950) An outline of general system theory. Br J Philos Sci 1:134–165. doi:10.1093/bjps/I.2.134

Weaver W (1948) Science and complexity. Am Sci 36:536–544

Wiener N (1948) Cybernetics of control and communication in the animal and the machine. Herman Editions, Paris

Chapter 7
Research in Two Case Studies: Irrigation and Land Use in the Fergana Valley and Water Management in the Lower Jordan Valley

Christine Bismuth, Sebastian Hoechstetter, and Oliver Bens

Abstract This chapter aims to present an overview of the two case studies "Irrigation and land use in the Fergana Valley" and "Water management in the Lower Jordan Valley". It names the main criteria and reasons for their selection as cases in the study of implications of major water engineering projects. The political and cultural conditions in these regions and the major issues and problems concerning water management are outlined in broad terms.

Keywords Fergana Valley • Lower Jordan Valley • Central Asia • Irrigation • Land use • Water management • Major water engineering projects • Red Sea • Dead Sea Conveyance Project • Study of alternatives • World Bank

7.1 Selection of the Case Studies: A Wide Spectrum of Socio-Economic and Ecological Framework Conditions

There are many examples and varying types of major water engineering projects (MWEPs) over the globe. The range of related technologies, resource challenges, impacts and the political and governance systems that such projects are embedded into is equally wide. In order to cover as many practical and theoretical aspects and framework settings of MWEPs, we have decided to select two case studies: the Uzbek

C. Bismuth (✉)
Interdisciplinary Research Group Society - Water - Technology,
Berlin-Brandenburg Academy of Sciences and Humanities,
Jägerstraße 22/23, 10117 Berlin, Germany

Helmholtz Centre Potsdam - GFZ German Research Centre for Geosciences,
Telegrafenberg, 14473 Potsdam, Germany
e-mail: bismuth@gfz-potsdam.de

S. Hoechstetter • O. Bens
Helmholtz Centre Potsdam - GFZ German Research Centre for Geosciences,
Telegrafenberg, 14473 Potsdam, Germany

© The Author(s) 2016
R.F. Hüttl et al. (eds.), *Society - Water - Technology*, Water Resources
Development and Management, DOI 10.1007/978-3-319-18971-0_7

parts of the Fergana Valley in Central Asia and the Lower Jordan Valley in the Middle East, which entail a broad range of issues that are relevant to the comprehensive evaluation of major water engineering projects. Furthermore, the selection of the case studies reflects the competences and experiences of the contributors to this volume, which has improved the accessibility to information and data.

The case studies cover the most prominent issues being discussed throughout the world when it comes to the implementation of MWEPs, including issues of Integrated Water Resource Management (IWRM), water governance, the "food-water-energy nexus", ecosystem services, issues of transboundary water management, the challenges posed by climate change, political/governance instruments and societal transformation.

Both case studies are located in geopolitical hot spots, where water appears both as a source of conflict and as an occasion for cooperation. Each of the case studies reflects the weaknesses of the international law and regulations with concern to transboundary management issues. Both regions are affected by climate change with mainly negative consequences on water availability. Societal issues, political transformations and the consequences of historical decisions characterise current water management systems in the selected regions. Very different approaches are pursued in the case studies with regard to solutions for overcoming water shortages, securing energy and producing food. The regions illustrate the significant impact of state and governmental failures on the water management. In addition, the lessons learned so far both in the Fergana Valley and in the Lower Jordan Valley reveal that many problems of water management cannot be solved only within the water management sector in its own narrow definitions, but that they have to be tackled from a more general societal and economic perspective.

The Red Sea–Dead Sea (RSDS) Conveyance Project in the Middle East offers the opportunity to study a major project in its planning phase. In many countries other than Israel and Jordan respectively, seawater desalination will be a major strategy in the future for overcoming water shortages. Therefore, the Lower Jordan Valley case study may be perceived as being exemplary for countries under similar conditions. The Fergana Valley case study focuses on past decisions and the resulting path dependencies; this shows how consequences of technically driven decisions have to be dealt with by following generations.

Information and material on the case studies was collected with comprehensive literature research and analysis. In 2014, a research journey to the Fergana Valley was carried out by several authors of this volume with the support of the German Academic Exchange Service (DAAD). The central Fergana Valley irrigation systems and the role of the water users associations (WUAs) were in the focus of this research visit. Discussions with local stakeholders and decision makers and the exchange with researchers from Central Asia provided valuable insight into the regional situation.

With regard to the case study of the Lower Jordan Valley, an international workshop on the RSDS project with Israeli, Jordanian and Palestinian experts was organised in 2013. A research trip to Jordan was carried out for the preparation of this workshop, its results eventually being presented to and discussed with members of the Israel Academy of Sciences and Humanities. In addition, a number of Israeli, Jordanian and Palestinian experts have been interviewed.

The outcome of these extensive discussions, studies and meetings will be presented in Parts III and IV of this volume. In the following, we will provide a brief overview of the situation in the two case study regions and illustrate their similarities and differences.

7.2   The Fergana Valley

The Fergana Valley is situated in the "heart" of Central Asia. In ancient times, branches of the famous Silk Road crossed the valley. The valley is shared between Uzbekistan, Tajikistan and Kyrgyzstan. The Syrdarya River and some of its major tributaries cross the valley on their way to the Aral Sea (see Fig. 7.1). Its extension is limited by the surrounding mountains.

It is one of the most fertile regions in Central Asia and one of the oldest irrigation cultures worldwide. Already in the Neolithic Age, wheat and barley were watered by means of gravity irrigation. The historical extension of the irrigation is compa-

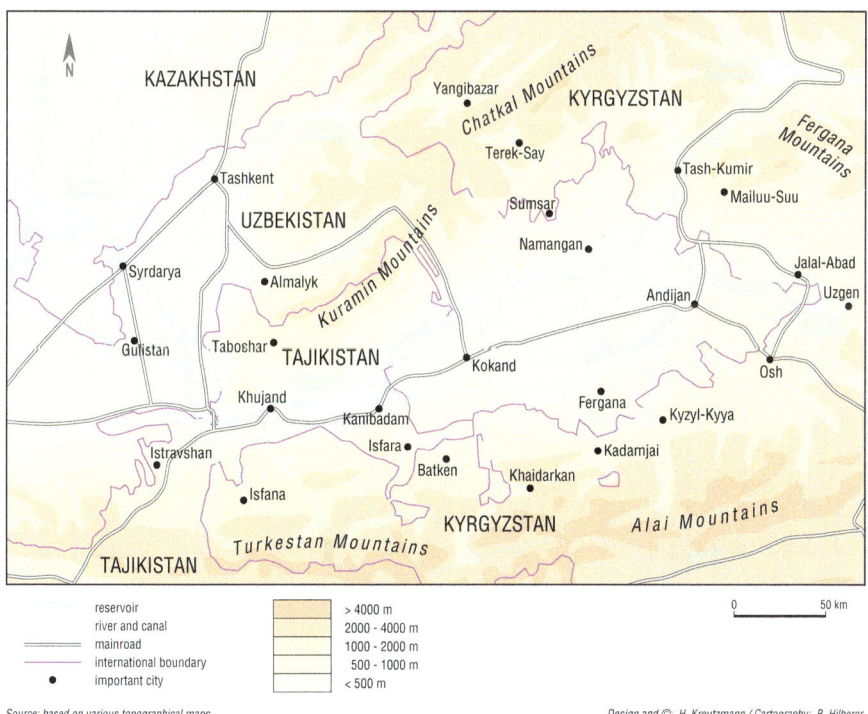

Fig. 7.1 Topographic map of the Fergana Valley (Source: Kreutzmann 2015, in this volume, pp. 113–127)

rable to the extension of present irrigation schemes (Francfort and Lecomte 2002). Before Russian colonisation, the valley was famous for its agricultural products such as melons, apricots, walnuts, wine and rice. Lucerne was the foundation for the valley's renowned horse breeding.

Russian colonisation in the nineteenth century is the starting point of radical changes and societal upheavals in the region referred to as "Turkestan". Turkestan, whose boarders were not precisely defined during the period of the Russian Empire, approximately comprises the current states of Turkmenistan, Kazakhstan, Kyrgyzstan, Uzbekistan and Tajikistan. The borders between these modern successors of Turkestan were established during the regime of Stalin on the basis of language and ethnicity. Today these borders are a source of conflicts and problems; as after all, the region remains intermingled, both in ethnical and economic terms.

The driving forces behind the Russian colonisation were Russia's rivalry with the British Empire and its need for cotton cultivation areas. Thus, the Central Asian agriculture was strategically relevant for the Russian Empire. From the early days of the colonisation period, cotton played a key role as an important feedstock. A central agrarian administration had been founded already in 1904. In the further course of time, a cotton production programme with a state-owned and controlled irrigation network was established. But cotton could only be successfully introduced in the Fergana Valley with the construction of railways for the transportation of the cotton to the production centres but also with the construction of factories for combing and pressing the cotton in large balls. From the remaining cotton grains, oil is produced for human consumption (Jozan 2012). It was the converging of local interests with those in the Russian textile production centres which led to the introduction of cotton in the Fergana Valley (*ibid*). But cotton could only be introduced by the extension of the irrigated area, as the other land was dedicated to support the needs of the local populations (*ibid*). Around 1910, the existing autochthone irrigation techniques could not use the waters of the Naryn and the Syrdarya which limited the extension of cotton (*ibid*). It was only in a later period with the construction of large-scale hydraulic works that the cotton production could be extended.

The agrarian production systems were collectivised in the 1930s. At later stages, radical changes were forced on the traditional oasis systems (Cariou 2004). Small agrarian units were merged into larger ones and traditional rural village communities were broken up. Old irrigation systems were levelled, and large irrigation channels were constructed for the benefit of a modern industrialised agricultural production with cotton as the main product.

As a consequence, the decades-long overuse of water resources, excessive cotton production and agrarian mismanagement have led to the drying up of the Aral Sea.

Along the Syrdarya, approximately 40 dams with a storage capacity of 37×10^9 m^3, representing 89 % of its flow capacity, were constructed between 1950 and 1980. During those years, cotton production was the main economic sector. Irrigation was more important than hydropower generation. The discrepancies already existing between hydropower and irrigation demands were moderated by the Soviet authorities (Jozan 2008).

With the collapse of the Soviet Union, the structural problems in the water management sector were aggravated, and the diverting interests of the individual states became evident. While the geopolitical setting of Central Asia has changed, the "old elites" still have considerable influence on the institutions of the water management sector.

The water balance, which was already fragile during the Soviet regime, has completely lost its equilibrium (Eschment 2011). Agreements that were supposed to regulate the management of water between the different states have not been respected and must be considered as failures (ICWC 2002). A line of division runs between the upstream states of the Syrdarya and Amudarya Rivers (Kyrgyzstan and Tajikistan) on the one hand and the downstream countries (Uzbekistan and Kazakhstan) on the other hand. The former are aiming at hydropower generation (particularly during the winter months) to cover their energy needs, while the latter require water mainly during the summer months to irrigate agricultural land. This relationship of dependence is illustrated in Fig. 7.2.

The hibernal water releases from the upstream dams impact the river banks and cause flooding (Jozan 2008). The climate change-induced retreat of the mountain glaciers will not only have considerable impact on water availability, but also on the seasonal distribution of the water resources. The dependency on precipitation as a source of (irrigation) water will increase. Climate change may therefore aggravate the already existing conflicts between the riparians.

The institutional settings, the role of the water users associations (WUAs) in Uzbekistan and the problems caused by the irrigation and drainage systems in the Fergana Valley constitute the focus of our analyses presented in Part III of this volume.

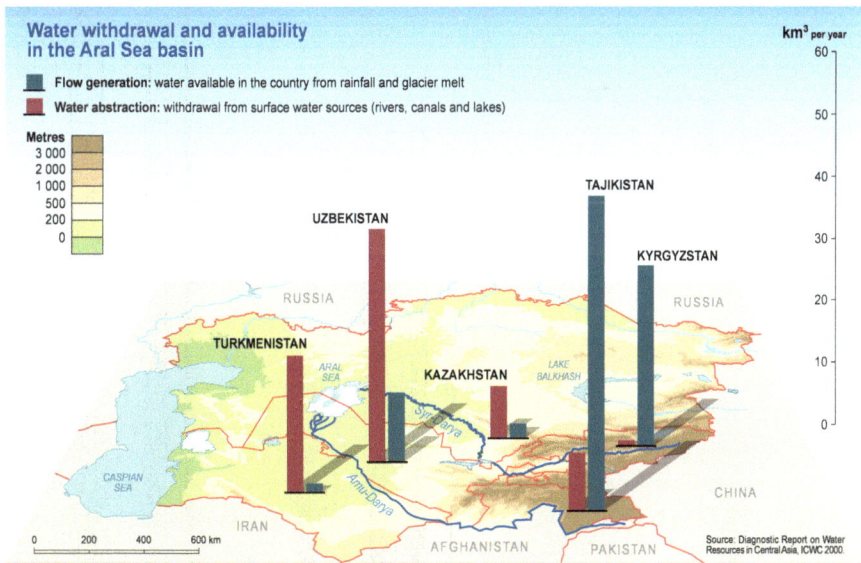

Fig. 7.2 Water withdrawal and availability in the Aral Sea Basin (Source: Rekacewicz and Novikov 2006, http://www.grida.no/graphicslib/detail/water-withdrawal-and-availability-in-aral-sea-basin_85d9)

The transformation processes and the resulting path dependencies are examined in Kreutzmann (2015, in this volume, pp. 113–127), while the institutional legacies, the reforms and the interstate cooperation after the collapse of the Soviet Union are studied in Moss and Dobner (2015, in this volume, pp. 101–111).

The complexity of the irrigation and drainage systems and the resulting limitations are focused on in Kenjabaev and Frede (2015, in this volume, pp. 129–148), while the socio-economic settings are analysed in Hansjürgens (2015, in this volume, pp. 169–186).

The water users associations are a central player in the valley's water management sector; therefore, their role and the reasons for their successes and failures are addressed in Moss and Hamidov (2015, in this volume, pp. 149–167).

7.3 The Lower Jordan River Basin and the Red Sea–Dead Sea Conveyance Project

The Lower Jordan River Basin stretches from Lake Tiberias (also known as Sea of Galilee, Kinneret or Lake of Gennesaret) to the Dead Sea, the final recipient of the Jordan River (Fig. 7.3). The Jordan River Basin is shared between Lebanon, Syria, Jordan, the Palestine Territories and Israel.

Water management is one of the major issues in the relationship between the people of this region. In the past, water has been both a source of conflict but also the opportunity for cooperation in one of the world's major political hot spots. All riparian states are confronted with an excessive use of the existing water resources, population increase and the impacts of climate change. The over-extraction of water – mainly for agricultural purposes – has led to a constant decline in the Dead Sea level of about 1 m/year. The consequences are the erosion of the coastal sea line, the formation of sinkholes and the salinisation of groundwater sources around the Dead Sea (see Fig. 7.4).

Lake Tiberias, the major water reservoir of the region, groundwater resources, and the River Jordan itself are also severely impacted by unsustainable water uses. And in the future, water stress will most probably increase.

In order to stop the further decline of the Dead Sea water level and to develop new water sources, Jordan and Israel have proposed that the World Bank finances an ambitious plan consisting of the desalination of water from the Red Sea and conveying the brine to the Dead Sea. The whole scheme is therefore referred to as the "Red Sea–Dead Sea (RSDS) Conveyance Project". The World Bank has agreed to carry out a feasibility study of the project – provided that the Palestine Territories become involved in the project. As a result of the political pressure of NGOs from the environmental sector, the "Study of Alternatives" (Allan et al. 2012) was conducted in addition to an environmental impact assessment.

The non-technical possibilities to restore the natural flow of the Jordan River as an alternative to the major technological solution pursued by the RSDS project, as well as the controversial role of the potash companies, which extract around 300×10^6 m³/year of water from the Dead Sea, were among the issues of greatest

Fig. 7.3 The Jordan River Basin (Source: UN-ESCWA and BGR 2013, p. 173)

Fig. 7.4 Pronounced erosion sinkholes (Photo: André Künzelmann, UFZ)

public concern and interest. There are environmental concerns regarding the likelihood of breakage or leakage of the conveying pipes, and the probable negative effects this would have on the Arrava groundwater layer. Further ecological threats could arise from the possible formation of gypsum and an algae blossom in the Dead Sea.

Shortly after the finalisation of the study of alternatives, Jordan, Israel and the Palestine Territories signed a memorandum of understanding (MoU) concerning a Red Sea–Dead Sea Conveyance Project as a first stage of a larger project. The MoU foresees a water swap between Israel and Jordan. In a first stage, a desalination plant at Aqaba (Jordan) will be built to produce 80×10^6 m³/year of freshwater, of which $30–50 \times 10^6$ m³/year will be sold to Israel, the remaining being used in Jordan. The brine is supposed to be pumped to the Dead Sea via a pipeline. Jordan will receive water from Beit Zera (Lake Tiberias) via the King Abdallah Canal to secure the water provision of Amman. The Palestine Territories will receive 30×10^6 m³/year from the desalination plants on the Israeli Mediterranean coast in return for money (Markel 2014).

Whatever solution will be chosen to overcome the water shortage in the region in the end, cooperation among the riparian states will be essential. With concern to the present political situation, the analysis in this volume will be restricted mainly to Jordan, the Palestine Territories and Israel, even though Syria and Lebanon have a stake in the region's water management as well. Bismuth (2015, in this volume, pp. 189–204) examines the institutional settings, the water management and the specific relationships between the main parties. The Red Sea–Dead Sea Conveyance Project and the planned water swap is studied by Malkawi and Tsur (2015, in this volume, pp. 205–225). Jordan's water management will be in the focus of the contribution of Yorke (2015, in this volume, pp. 227–251). Here, the social and institutional obstacles to an improved water management in Jordan are analysed. Bismuth et al. (2015, in this volume, pp. 253–275) focus their analysis on the technical and financial instruments which Israel has introduced to face the water crisis during the last decade.

7.4 Significance and Exemplary Importance of the Case Studies

The fates of the Aral Sea and the Dead Sea are both exemplary for the slow decline or even disappearance of many inland lakes under arid climate conditions. Both case studies are relevant exemplary cases of the most common water problems and the role of MWEPs – either as an instrument to solve some of those problems or as a cause of additional problems. The case studies are also examples for the effects of water projects that are mainly based on the advancements of hydrological engineering and on more or less sophisticated technological measures. Furthermore, the river ecosystems affected by these measures are fragile, complex and valuable, and are embedded into a many-faceted societal context characterised by opposed interests. We will focus not only on the frequently unpredictable effects of "technologies" on complex natural systems but also on the roles of the different stakeholders in the decision-making process. The case studies illustrate the insufficient dialogue and exchange between the different scientific disciplines involved, namely the social sciences on the one hand and the technical natural-science-based disciplines on the other. Stimulating such a dialogue and presenting new views on solutions and perspectives is one of the major objectives of our research.

This entire discourse is not only of relevance for water management projects – but rather, these findings can be transferred to all major engineering projects at the interface between ecosystems and societies. In view of the increasing numbers of planned MWEPs, some at a larger scale and extent than any other project to date, new approaches are needed, especially when it comes to secure future resources of food, water and energy.

References

Allan JA, Malkawi AIH, Tsur Y (2012) Red Sea – Dead Sea water conveyance study program. Study of alternatives. Preliminary draft report. World Bank Publications, Washington, DC

Bismuth C (2015) Cooperation and power asymetries in the water management of the Lower Jordan alley – the situation today and the path that has led there. In: Huettl RF, Bens O, Bismuth C, Hoechstetter S (eds) Society – water – technology: a critical appraisal of major water engineering projects. Springer, Dordrecht, pp 189–204

Bismuth C, Hansjürgens B, Yaari I (2015) Technologies, incentives and cost recovery: is there an Israeli role model? In: Huettl RF, Bens O, Bismuth C, Hoechstetter S (eds) Society – water – technology: a critical appraisal of major water engineering projects. Springer, Dordrecht, pp 253–275

Cariou A (2004) Le jardin saccagé. Anciennes oasis et nouvelles campagnes d'Ouzbékistan. Ann Geogr 113:51–73

Eschment B (2011) Wasserverteilung in Zentralasien. Ein unlösbares Problem? Friedrich Ebert Stiftung, Berlin

Francfort H-P, Lecomte O (2002) Irrigation et société en Asie centrale des origines à l'époque achéménide. Ann Hist Sci Soc 57:626–663. doi:10.3406/ahess.2002.280068

Hansjürgens B (2015) Theory, market and the state: agricultural reforms in post socialist Uzbekistan between economic incentives and institutional obstacles. In: Huettl RF, Bens O, Bismuth C, Hoechstetter S (eds) Society – water – technology: a critical appraisal of major water engineering projects. Springer, Berlin/Dordrecht, pp 169–186

ICWC (2002) II. Diagnostic report on water resources in Central Asia. http://tajikwater.net/docs/diagnostic_reportwater_080719.pdf. Accessed 17 Feb 2015

Jozan R (2008) "État délinquant" ou modèle déviant? Retour sur le non-respect du traité international de partage de la ressource en eau du Syr Darya. Flux 1:46–60

Jozan R (2012) Les débordements de la Mer d'Aral. Presses Universitaires de France, Paris

Kenjabaev SM, Frede H-G (2015) Irrigation infrastructure in Fergana today: ecological implications – economic necessities. In: Huettl RF, Bens O, Bismuth C, Hoechstetter S (eds) Society – water – technology: a critical appraisal of major water engineering projects. Springer, Dordrecht, pp 129–148

Kreutzmann H (2015) From upscaling to rescaling – the Fergana Basin's tranformation from Tsarist irrigation to water management for an independent Uzbekistan. In: Huettl RF, Bens O, Bismuth C, Hoechstetter S (eds) Society – water – technology: a critical appraisal of major water engineering projects. Springer, Dordrecht, pp 113–127

Malkawi A, Tsur Y (2015) Reclaiming the Dead Sea: alternatives for action. In: Huettl RF, Bens O, Bismuth C, Hoechstetter S (eds) Society – water – technology: a critical appraisal of major water engineering projects. Springer, Dordrecht, pp 205–225

Markel D (2014) Viewpoint: the best alternative. In: The Jerusalem Post. 13. Januar. http://www.jpost.com/Jerusalem-Report/The-Region/Viewpoint-The-best-alternative-337129. Accessed 9 Apr 2015

Moss T, Dobner P (2015) Between multiple transformations and systemic path dependencies. In: Huettl RF, Bens O, Bismuth C, Hoechstetter S (eds) Society – water – technology: a critical appraisal of major water engineering projects. Springer, Dordrecht, pp 101–111

Moss T, Hamidov A (2015) Where water meets agriculture: the ambivalent role of the water users associations (WUAs). In: Huettl RF, Bens O, Bismuth C, Hoechstetter S (eds) Society – water – technology: a critical appraisal of major water engineering projects. Springer, Dordrecht, pp 149–167

Rekacewicz P, Novikov V (2006) Water withdrawal and availability in the Aral Sea Basin. In: UNEP/GRID Arendal. http://www.grida.no/graphicslib/detail/water-withdrawal-and-availability-in-aral-sea-basin_85d9. Accessed 3 Mar 2015

UN-ESCWA, BGR (2013) Inventory of shared water resources in Western Asia: Chapter 6 Jordan River Basin. United Nations Economic and Social Commission for Western Asia, Federal Institute for Geosciences and Natural Resources, Beirut

Yorke V (2015) Jordan's shadow state and water management: prospects for water security will depend on politics and regional cooperation. In: Huettl RF, Bens O, Bismuth C, Hoechstetter S (eds) Society – water – technology: a critical appraisal of major water engineering projects. Springer, Dordrecht, pp 227–251

Part III
The Fergana Valley – Uzbekistan's Hydro-Agricultural System Between Inertia and Change

Chapter 8
Between Multiple Transformations and Systemic Path Dependencies

Timothy Moss and Petra Dobner

Abstract This introductory section on water management in the Fergana Valley makes the case for viewing this major water engineering project (MWEP) in terms of two core positions: the interdependency of complex factors at play and the coexistence of forces for change and obduracy. We argue, firstly, that water management in the Fergana Valley is inextricably tied up with agriculture policy and practice, outlining how the region's irrigation system is predicated upon post-Soviet agriculture. We illustrate, secondly, how this relationship is shaped not only by powerful path dependencies – in the shape of physical structures, sunk costs and institutional arrangements – but also by changes, both radical and incremental, in response to system failure, shifting political preferences or the emergence of viable alternatives. We conclude by setting the stage for the subsequent detailed analyses of selected arenas critical to the development of Fergana Valley's irrigation system.

Keywords Fergana Valley • Syrdarya River • Uzbekistan • Aral Sea • Irrigation • Post-Soviet agriculture • Cash crop • Path dependency • Transformation • Major water engineering projects

8.1 Post-Soviet Transformation as a Multi-dimensional, Long-Term Process

Any study of water management in the Fergana Valley begins with a litany of problems associated with its immense and intricate irrigation system. Which problems are selected, how they are presented and what solutions are proffered varies significantly,

T. Moss (✉)
Leibniz Institute for Regional Development and Structural Planning (IRS),
Flakenstraße 28-31, 15537 Erkner, Germany
e-mail: MossT@irs-net.de

P. Dobner
Institute of Political Science and Japanese Studies, Martin-Luther-Universität
Halle-Wittenberg, Emil-Aberhalden-Str. 7, 06099 Halle/Saale, Germany
e-mail: petra.dobner@politik.uni-halle.de

© The Author(s) 2016
R.F. Hüttl et al. (eds.), *Society - Water - Technology*, Water Resources
Development and Management, DOI 10.1007/978-3-319-18971-0_8

101

however, according to the author's perspective. For some, the main problem is how to keep the existing irrigation system operating in order to maintain current modes of intensive agricultural production, focussing on the collapse of effective irrigation management on farms following de-collectivisation and engineering solutions to improve water use efficiency (Dukhovny et al. 2009; Karimov et al. 2012). For others, the main problem is the transnational dispute between the riparian states of the Syrdarya River surrounding the allocation of water from the upstream reservoirs which, after over 20 years of negotiation, has reached an impasse, prompting each nation state to pursue its own, second-best option (Weinthal 2002; Megoran 2004; Wegerich et al. 2012). For others still, the core issue is about strengthening the hand of water users – i.e. farmers – in managing irrigation systems on the ground as a means of improving water-use practices and challenging the authoritarian and bureaucratic procedures of current decision-making (Abdullaev et al. 2009; Gunchinmaa and Yakubov 2010; Abdullaev et al. 2010; Dukhovny et al. 2013). A fourth group of authors targets the dire ecological impacts of the existing irrigation system and its management, notably the Aral Sea catastrophe, deploring the absence of environmental issues in debates on the future of the irrigation system in the region (Spoor 1998; White 2013). This categorisation is by no means mutually exclusive or exhaustive. Further problems confronting the Fergana Valley irrigation system include rapid population growth, negative impacts of climate change, land degradation, infrastructure disrepair, increased crop competition, dependence on donor support, the emigration of specialists and an authoritarian political regime (Dukhovny et al. 2009; Abdullaev et al. 2009; Gunchinmaa and Yakubov 2010; Abdullaev and Atabaeva 2012).

None of these problems exists in isolation. They are each – to a greater or lesser degree – connected with one another. Local decision-making on water allocation is, for instance, powerfully shaped by crop production quotas set by central government. To take another example, the huge scale of the valley's irrigation system has a significant bearing on its hierarchical governance structures. Any attempt to analyse the challenges confronting the management of this major water engineering project (MWEP) must address the interdependency of the complex factors at play. These cannot be limited to the technical or physical aspects of hydrological engineering or plant science but must consider also the social, economic, political and cultural dimensions of the land-water-food nexus, as well as its environmental aspects, notably biodiversity and ecosystem services. This is why the authors of this book propose to conceive of such MWEPs as complex socio-technical, social-ecological and political-economic systems or configurations. Identifying the interdependencies between the components of such systems and how they influence one another is key, we believe, to revealing potential responses to the challenges posed. This is particularly pertinent regarding the connectivity between the water management regime, land reforms and agricultural production in Uzbekistan. Attempting to resolve irrigation problems by focusing solely on water management issues and relegating agricultural policy and practices to background factors misses the point. Appreciating the interconnectivity of the multiple components constituting a MWEP such as the Fergana Valley irrigation system is, therefore, the first core position we take in this chapter.

Our second core position is that a MWEP of this kind is shaped by a combination of forces for both change and obduracy. On the one hand, large technical systems, such as irrigation networks, are renowned for their path dependency (Pierson 2000; Melosi 2005; Sehring 2009). Not only their physical structures, with their high sunk costs and material embeddedness, limit future options for adaptation; also the institutional arrangements developed over time to manage them reveal typically a high degree of persistence. On the other hand, large technical systems can and do change, sometimes quite dramatically, for instance in response to system failure, shifting political preferences or the emergence of viable alternatives (Summerton 1994). What this means for the MWEP in the Fergana Valley is that we need to identify the path dependencies (technical, environmental, organisational, economic, etc.) working to sustain the irrigation system as it is, but also the forces at play which are pushing for change, whether radical or incremental. For, as a historical perspective on the MWEP reveals, Fergana Valley's irrigation system may have been in many ways path dependent, but it was never static, adapting constantly to shifts in political regime, economic strategy, environmental limitations and geopolitical disputes.

These shifts became more sudden and disruptive with the collapse of the Soviet Union and emergence of independent Central Asian states out of the former socialist republics. Here, though, we need to be wary of over-interpreting the significance of this single act of nation-building for subsequent water and agricultural policy. The sudden switch of political regime in 1991 is deceptively suggestive of an abrupt system transformation reaching down into all walks of life. In reality, the radical transformation of some dimensions of society was accompanied by the strong path dependence of others. The prior configuration of technical, economic, organisational and social components of Fergana Valley's irrigation system was certainly destabilised after 1991, but not discarded. What we have been witnessing ever since is the continuous process of reordering this relationship. This process is not happening automatically, following some hidden hand of economic rationality, but in a non-linear, unpredictable and context-dependent way in response to shifting conditions and human agency. The multiple stakeholders involved – from senior water officials in central government agencies to household farmers living off secondary crops – are each trying to cope with this combination of transformation and path dependency in their own ways to advance their own interests. Given the stakes involved, this process is highly contested.

In the remainder of this section, we outline the principal dimensions of transformation and path dependency for first the agriculture sector and then the water sector, highlighting their interdependency. We conclude by setting the stage for the subsequent detailed analyses of selected arenas critical to the development of Fergana Valley's irrigation system.

8.2 Transformation and Path Dependencies in Fergana Valley's Post-Soviet Agriculture

The principal transformation of agriculture in Uzbekistan since independence relates to land ownership on the one hand. Land reform was not immediate after 1991, as in the case of the Kyrgyz Republic, but more gradual (Hamidov 2007). Nonetheless, its effects were far-reaching, including on the irrigation system (see below). Land reform in Uzbekistan began during the 1990s with the dissolution of the large-scale collective farms (*kolkhozes*) and state farms (*sovkhozes*) following their bankruptcy as a result of declining state subsidies and investment (Spoor 2004). They were replaced at first by smaller semi-cooperative farms (*shirkats*), still maintaining Soviet-style management. In the early 2000s, a further phase of land reform replaced the *shirkats* with individual farms, a process completed by 2006 (Gunchinmaa and Yakubov 2010). As land was transferred by long-term lease to individuals, a situation was created in which a huge number of people became small landowners and farmers who often had little experience or expertise in agriculture. The resulting problems for agricultural production prompted a fresh round of land reform with the aim of reducing the number of small, inefficient farms. Within the space of just 1 year, between 2006 and 2007, the number of individual farmers in Uzbekistan was reduced from 200,000 to 100,000.

On the other hand, today's agriculture system in Uzbekistan is marked by deep-rooted path dependency relating in particular to state planning of agricultural production and the prioritisation of cash crops. Despite land reforms, state production quotas for cotton and wheat continue to dominate the system (Abdullaev et al. 2009). For these two cash crops, regarded by the government as essential for national (food) security, the state sets quotas for production, just as in Soviet times. The national targets are sub-divided into targets for the individual provinces and districts, with the head of each administrative district being held personally responsible for meeting his/her target. This creates enormous pressure to fulfil the state plan, pressure which is passed on down the line to the individual farmer. For these cash crops, the farmer is further bound by the obligation to buy seed, pesticides and fertiliser from the state at a certain time and to sell cotton and wheat he/she produces to the state at a fixed price below market prices. From these production quotas, water needs are calculated and a plan for water allocation from national to on-farm level devised (see 8.3). Beyond the production of cotton and wheat, farmers are permitted to grow other crops, such as rice, vegetables or fruit, but this is only feasible if it does not endanger meeting the cash crop quotas. In practice, there is generally too little time and too little water left after harvesting the cash crops, rendering a second harvest difficult. Those private farms that specialise in products beyond cotton and wheat are responsible for financing all investments and marketing their goods, yet often receive only poor quality water or none at all (Gunchinmaa and Yakubov 2010).

In effect, there exists a dual agricultural structure characterised by a dominant, yet inefficient state-driven system of cash-crop production on the one hand and a much smaller, but emergent system of household plots and secondary harvests producing crops like maize, sorghum, rice, sunflower, vegetables and fruit on the other. In recognition of the higher productivity and market value of these products, there has recently been a modest shift in agricultural policy away from cotton not only to wheat but also to rice and vegetables. These products require different cropping patterns and, by consequence, different irrigation practices. This is creating new conflicts over water distribution between the harvests of cotton, winter wheat and secondary crops, revealing also serious limitations to the existing irrigation system (Kenjabaev 2014).

8.3 Transformation and Path Dependencies in Fergana Valley's Post-Soviet Irrigation System

Water management in Uzbekistan has undergone transformation in a number of ways since independence, most of them in the form of organisational restructuring to meet new challenges. The immediate change following the dissolution of the USSR was to set up a trans-boundary central organisation – the Interstate Coordination Water Commission (ICWC) – in 1992 to coordinate water management between the new riparian states. This body replaced the central Soviet government as the head of a hierarchy of water authorities reaching from the national to the local level. These comprise today, for the Fergana Valley, the transnational Basin Water Organisation Syrdarya responsible for implementing ICWC policy, the national government with – above all – the Ministry of Agriculture and Water Resources, sub-basin water management organisations at the regional level and water users associations (WUAs) responsible for on-farm irrigation.

Originally, irrigation was organised around the territorial jurisdictions of provinces, districts and local authorities. In 2003, in accordance with a core principle of Integrated Water Resources Management (IWRM), the organisation structure of irrigation was reordered spatially around hydrographic boundaries, i.e. corresponding to the spatial reach of an irrigation system or sub-system (Dukhovny et al. 2009). In Uzbekistan as a whole, ten basin irrigation systems authorities were established and, under their authority, 63 irrigation system authorities and main canal authorities. In the Fergana Valley basin and sub-basin, authorities were created for the Fergana Canal and its sub-canals. This organisational innovation is judged to have proved remarkably successful in minimising conflicts in the region over water allocation (Abdullaev et al. 2009). In particular, disputes between the upstream province of Andijan and the downstream province of Fergana relating to water use along the South Fergana Canal have been resolved earlier and more effectively since the replacement of provincial with irrigation system-based authorities. The new hydrographic boundaries are, though, only one reason for this improvement in inter-provincial relations. A second reason has been the creation of a new participatory governance structure, comprising nine Unions of Water Users along the South Fergana Canal, which represent the interests of water users associations in the Canal Water Committee (Abdullaev et al. 2009).

The creation of water users associations (WUAs) across the country is the third significant organisational transition after the creation of the ICWC and the reordering of irrigation management around hydrographic boundaries. Given the strong association of WUAs globally with the IWRM paradigm, it is tempting to interpret their creation in Uzbekistan in the early 2000s primarily as an instrument of IWRM. In reality, they were introduced by central government decree in response to an overwhelming crisis in managing on-farm irrigation systems following the collapse of the collective farms. Here we note a dramatic example of connectivity between land reform and irrigation management and, on a broader level, of the embedment of agricultural structures, problems and development potentialities in political power relations, financials needs of the state and modes of political decision making. In the initial phases of land reform, little thought was given to the impacts it might have on irrigation. On-farm irrigation and drainage infrastructure, formerly managed and maintained by staff of the collective farms, had no one responsible for them once these were disbanded (Abdullaev et al. 2010). Water distribution became an issue of contestation and competition amongst the new land-owning units. Many trained and experienced engineers, irrigators and agronomists left for Russia or Kazakhstan. The combined effect of this institutional void and exodus of skilled personnel was a catastrophic drop in the water efficiency of irrigations systems. In the Fergana Valley, losses were reported of up to 55 % of irrigation water supply (Dukhovny et al. 2009). This was the principal reason for the sudden creation of WUAs: to fill an institutional gap left by land reform. Within the space of a few months, on the back of a central government decree, thousands of WUAs emerged. The nature of their emergence is indicative, however, of many of their subsequent shortcomings. Firstly, the government was more interested in creating new organisations than making them viable or fit-for-purpose, with the result that, without adequate funding, most WUAs exist on paper only (Wegerich 2006; Abdullaev et al. 2010). Secondly, being the product of top-down bureaucratic policy, most WUAs do not enjoy the support of their members (see Sehring 2009; Abdullaev et al. 2010). Most water users regard their WUA as the extended arm of state water authorities rather than the self-governing bodies to which they normally ascribe. Many commentators see WUAs in Central Asia as inactive and not financially viable (Wegerich 2000, 2006; Abdullaev et al. 2010).

A fourth important source of innovation in the field of irrigation is undoubtedly the IWRM Fergana Project run by the Scientific and Information Centre of the ICWC and the International Water Management Institute (IWMI) and funded by the Swiss Development Cooperation (SDC). This transnational project, covering the Fergana River Basin in three countries (Kyrgyz Republic, Uzbekistan and Tajikistan), ran from 2001 to 2012 with the aim of improving the efficiency of water resources management by introducing IWRM principles and practice (Dukhovny et al. 2009, 2013). The measures promoted include supporting the reordering of irrigation management around hydrographic boundaries (see above), the increased participation of water users in decision-making bodies (as via the Unions of Canal Users, cited above), training and education for hydro-engineers and farmers (see Sect. 8.4) and research on improved irrigation techniques. Those involved in the project credit it with helping to reduce water used in the region by 12–25 %, increase the equity and stability of water delivery in South Fergana and improve allocation between upstream and downstream users, thereby dramatically reducing the number of conflicts between canal administrations and WUAs, between WUAs and between WUAs and farmers (Dukhovny et al. 2013). They concede, however, that the IWRM Fergana project did not pay sufficient attention to issues of governance, ecology and climate change (Dukhovny et al. 2014). The IWRM Fergana project can be credited, though, with setting an important example of what can be changed with the necessary political will and financial resources.

Parallel to these transformative factors, the irrigation system of the Fergana Valley is characterised by multiple dimensions of path dependency. First, and visually most obvious, is the path dependency of the physical infrastructures which transport water from massive reservoirs upstream to individual farms in the valley. This complex and hierarchical network of rivers, reservoirs, weirs, major canals, sub-canals, pipes, valves and furrows has been built up over the past 70 years and is deeply embedded – literally and metaphorically – in the landscape of the Fergana Valley. Its physical existence and size alone limits the options for using water in the region because alternative distribution systems of such dimensions are simply not viable. For example, the existing irrigation system is proving ill-equipped to adapt to the increasing diversity of crops beyond cotton, such as rice, wheat and vegetables, which each require different irrigation regimes (Abdullaev et al. 2010). Alterations to the infrastructure system do not challenge its underlying logic of securing adequate water supply for agriculture, but merely equip it to meet new challenges and adapt to technological advances. The automated management system at Ush Qoron is an illustration of how a major weir used for directing water along different canals has been upgraded constantly since the 1970s to address shifting demands and respond to unpredictable water restrictions from the upstream Toktogul reservoir in the Kyrgyz Republic.

The second dimension of path dependency relates to organisational procedures of water allocation. Water for irrigation is distributed, as during Soviet times, according to a central planning system. Data on water needs may be collected at the local level and fed up the system hierarchy, but it is the transnational ICWC which determines overall water quotas for the riparian states. These are then passed down the multilevel structure of the water bureaucracy, from the national Ministry of Agriculture and Water Resources, via the Central Water Management Administration, the 10 basin irrigation authorities, the 63 irrigation system authorities and main canal organisations down to the WUAs and the farmers (Dukhovny et al. 2013). This system allows for continuous intervention by national water authorities in water allocation at subordinate levels, not only under "normal" circumstances but also in emergency situations of water shortage when, for instance, a basin irrigation organisation can set limits to the production of cotton and wheat or prohibit a second harvest (Dukhovny et al. 2013). Such emergency measures are unusual in that they are a rare example of water management taking a priority over agricultural production.

The top-down procedure for allocating water for irrigation is indicative of more deep-rooted, institutional dimensions of path dependency of water management institutions in Uzbekistan and Central Asia generally (Sehring 2009; Abdullaev et al. 2009, 2010). Soviet and even pre-Soviet patterns of behaviour, professional norms and routinised practices are still highly influential in shaping the responses of water authorities to new challenges (Sehring 2009). Many new regulations and organisations created in the name of IWRM are undermined in practice by the prevalence and persistence of such informal institutions. Representatives of new organisations are often not integrated into decision-making processes and have weak mandates (Schlüter et al. 2010). There is also considerable continuity of leading personnel in water authorities at different levels, from the national ministries via the basin water organisations to provincial departments, whose biographies stretch back to the Soviet era. This "hydrocracy", or hydraulic bureaucracy, has remained a constant across regime change and organisational restructuring. Water authorities in the Fergana Valley alone can count on 7000 employees – many of them trained in the Soviet era – to implement state water policy (Interview Swiss Agency for Development and Cooperation). The number of water bureaucrats has, however, been substantially reduced since the Soviet period. Indeed, Abdullaev argues that, as a result of severe budget cuts, reduced staff levels and poor on-farm policy implementation, water officials no longer command the respect or play the role they once did in the heyday of Soviet irrigation engineering, when they were the celebrated heroes of the hydraulic mission of modernisation (2012, p. 103). There is, therefore, ambivalence to the modern "hydrocracy": whilst striving to improve the efficiency of irrigation systems it is reinforcing the path dependency of not only the physical infrastructure but also its own raison d'être.

A fourth dimension of path dependency to water management is distinctive by its absence: consideration of environmental impacts. Ecology was never an issue of great concern to irrigation and agricultural production in Soviet times and this has changed little since independence. International shock at the ecological collapse of the Aral Sea is barely reflected in the national debate on the future of agriculture and irrigation. Restoring the Aral Sea has *de facto* been given up by the authorities as a lost cause. In face of overwhelming environmental problems and limited prospects for transnational solutions, resignation has set in (White 2013). Even on a local or regional scale there is little discussion, for instance of the need for minimum environmental flows to sustain ecosystem functions. There is no sign that ecosystem water needs are playing an integral part in water allocation planning in Uzbekistan (Schlüter et al. 2010).

8.4 Researching the Fergana Valley: Tasks and Topics

What role can research play in helping to resolve, or – at least – cope with, these complex challenges? Given the legacy of scientific endeavour in this field since Soviet days and the wealth of scholarly attention paid to water management in Central Asia in the years since independence, this is not a trivial question. In the face of what to many observers appears an intractable situation research can first of all raise understanding of the processes outlined above, querying assumptions and revealing inconsistencies of existing knowledge. On this basis, it can devise alternative options, mapping out potential futures and extending the boundaries of debate. From different disciplinary perspectives, it can support initiatives to shape change, identifying promising points of entry and potential pitfalls as well as accompanying projects through implementation with expertise and critical reflection.

The sections in this chapter all represent attempts to make sense out of the past, present and potential future of the MWEP that is Fergana Valley's irrigation system. They share a common concern to explore the interdependencies of the region's regimes for agricultural production and water management, the first core position of this chapter. Here, the interest lies in revealing not only how each policy field affects the other, but also what institutional misfits between the two – in terms of their differing organisational structures, incentive mechanisms and power relations – make it so difficult to work across them effectively. In line with our second core position, the sections all address the relationship between transformation and path dependency. Transformation is treated as a continuous process of radical and incremental change, rather than a sudden shift. Path dependency is deprived of any deterministic assumptions. Unpacking the emergent hybrids of the old and the new is the task at hand.

The topics of these in-depth analyses have been selected by virtue of what we regard as their critical importance to the future of Fergana Valley's irrigation system and to prospects for making this future more sustainable. In the first piece, Hermann Kreutzmann explores issues of geographical interdependence and rescaling emerging from attempts at inter-state cooperation since 1991 and how the riparian states are responding to the geopolitical impasse. In the following section, Shavkat Kenjabaev and Hans-Georg Frede study the region's irrigation infrastructures in terms of the environmental impacts they are having, the economic conditions they require and the political constraints they are facing. Timothy Moss and Ahmad Hamidov then focus in on the region's WUAs to analyse how far they are fulfilling the multiple expectations vested in them and what the prospects are for giving water users a greater say in the management of Fergana Valley's irrigation system. Finally, Bernd Hansjürgens investigates the potential for economic incentives to encourage more efficient use of irrigation water and crop production and the institutional obstacles to implementing them in the region.

Field Trips and Discussions

1. Central Office of Fergana Irrigation District, Fergana, 4 May 2014
2. Office of South Fergana Canal Management, 5 May 2014
3. WUA Kadyrjon Azamjon, 5 May 2014
4. Swiss Agency for Development and Cooperation, Tashkent, 8 May 2014
5. Friedrich-Ebert-Stiftung, 8 May 2014

References

Abdullaev I, Atabaeva S (2012) Water sector in Central Asia: slow transformation and potential for cooperation. Int J Sustain Soc 4:103–112. doi:10.1504/IJSSOC.2012.044668

Abdullaev I, Kazbekov J, Manthritilake H et al (2009) Participatory water management at the main canal: a case from South Ferghana canal in Uzbekistan. Agric Water Manag 96:317–329. doi:10.1016/j.agwat.2008.08.013

Abdullaev I, Kazbekov J, Manthritilake H et al (2010) Water user groups in Central Asia: emerging form of collective action in irrigation water management. Water Resour Manag 24:1029–1043. doi:10.1007/s11269-009-9484-4

Dukhovny V, Sokolov V, Manthrithilake H (2009) Integrated water resources management: putting good theory into real practice. Central Asian Experience. Scientific Information Center of the Interstate Commission for Water Coordination (SIC ICWC), Tashkent

Dukhovny VA, Sokolov VI, Ziganshina DR (2013) Integrated water resources management in Central Asia, as a way of survival in conditions of water scarcity. Quat Int 311:181–188. doi:http://dx.doi.org/10.1016/j.quaint.2013.07.003

Dukhovny V, Yakubov S, Kenjabaev SM (2014) Role of drainage water management in irrigated agriculture of Central Asia. In: Proceedings of the 12th ICID international drainage workshop, "drainage on waterlogged agricultural areas", St. Petersburg, 23–26 June 2014

Gunchinmaa T, Yakubov M (2010) Institutions and transition: does a better institutional environment make water users associations more effective in Central Asia? Water Policy 12:165–185. doi:10.2166/wp.2009.047

Hamidov A (2007) Water user associations in Uzbekistan: review of conditions for sustainable development. Working paper for the World Bank Institute (WBI). Washington, DC

Karimov A, Molden D, Khamzina T et al (2012) A water accounting procedure to determine the water savings potential of the Fergana Valley. Agric Water Manag 108:61–72. doi:http://dx.doi.org/10.1016/j.agwat.2011.11.010

Kenjabaev SM (2014) Ecohydrology in a changing environment. Dissertation, Justus-Liebig-Universität Gießen

Megoran N (2004) The critical geopolitics of the Uzbekistan–Kyrgyzstan Ferghana Valley boundary dispute, 1999–2000. Pol Geogr 23:731–764. doi:http://dx.doi.org/10.1016/j.polgeo.2004.03.004

Melosi MV (2005) Path dependence and urban history: is a marriage possible? In: Schott D, Luckin B, Massard-Guilbaud G (eds) Resources of the city: contributions to an environmental history of modern Europe. Ashgate Publishing Limited, Aldershot, pp 262–275

Pierson P (2000) Increasing returns, path dependence, and the study of politics. Am Polit Sci Rev 94:251–267. doi:10.2307/2586011

Schlüter M, Hirsch D, Pahl-Wostl C (2010) Coping with change: responses of the Uzbek water management regime to socio-economic transition and global change. Environ Sci Pol 13:620–636. doi:10.1016/j.envsci.2010.09.001

Sehring J (2009) Path dependencies and institutional bricolage in post-soviet water governance. Water Altern 2:61–81

Spoor M (1998) The Aral Sea Basin crisis: transition and environment in former soviet Central Asia. Dev Chang 29:409–435. doi:10.1111/1467-7660.00084

Spoor M (2004) Agricultural restructuring and trends in rural inequalities in Central Asia. A socio-statistical survey. United Nations Research Institute for Social Development, Geneva

Summerton J (1994) Introductory essay: the systems approach to technological change. In: Summerton J (ed) Changing large technical systems. Westview Press, Boulder, pp 1–21

Wegerich K (2000) Water user associations in Uzbekistan and Kyrgyzstan: study on conditions for sustainable development, Occasional paper no. 32. Water Issues Study Group, London

Wegerich K (2006) "Illicit" water: un-accounted but paid for. Observations on rent-seeking as causes of drainage floods in the Lower Amu Darya. Paper presented at the CERES-IWE research seminar "News from the water front alternative views on water reforms", Utrecht, 8 November 2006

Wegerich K, Kazbekov J, Lautze J et al (2012) From monocentric ideal to polycentric pragmatism in the Syr Darya: searching for second best approaches. Int J Sustain Soc 4:113–130. doi:10.1504/IJSSOC.2012.044669

Weinthal E (2002) State making and environmental cooperation: linking domestic and international politics in Central Asia. MIT Press, Cambridge

White KD (2013) Nature–society linkages in the Aral Sea region. J Eur Stud 4:18–33. doi:http://dx.doi.org/10.1016/j.euras.2012.10.003

Chapter 9
From Upscaling to Rescaling: Transforming the Fergana Basin from Tsarist Irrigation to Water Management for an Independent Uzbekistan

Hermann Kreutzmann

Abstract The Fergana Valley is regarded as one of the most fertile irrigated oases in Asia. The genesis of these highly productive agricultural lands is the result of a lengthy process that originated long before the Kokand Khanate controlled most of the valley. Major transformations occurred during Tsarist Russian rule when the upscaling of this irrigated land commenced and when Fergana was integrated into long-distance exchange networks. Major water works were planned, but only implemented in a massive fashion during Soviet rule with its major campaigns for modernisation and planning on a large scale. Cross-border management was established in the highly integrated water scheme at the same time. Rescaling is a challenge of contemporary times with the pressures of globalisation, independent Uzbekistan's dependence on water supplies from its neighbours who share the Fergana Valley's hydraulic resources and the conditions governing the cotton world market.

Keywords Post-socialist transformation • Modernisation • Path-dependent development • Central Asia • Aral Sea • Fergana Valley • Water management • Water conflicts • Major water engineering projects

9.1   Introduction

> The waters of the Oxus and Yaxartes still flow into Lake Aral. The great task of the future will provide work until the last drops of these rivers have been diverted to agriculture. Realize this: not until Lake Aral is dry need we expect an end to the development of the country between the rivers; until then we can expect a steady increase of produce and population. Realize this, and you then know what it means to speak of the future of the Duab. (Read on 27 March 1907 by Willi Rickmers (1907, p. 8) at the Central Asian Society in London)

H. Kreutzmann (✉)
Department of Geography, Freie Universität Berlin,
Malteserstraße 74, 12249 Berlin, Germany
e-mail: h.kreutzmann@fu-berlin.de

© The Author(s) 2016   113
R.F. Hüttl et al. (eds.), *Society - Water - Technology*, Water Resources
Development and Management, DOI 10.1007/978-3-319-18971-0_9

Major water engineering projects (MWEPs) have been held up as significant symbols of modernisation. Technological approaches address the power and potential of ecological and societal transformation. Most of the major projects are located in areas with antecedents. Some of these have been irrigation hubs since ancient times. The Fergana Valley is no exception to this widespread phenomenon (Fourniau 2000; Francfort and Lecomte 2002; Cariou 2004). Known for its fertility in the midst of an arid environment, the utilisation of the Syrdarya waters formed the basis of a success story of the upscaling, expansion and extension of irrigation which, in turn, lead to the drying-up and shrinking of Aral Sea with all its consequences. The nineteenth-century Russian administrator and scholar Aleksandr Fedorovich Middendorf said that

> [...] over thousands of years the populace had constructed huge water channels, carried out large-scale fertilization and planted whole forests of shade-giving trees for fruits and wood, with 'each individual tree being in need of life-giving water'. The Kokandis planted fields of wheat, barley, millet sorghum, corn, rice beans, sesame, flax, hemp, cotton, and alfalfa while their gardens included melons, water melons, cucumbers, pumpkins, grapes, apricots, peaches, apples, pears, quinces, nuts, plums, cherries, not to mention onions, carrots, beets and other produce. The main grain crop was wheat, which Kirghiz cattle ranchers raised on the lower slopes of the Alai range as a kind of side business. [...] The expansion of irrigation after the early eighteenth century increased the number of villages and reduced the area available for grazing. Cotton growing always had held a special place throughout the Kokand Khanate, but in the nineteenth century farmers also began cultivating American long-fibred hybrids. (Middendorf 1882, pp. 11–12)[1]

This stated diversity of cultivars hints at the enormous fertility attributed to the Fergana Valley, the tradition of cotton cultivation prior to Russian colonisation and an early inclination to produce market-oriented cash crops. In the contemporary context, only cotton and wheat are discussed as the leading cash and subsistence crops of independent Uzbekistan. Cotton has been identified as the most controversial crop, the importance of which Max Spoor (2007) questioned as either a "curse" or "foundation for development".

Through several transformations, the Fergana Valley has come a long way from the fertile hub of the Kokand Khanate (Fig. 9.1) to a tripartite bone of contention shared by Kyrgyzstan, Tajikistan and Uzbekistan. Looking at the path-dependent development of the growing irrigation systems in the Fergana Valley, the impact of modernisation as the guiding principle for transforming economy and society demands further analysis. The nexus of colonial/imperial domination and the experiment of modernising and expanding the water sector need to be explored to shed some light on the effects of upscaling an irrigation network and the challenges of rescaling in recent times. Imperial attitudes and boundary making during the latter half of the nineteenth century shifted borders as the result of the *Great Game* and have changed scale over time (Kreutzmann 2013a). The reconfiguration of territoriality poses an interesting challenge in understanding the impact on a "fixed infrastructure" such as the MEWP in the Fergana Valley. Borders have been shifted over time; the extent of controlled catchment areas, the size of the water management

---

[1] Middendorf (1882, pp. 11–12); quoted according to the translation of Dubovitskii and Bababekov (2011, p. 58). See as well Joffe (1995, p. 369).

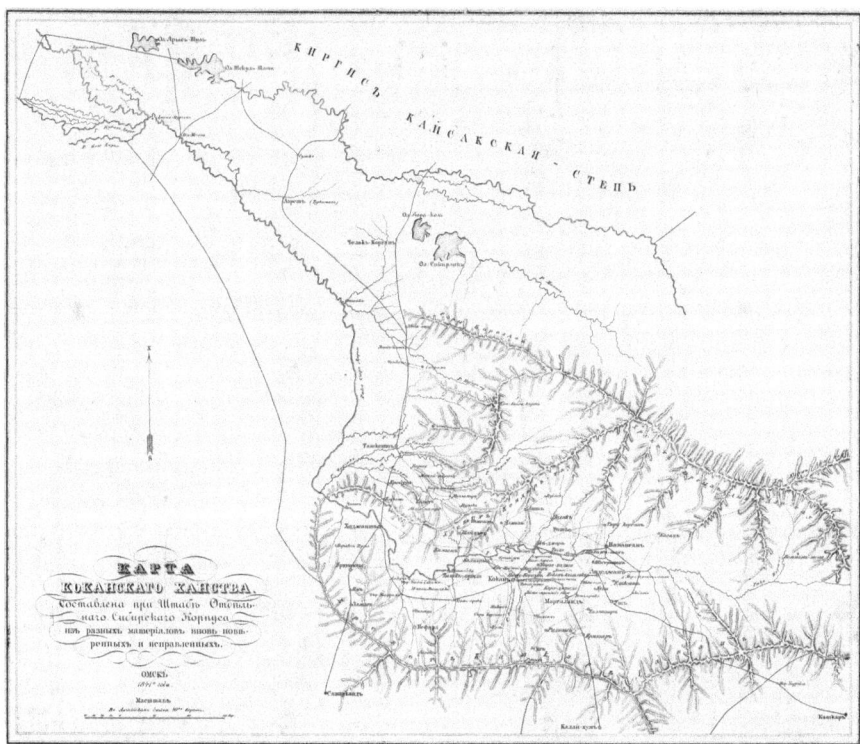

Fig. 9.1 One of the earliest maps of the Fergana Valley was produced in Omsk in 1841 and published in St. Petersburg in 1849. It shows the Fergana Valley as a unit and as dominated by the Kokand Khanate (Source: Zapiski Russkago Geograficheskago Obshchestva 1849)

system and the socio-economic environment have experienced several transformations. The present-day challenge of adjusting the irrigation network's functioning to new borders, conflicts among neighbours and global relations needs further reflection when taking Neil Brenner's statement into context emphasising the interrelationship of these processes characterised as "[…] the deterritorialisation of social relations on a global scale [that] hinges intrinsically upon their simultaneous reterritorialisation on sub-global scales within relatively fixed and immobile configurations of territorial organisation" (Brenner 1999, p. 62). We will trace and link the effects of globalisation and modernisation to path-dependent developments in the Fergana Valley with its creation of a complex network of interconnected rivers and main canals. We will look at the existing system as a historico-genetic product faced with contemporary challenges differing significantly from the frame conditions and traits they initiated. And based upon the following overview of salient features that frame the modes of operation of this MWEP, we will highlight three aspects to assess the challenges Uzbekistan is confronted with in Fergana. First, the often neglected Tsarist strategy of modernisation, its emphasis on cotton cultivation and its transition and translation into the Soviet cotton-based growth model for "white gold" in the Fergana Valley; second, the challenges of spatial rescaling following Uzbekistan's independence; and, third, the constraints of operating a system that is presently heavily dependent on good relations with neighbourly and favourable global market conditions.

9.2 Salient Features of Fergana's Irrigation System

Among the six major Asian river basins – Brahmaputra, Indus, Ganges, Mekong, Amudarya – the Syrdarya with a catchment of 402,760 km^2 with a mean annual flow of 39 km^3 is by far the smallest, but with operational challenges in maintenance and constraints similar to the others (Savoskul and Smakhtin 2013, p. 2). During the last half century, the flow characteristics of the Syrdarya have significantly changed: snowmelt contribution decreased by one-fifth, while the share of glacier runoff increased. Various calculations suggest a marginal impact on future annual water flows and quantities available, but changes to seasonal irrigation water availability with less water in spring and summer and higher shares in autumn and winter (Savoskul and Smakhtin 2013, p. vii, 35). Such a shift will detrimentally affect the cultivation patterns that depend on summer cultivation and winter flooding for drainage.

A century ago the Fergana Valley (Fig. 9.2) constituted one-third of all irrigated lands in the territories that are now Uzbekistan and provided nearly two-thirds of all Central Asian cotton for the Russian textile industry (Pierce 1960, p. 169; Thurman 1999, pp. 11–12).[2] The expansion of cultivated land has been substantial: in 1930 the irrigated land in Fergana accounted for 530,000 ha and was increased to 650,000 ha by 1950. In the post-Stalin era, a major boost during the *Virgin Lands Campaign* and subsequent developments led to an irrigated area of 1.5 million ha by 2005. The share of Uzbek Fergana alone doubled between 1924 and 1985 to 800,000 ha (Bichsel and Mukhabbatov 2011, p. 254; Wegerich et al. 2012, p. 550).[3] While the share of cotton was only 9.6 % in the Fergana Oblast' in 1885, it increased to 42.0 % by the end of the Tsarist rule. Within the Fergana Districts of the Uzbekistan Socialist Soviet Republic, its share was more than three quarters in 1930 and fell to 62.3 % by the end of the Soviet rule (Bichsel 2009, p. 17). Overall, Fergana mirrors the trend in cotton production which itself seems to have been defined as the backbone of Soviet agrarian science and proof of technological development. For the time between 1913 and 1990, Uzbekistan's record contains a four-fold expansion in the area of cotton cultivation, an increase in output by a factor of ten, and the yield per hectare rose from 1.2 to 2.8 t/ha (Spoor 2007, pp. 58–59). In contrast, wheat production decreased during the Soviet period. Emphasis was put

---

[2] By the turn to the twentieth century, the Russian textile industry imported 36 % of its raw cotton from the three Central Asian cultivation areas in Bukhara, Fergana, and Khiva (Joffe 1995, p. 369).

[3] For the earlier periods a growth in irrigated lands took place in Fergana Oblast from 593,246 ha in 1885 to 833,850 ha in 1916; during the transition a decrease to 322,640 ha from a share of 42.0–5.7 % of cotton cultivation occurred until 1922, and a subsequent recovery with high growth rates was on record (Thurman 1999, pp. 264–265).

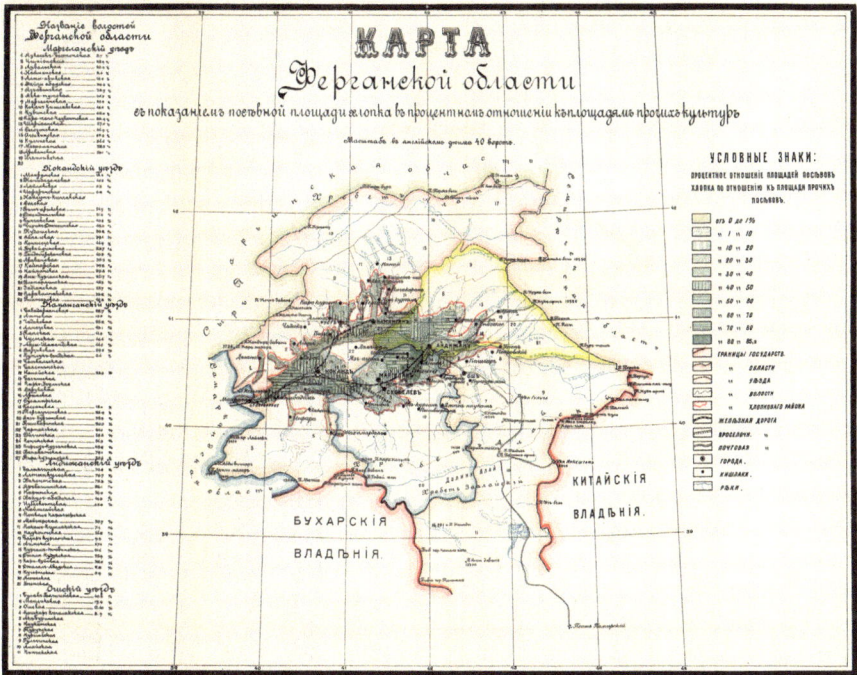

Fig. 9.2 Map of Fergana Oblast showing the percentage shares of cotton (*green* shading in the legend) in comparison to other crops in the irrigated lands in 1913. The urban centres within the cotton-growing areas were Andijan, Kokand, Margilan, Namangan and Skobelev (Source: Ponyatovsky 1913)

on cotton cultivation, which has remained the leading crop under state conditions, despite wheat production share significantly increasing in recent years. Uzbekistan is setting an impressive example by presently generating 82.9 % of its wheat consumption from domestic production, while this share from own production was only 14.3 % when it became independent. In 2011 Uzbekistan remained ranked sixth among world cotton suppliers with 983,000 t (Perekhozhuk et al. 2013). Fergana has remained one of the prime suppliers of raw cotton as the processing industries were mainly located far away in the industrial centres of the West.

9.3 Historical Setting of Tsarist Plans to Utilise Central Asian Water Resources to Large-Scale Soviet Irrigation in the Fergana Valley

With the conquest of the Kokand Khanate in 1876, the Fergana Oblast of the General Governorate Turkestan was established and promoted cotton production for the up-and-coming Russian industry (Thurman 1999). The establishment of major irrigation systems in Fergana during the Tsarist rule can be interpreted as part of the civilising mission which was so aptly commented upon by Fyodor M. Dostojevski who compared Russian interest in Central Asia with the same mission of European powers in North America (Hauner 1992). The control of regional resources and experiments with new technologies in the Central Asian "laboratory" enabled the colonial power to envisage an interconnected system of producing raw materials in Central Asia and processing them in Russia. The establishment of an extensive engineered water management system with major rivers connected supported the construction of huge reservoirs and the greening of the desert and steppe. Thus, the conversion of the Fergana Valley into a fertile, irrigated, cotton-producing oasis of major dimensions was in tune with the Tsarist modernisation and civilisation programme (Fig. 9.3). A scientific approach in understanding system properties accompanied with measurements and control mechanisms for the implementation of a major scheme symbolises the holistic imperial approach during the Tsarist imperial regime. The Fergana Valley was integrated into an empire where it fulfilled a specific function. The irrigation schemes that were planned incorporated the basic grid of the entire present-day irrigation network; even the Toktogul Reservoir was envisaged during Tsarist times, although it was eventually built and opened in 1975 (Fig. 9.4).

Fig. 9.3 Schematic irrigation map of Turkestan, depicting the state of 1913 entitled "Skhematiceskaya irrigatsionnaya karta Turkestana". It shows the linkages between the Fergana Valley and the Khanates of Bokhara and Khiva. The grand plan of future irrigation (*red colour*) is outlined here starting from a reservoir in Toktogul via Fergana to the oasis of Tashkent. A scientific approach is shown with several measurement stations (Source: Glavnoe Upravlenie Zemleustroistva i Zemledeliya, Otdel Zemel'nykh Uluchshenii 1914)

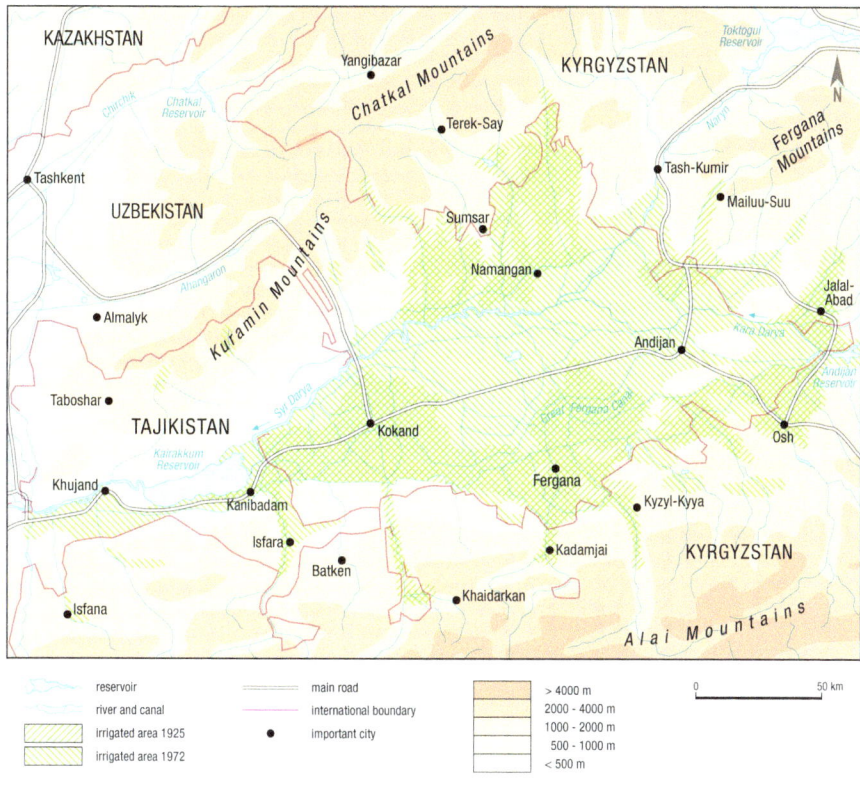

Fig. 9.4 Comparison of irrigated areas in the Fergana Valley 1925 and 1972 (Source: own design based on Bulaevskii 1925 and Benyaminovich and Tersitskii 1975)

Soviet rule divided existing territorial units, created new republics and, at the same time, integrated them under the roof of a political union. The creation of eponymous republics – meant to respect and to reflect the ethnic divide in Central Asia – and the implementation of autonomy policies that lacked any sense of autonomous decision-making (Kreutzmann 2013b) provided the blueprint for profound interference to the socio-economic structures of the formerly independent khanates and administratively divided the Fergana Valley along newly created and delineated boundaries. Other reforms such as collectivisation of productive resources, the introduction of central planning, and expansion of communication infrastructure on the Soviet Union's scale and the supply of citizens with similar sets of low-cost consumer goods indicate the centralised approach in state organisation.

Plans to expand the irrigation system and to integrate the long-distance water management in a scientific approach to agrobusiness and agricultural technologies, as well as raising it to become part of the industrial processing of agricultural goods, gained significant momentum after 1930 when major economic reforms and social transformations had been accomplished. The construction and maintenance of these complex major irrigation systems were made possible at the highest level of planning and implementation (Fig. 9.4). The Soviet Ministry of Land Reclamation and Water Resources (*minvodkhoz*) was the highest hydraulic authority with regional water resources management departments (*oblvodkhoz*) with several name changes over time: *Turkvodkhoz* (Turkestan Water Management Department of the People's Commissariat of Agriculture 1918–1924), *Uzvodkhoz* (Uzbekistan Water Management Department of the People's Commissariat of Agriculture 1925–1931), *Glavvodkhoz* or *TsJUPR* (Chief Administration of Water Management 1931–1938), *Uzvodkhozi* (Uzbekistan People's Commissariat of Water Management, part of *Narkomvodkhoz* (People's Commissariat of Water Management 1938–1946)) and *Minvodkhoz* (USSR Ministry of Water Management 1946–1991, containing *Uzminvodkhoz* (Uzbekistan Ministry of Water Management, since 1946)). The lower level was structured by district water resources management departments (*rayzemvodkhoz* (1930–1938); *rayvodkhoz*, since 1938) and organised the supply for the collective farms (*kolkhoz, sovkhoz*) (Thurman 1999, p. 263). Within the agricultural sector the professional decision-making capacity of the farm worker (*dehqan*) continuously declined, a phenomenon that served the needs of superordinate administrators and made them powerful controllers. Generic bureaucrats made "expert" decisions and instructed the farm workers what they should do on the fields. This approach has persisted until today in the guise of water user associations.

The principle of water management was governed by a hierarchical territorial approach rather than by a hydraulic or catchment area-based one. The centralised economic planning fulfilled two objectives. First, it fitted in the grand plan for the Soviet Union, by attributing certain tasks to individual republics, and even districts, by setting production standards and objectives for resource allocation and production; second, it took into consideration the specific needs of different republics and created a certain degree of lower-level authority. Many statistics and accounts were produced on the administrative level of republics although central rules and regulations were applied. In the case of Fergana, water management followed a holistic approach by integrating territories from three neighbouring republics into one system, but attributing production figures to republics and recording on state levels (Fig. 9.5). The success of Uzbekistan's economy was solely judged upon cotton production (Obertreis 2007, p. 171). Indigenous knowledge and local expertise were completely refuted; modernisation strategies based on science and technology were expected to replace and surpass them. Jonathan Thurman is explicit in stating about Fergana: "[...] the 'cradle' of irrigated cotton growing in Central Asia. Indigenous modes of organisation that effectively fostered farmer participation here were undermined by a colonising power, which transformed them into state-dominated organisations incapable of effective management" (Thurman 1999, p. iv). The quest for modernisation was the same in Tsarist Russia and the Soviet Union. Both regimes utilised Central Asia as supplier of raw materials needed for its

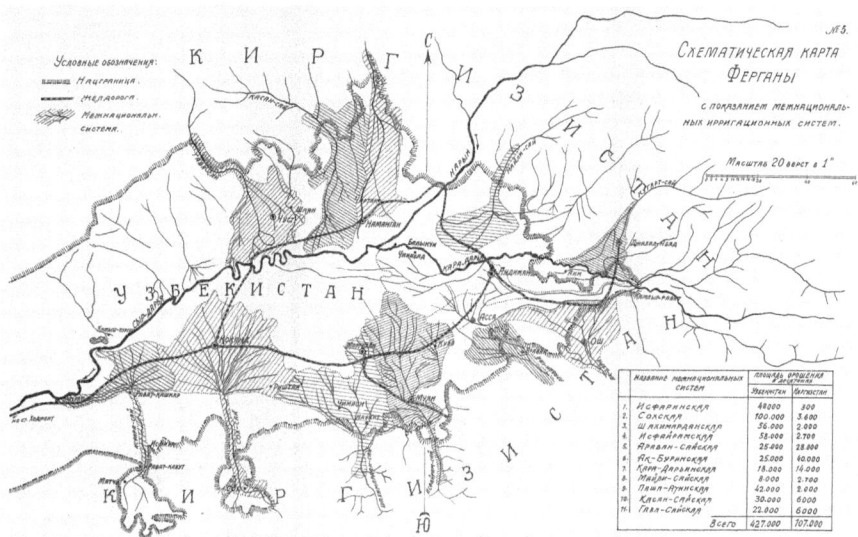

Fig. 9.5 Schematic map of the Fergana Valley [Skhematiceskaya karta Fergany s pokazaniem mezhnatsionalnykh irrigatsionnykh sistem] showing the Soviet Union's trans-boundary irrigation system and the irrigated areas in Kyrgyzstan (107,000 dessiatines, one dessiatine equals 1.0925 ha) and Uzbekistan (427,000 dessiatines) in 1925 (Source: Bulaevskii 1925)

industries in the western parts of the empire.[4] The share of regional processing of cotton never reached significant proportions. The newly introduced boundaries did not significantly affect the everyday life of people, their communication and travel. The effect of these borders became severely felt after the independence of the Central Asian republics in 1991.

9.4 Geographical Rescaling After Independence

Independence in the Central Asian republics in 1991 brought about functioning borders, something hitherto unknown. Conflicting interests of independent countries resulted in a multitude of unsolved problems. One was the delineation of international boundaries. In the aftermath of the dissolution of the Soviet Union, the

---

[4] In his assessment of the Fergana experiment, Jonathan Thurman (1999, p. 222) is highly critical about its function: "Under Stalin, the Ferghana Valley had the misfortune of becoming the Soviet model for an irrigated agricultural zone in the East. It was here that planned delivery of water in Central Asia was first applied and tested, and that cotton cultivation reached the most absurd proportions. Construction and maintenance programs focused on Ferghana as if it were another republic, separate from the rest of Uzbekistan. Throughout the Stalin era, the Soviet media commonly spoke of 'transferring the Ferghana experience' in irrigation to other areas of Central Asia. This would be done after World War II, with disastrous results".

Fig. 9.6 Enclaves and exclaves in the borderlands of Tajikistan, Kyrgyzstan and Uzbekistan after independence (Source: modified after Kreutzmann 2013a, p. 18)

Moscow Institute of Political Geography conducted a survey in 1992 and recorded 180 border and territorial disputes among new neighbours (Halbach 1992, p. 5). A few cases illustrate the range of conditions and demands in Central Asia: irredentist movements in Turkmenistan expect Uzbekistan to "return" the territory of the Khanates of Khiva and Khorezm. The long-standing demand of Tajik nationalists has been reiterated that Samarkand and Bukhara as the centres of Tajik culture must be "returned". The divide of Fergana into three sections is questioned, and the present-day economic and commercial centre of Southern Kyrgyzstan – Osh Oblast' – is claimed by Uzbekistan. A multitude of legal documents about boundary decisions taken in the 1920s concerning territorial issues in Central Asia are archived in Tashkent. The Uzbekistan government blocks access to researchers from neighbouring countries wanting to consult the archival material.

The Fergana and Alai Valleys alone contain seven enclaves and exclaves through which major traffic routes lead and which are valuable resource-rich areas (Fig. 9.6). Regularly, the freedom of travel is affected, especially when diplomatic relations worsen. The inhabitants of these enclaves have become pawns in bilateral negotiations. Some of the border closures have been justified in the aftermath of attacks from Afghanistan-trained rebels, which plundered Tajik and Kyrgyz villages on their way to the Fergana Valley in 1999 and 2000. Territorial rescaling has been one of the major challenges for post-Soviet societies far beyond the unsolved issues of commanding a compact state territory, mutually accepted international boundaries and respective sovereignty.[5]

---

[5] The Central Asian republics are facing challenges that are well known from other irrigation networks that had to be divided such as the Indus waters between India and Pakistan in the process of decolonisation (Kreutzmann 2011).

Following independence, Fergana's water management system faced similar challenges. The irrigation network was implemented at a time when no boundaries affected the decision-making of concerned engineers and politicians. The meshed system of canals and reservoirs had emerged from a genetic expansion of irrigation from Tsarist to Soviet times (Fig. 9.7). Previously, abundant quantities of water were available; post-Soviet management had to cope with the challenges of importing water from neighbours and sharing the network across borders.

The legacy of past decisions and constructions is an in-built factor of the Fergana system that is represented in the structural pattern of the existing meshed system, the incorporation of new additions to the system by sharing the available water and the organisation seasonal shifts in a decentralised balancing of flow characteristics. Kai Wegerich et al. (2012) have analysed the hydrological constraints and its consequences:

> In the past, a management regime was implemented that fitted the state objective (expansion of the state order on cotton production) regardless of sources, distances, inter-linkages, and

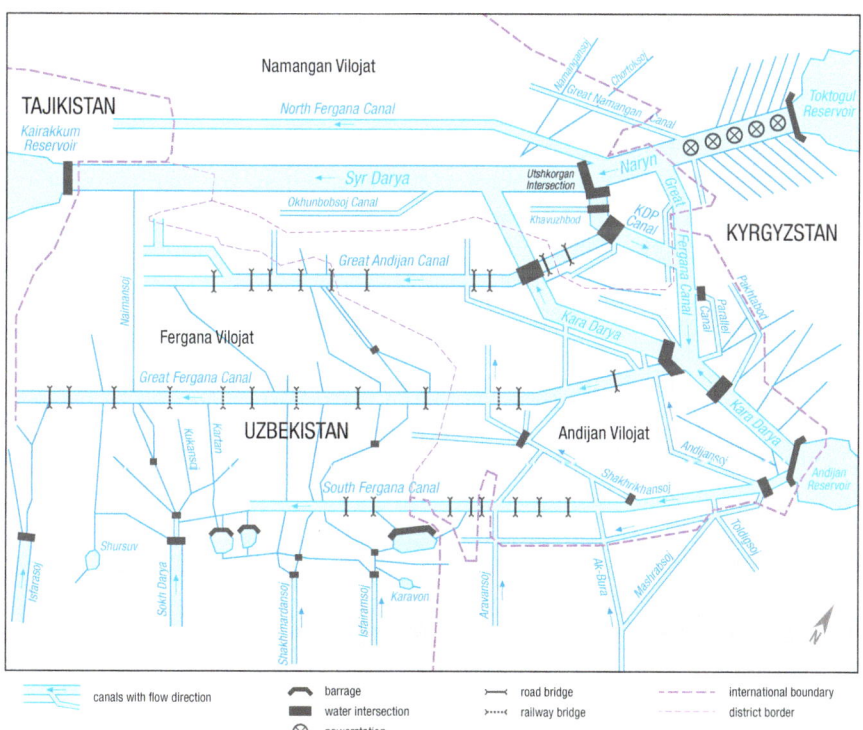

Fig. 9.7 The present state of Uzbekistan's irrigation network in the three bordering provinces (Viloyat Andijan, Fergana and Namangan) with Kyrgyzstan's share in the East and Tajikistan's share in the West of the Fergana Valley. All major reservoirs are located outside of Uzbekistan's direct control and influence (Source: own design based on management plans displayed in various offices of Uzbekistan's irrigation authorities in the Fergana Valley in 2014)

costs. Obviously, the past objectives are reflected in the current design of the system; consequently, it may not be possible to implement new objectives regarding water management and governance at the sub-catchment and irrigation system levels without adjusting and modifying the original design. (Wegerich et al. 2012, pp. 562–563)

In their opinion any institutional and/or technical restructuring requires a thorough analysis of the complexities of the existing meshed system in order to identify the spatial and organisational units where reforms could be implemented (Wegerich et al. 2012, p. 563). The authors conclude that only a far-reaching reconstruction and adaptation of the water management system would be feasible for smooth future operations. Beyond hydrological considerations, the question of the water source occurs. The Fergana Valley taps most of its irrigation water from the Syrdarya via Toktogul Reservoir and the Kara Darya via the Andijan Reservoir, both of which are located in Kyrgyzstan. Since the Kyrgyz government has adopted its authority over the management of their water resources, the dispute about the allocation of river water is regularly revived when it comes to compensation measures and contribution to the actual costs for providing network maintenance and supply of water. A water treaty was approved by the four Central Asian neighbours in Almaty in 1992 which allocated fixed shares of the Syrdarya water: Kyrgyzstan 1 %, Tajikistan 9.2 %, Kazakhstan 38.1 % and Uzbekistan 51.7 % (Bichsel and Mukhabbatov 2011, pp. 261–263). The water key will need to be renegotiated as Kyrgyzstan is in the process of expanding its energy-producing sector and its irrigated agriculture. The recent signing of the Central Asia South Asia Electricity Transmission and Trade Project (CASA-1000)[6] project treaty and its financing by the World Bank for the transmission of hydroelectricity from Kyrgyzstan's "surplus" production to Afghanistan and Pakistan via Tajikistan affects the weak equilibrium between Uzbekistan and Kyrgyzstan and changes the latter's bargaining power. Consequently, the political dimension of negotiating exchange of water for monetary funds or energy will gain momentum in the future. Ecological constraints such as growing water logging and salinisation, as well as disputes in times of scarce seasonal supply, have aggravated the situation between competitors along the Syrdarya (UNEP et al. 2005, p. 24; Bichsel and Mukhabbatov 2011, p. 264). It appears that the quality of water and land in terms of salt content and contaminations is significantly deteriorating, although the quantity of available irrigation water is estimated and projected not to dwindle.

9.5 Persistence and Change Within Uzbekistan's Water Management

Given these major challenges, it is surprising how many elements of a powerful reluctance to change can be observed in the Fergana Valley. The institutional set-up has maintained its overall control over all operations despite opening up to world

---

[6] See for details and maps www.casa-1000.org.

market conditions for the export of cotton and its processed products and in spite of a quest for privatisation in land holdings. The farmer seems to be the least involved actor in making any decisions about cultivation patterns, timing of water allocation and selection of agricultural inputs. Water user associations have given a new name to the old control mechanism. The water management system in Fergana seems to be the least in line with the changed conditions and with the changing conditions of a growing population and expanding demands of the up-and-coming generation. The latest report by the International Crisis Group (ICG) (2011) on "water pressures in Central Asia" states that "[t]he concerns Crisis Group identified in 2002 – inadequate infrastructure, poor water management and outdated irrigation methods – remain unaddressed, while the security environment is bleaker". The ICG recommends to the governments sharing the Fergana Valley: "Commit to resolving border demarcation problems without using water or energy as a coercive factor; facilitate cross-border cooperation between police forces and form a tripartite intraregional council to oversee day-to-day management of water and land resources parallel to high-level border delimitation negotiations".[7] The hydraulic constraints that were imposed on Kyrgyzstan, Tajikistan and Uzbekistan after independence remain to be the result of path-dependent developments from Tsarist and Soviet legacies. Rescaling could become a necessity if no consensus among neighbours can be accomplished that serves the purpose. Rescaling is a major challenge in changing the rules of the game within respective countries. It seems that practical measures to address the challenges of rescaling the water management system require a transformative spirit from both inside and outside of the Fergana Valley.

References

Benyaminovich ZM, Tersitskii DK (1975) Irrigation in Uzbekistan II. Fan Publishing House, Tashkent (in Russian)

Bichsel C (ed) (2009) Conflict transformation in Central Asia. Irrigation disputes in the Ferghana Valley. Routledge, Abingdon

Bichsel C, Mukhabbatov K (2011) Land, water, ecology. In: Starr FS, Beshimov B, Bobokulov II, Shozimov P (eds) Ferghana valley. The heart of Central Asia. M. E. Sharpe, Armonk/London, pp 253–277

Brenner N (1999) Beyond state-centrism? Space, territoriality, and geographical scale in globalization studies. Theory Soc 28:39–78. doi:10.1023/A:1006996806674

Cariou A (2004) Le jardin saccagé. Anciennes oasis et nouvelles campagnes d'Ouzbékistan. Ann Geogr 113:51–73

---

[7] See the detailed Europe and Central Asia Report No. 233, released on 9 September 2014 at http://www.crisisgroup.org/en/regions/asia/central-asia/233-water-pressures-in-central-asia.aspx?utm_source=central-asia-report&utm_medium=1&utm_campaign=mremail   (=http://tinyurl.com/ocbbkln) and accessed on 12 September 2014.

Dubovitskii V, Bababekov K (2011) The rise and fall of the Kokand Khanate. In: Starr FS, Beshimov B, Bobokulov II, Shozimov P (eds) Ferghana valley. The heart of Central Asia. M. E. Sharpe, Armonk, pp 29–68

Fourniau V (2000) Some notes on the contribution of the study of irrigation to the history of Central Asia. In: Kreutzmann H (ed) Sharing water. Irrigation and water management in the Hindukush-Karakoram-Himalaya. Oxford University Press, Oxford/New York, pp 32–54

Francfort H-P, Lecomte O (2002) Irrigation et société en Asie centrale des origines à l'époque achéménide. Ann Hist Sci Soc 57:626–663. doi:10.3406/ahess.2002.280068

Halbach U (1992) Ethno-territoriale Konflikte in der GUS. Bundesinstitut für ostwissenschaftliche internationale Studien, Köln

Hauner M (1992) What is Asia to us? Russia's heartland yesterday and today. Routledge, London

International Crisis Group (2011) Water pressures in Central Asia. International Crisis Group, Brussels

Joffe M (1995) Autocracy, capitalism and empire: the politics of irrigation. Russ Rev 54:365–388

Kreutzmann H (2011) Scarcity within opulence: water management in the Karakoram Mountains revisited. J Mt Sci 8:525–534. doi:10.1007/s11629-011-2213-5

Kreutzmann H (2013a) The significance of geopolitical issues for internal development and intervention in mountainous areas of Crossroads Asia. http://www.geo.fu-berlin.de/geog/fachrichtungen/anthrogeog/zelf/Medien/download/Kreutzmann_PDFs/HK_Crossroads_Asia_WP07_Jan2013.pdf. Accessed 4 Sept 2014

Kreutzmann H (2013b) Boundary-making as a strategy for risk reduction in conflict-prone spaces. In: Müller-Mahn D (ed) The spatial dimension of risk. How geography shapes the emergence of riskscapes. Routledge, Abington/New York, pp 154–171

Middendorf AF (1882) Ocherki Ferganskoi doliny. Academy of Science Publishing, St. Petersburg (in Russian)

Obertreis J (2007) Infrastrukturen im Sozialismus. Das Beispiel der Bewässerungssysteme im sowjetischen Zentralasien. Saeculum 58:151–182

Perekhozhuk O, Bobojonov I, Glauben T (2013) Immer mehr, aber nicht genug. Über die wachsende Bedeutung von Weizen in Zentralasien. Zentralasien-Analysen 72:9–17

Pierce RA (1960) Russian Central Asia, 1867–1917: a study in colonial rule. University of California Press, Berkeley

Rickmers W (1907) Impressions of the Duab (Russian Turkestan). Central Asian Society, London

Savoskul OS, Smakhtin V (2013) Glacier systems and seasonal snow cover in six major Asian river basins: hydrological role under changing climate. International Water Management Institute (IWMI), Colombo

Spoor M (2007) Cotton in Central Asia "curse" or "foundation for development". In: Kandiyoti D (ed) The cotton sector in Central Asia economic policy and development challenges. The School of Oriental and African Studies, London, pp 54–74

Thurman J (1999) Modes of organization in Central Asian irrigation: the Ferghana Valley, 1876 to present. Dissertation, Indiana University

UNEP, UNDP, OSCE, NATO (2005) Environment and security. Transforming risks into cooperation. Central Asia: Ferghana/Osh/Khujand area. United Nations Environment Programme (UNEP), Geneva

Wegerich K, Kazbekov J, Mukhamedova N, Musayev S (2012) Is it possible to shift to hydrological boundaries? The Ferghana Valley meshed system. Int J Water Res Dev 28:545–564. doi: 10.1080/07900627.2012.684316

Map References[8]

Bulaevskii VF [Булаевский В.Ф.] (1925) Interethnic systems of Central Asia and form of their exploitation. In: Messenger of irrigation. No. 11. November 1925. Tashkent [Межнациональные системы Средней Азии и формы их эксплоатации. Вестник ирригации. № 11. Ноябрь. 1925 г. Ташкент]

Glavnoe Upravlenie Zemleustroistva i Zemledeliya Otdel Zemel'nykh Uluchshenii (1914) Report of the hydro-module division for 1913. Moscow [Отчет гидромодульной части за 1913 год. Главное Управление Землеустройства и Земледелия Отдел Земельных Улучшений. Москва]

Ponyatovsky SV [Понятовский С. В.] (1913) Experience of studying cotton cultivation in Turkestan and Trans-Caspian Province. V. F. Kirshbaum's printing house. St. Petersburg [Станислав Валентинович Понятовский (1913) Опыт изучения хлопководства в Туркестане и Закаспийской области. Типография В.Ф. Киршбаума, СПб]

Zapiski Russkago Geograficheskago Obshchestva (1849) Map of the Khanate of Kokand prepared at the headquarters of the Separate/Special Siberian Corps on the basis of diverse checked and corrected materials, prepared in Omsk 1841. Scale: In one English inch 40 verst. St. Petersburg [Карта Коканскаго Ханства составлена при Штабе Отдельнаго Сибирскаго Корпуса из разных материалов вновь поверенных и исправленных, Омск 1841. Масштаб: В Англиском дюйме 40 верст. СПб]

---

[8] The historical maps reproduced here were provided by the Pamir Archive Collection of Markus Hauser, Winterthur, who kindly permitted their publication. All other illustrations are copyrighted by Hermann Kreutzmann.

Chapter 10
Irrigation Infrastructure in Fergana Today: Ecological Implications – Economic Necessities

Shavkat Kenjabaev and Hans-Georg Frede

Abstract Managing water sustainably and efficiently is important for the Fergana Valley's (FV) irrigation-dominated agricultural system and, subsequently, for its rural population and environment. During the past decade, national water legislation and the organisation of integrated water resources management have been reformed in FV and this development continues. Nevertheless, their implementation has been limited by the lack of resources and the weakness of the institutions. Moreover, the future challenges water management faces in the region's agriculture are increasing all the time. These challenges include low water-use efficiency, fewer incentives for water users to increase land and water productivity, water shortages within the system, salinity and declines in key crop yields. Current irrigation strategies in the region are not adaptable enough to cope with variations in water supply and crop water requirements caused by land use and climate change. The objective of this chapter is to provide an overview of the irrigation water management in the region and to lay down some of the concepts and complexities in maintaining the existing irrigation infrastructure with the aim of increasing water productivity and environmental sustainability. We hope that this will help set the stage for productive discussions and to identify research needs.

Keywords Irrigation infrastructure • Water management • Water resources • Fergana Valley • Syrdarya River • Agricultural water use • Ecological impact

S. Kenjabaev (✉)
Scientific-Information Center of the Interstate Coordination Water Commission (SIC ICWC),
h.11, Karasu-4 Tashkent 100187, Uzbekistan
e-mail: kenjabaev@yahoo.com

H.-G. Frede
Institute of Landscape Ecology and Resource Management (ILR), Justus-Liebig-Universität
Gießen, Heinrich-Buff-Ring 26-32, 35392 Gießen, Germany

© The Author(s) 2016
R.F. Hüttl et al. (eds.), *Society - Water - Technology*, Water Resources
Development and Management, DOI 10.1007/978-3-319-18971-0_10

10.1   Introduction

This chapter outlines the problems existing in the management of the irrigation infrastructures in the FV and evaluates the benefits of irrigated agriculture and also the costs of environmental degradation and pollution caused by changes in water management.

The Fergana Valley, one of the oldest *world* oases, is unique with its highly fertile lands (Irrigation of Uzbekistan 1975), manageable river waters and adequate natural (hilly areas or *adyrs* in local language) and artificial (central plain part of Fergana) drainage. Combine the above with the local climatic conditions and we have the reason for the area's agricultural importance.

Despite this, there are obstacles to the well-being of the region's population, the key ones being the high rates of population growth, interdependence of water and energy (Baker 2011) and limited land resources (Dukhovny and Stulina 2012).

The geographical location of the FV is also a peculiarity. The valley is blocked in by high mountains in Tajikistan and Kyrgyzstan, while the valley itself is rather plain and arid. The specific climate and topography conditions limit the further development of land and water resources. Agriculture, accounting for about 90 % of withdrawal from the total water resources available in the region (CAWATERinfo 2012), heavily depends on irrigation. Moreover, agricultural production is highly vulnerable to climate change (Lioubimtseva and Henebry 2009). High fluctuations in precipitation and temperature increases may influence land use in irrigated lands, create difficulties in water management at regional and local levels and increase competition for scarce water resources amongst water users in various sectors.

Demographic pressure will further limit the availability of the water and land resources. Population density in the valley increased from an average 135.5 people per km^2 in 1980 to 229.1 people per km^2 in 2010 (CAWATERinfo 2012). Consequently, food demand has increased and the area of irrigated land has expanded, leading to further degradation of the surface and groundwater quality as agricultural activities have increased. The expansion of irrigated land has resulted in increasing water withdrawals from two main rivers, Amudarya and Syrdarya, and induced the problems of land degradation and water resource contamination and, as a final consequence, has led to the Aral Sea environmental disaster.

In their natural state, many of these lands, especially the plain lands within the Uzbek part of the valley, are poorly drained. Therefore, the irrigated lands were intensively developed in the second half of the twentieth century increasing the artificially drained area to about 682×10^3 ha in the valley (CAWATERinfo 2012). There are no doubts that the development of the collector-drainage network (CDN) had a positive impact on irrigated lands by preventing salt accumulation in soils and creating favourable soil aeration conditions. However, as a consequence of the development of the irrigation and drainage systems, a steady rise of return water was observed. This increase was caused by low irrigation and drainage efficiencies on the one hand and misuse (over-irrigation) of water resources on the other. At the same time, the collector-drainage infrastructure in the region is designed, naturally and artificially, to discharge most of its effluent back into the rivers (SIC ICWC 2004). Thus, a gradual increase of water salinity in the Syrdarya and Amudarya Rivers was observed since the 1950s (SIC ICWC 2004). As a result, environmental and epidemiological fragility and hygienic and sanitary conditions are emerging as a major problem, especially in the lower course provinces of the river basins (SC RUz 2006).

10.2   General Overview of Fergana Valley

10.2.1   Geography (Geomorphological Structure)

The Fergana Valley is one of the water-rich upstream subregions of the Syrdarya River basin. The valley is 300 km in length and 170 km in width with floor elevation of about 450 m above sea level (aSL) (Irrigation of Uzbekistan 1975). Some literature sources quote the administrative territory of the valley as 122,100 km² (Table 10.1). However, according to recent publications (UNECE 2011), the valley's basin area is 142,200 km². It is characterised as a locked intermountain depression, embosomed in the spurs of the Ala-Tau Range in the north, the Tian Shan Mountains in the east and the Alay Mountains in the south, with only a narrow mouth to the west through which the Syrdarya River drains the valley (see Fig. 10.4). The uppermost part of the geological profile (100 m or thicker) in and around the central valley varies from location to location, and it is primarily of Quaternary origin. The central part merges with outwash fans near the valley foothills.

Table 10.1 Administrative-territorial and demographic data in provinces of the Fergana Valley

Provinces	Established in	Number of districts	Territory (km²)	Population density (people per km²)			% rural population in 2010	Annual increase 2000–2010 (%)
				2000	2005	2010		
Andijan	1941	14	4303	515	552	621	72	1.7
Fergana	1876	15	6759	399	426	478	74	1.5
Namangan	1941	11	7440	263	283	320	64	1.7
Osh	1939	7	28,934	41	36	39	92	1.6
Djalalabad	1939	8	32,418	27	29	32	77	1.5
Batkent[a]	1999	3	17,049	22	24	25	76	1.2
Sogd[b]	2000	14	25,200	75	82	89	75	1.2

Author's compilation from different sources
[a]Batkent was established in 1999
[b]Sogd was named Leninabad from 1936 till 2000

10.2.2   Climate

The average air temperature in the valley is 13.1 °C, ranging from −8 °C to 3 °C in winters (January) and 17 °C to 36 °C in summers (July). Annual precipitation ranges from 109 to 502 mm, whereas evaporation ranges from 1133 to 1294 mm throughout the valley (Abdullaev et al. 2010; Reddy et al. 2012). The limited precipitation on the lowland plain coupled with high temperatures, low humidity and high degree of solar radiation causes a greater level of potential evapotranspiration. A steady transfer of air temperature from 0 °C starts in the second half of February. This is the time when the early fruit, such as apricot and almond, growing period begins. The autumn transfer from 0 °C is observed in the second half of December. The duration of the period with $T \geq 0$ °C is 280–310 days per annum. When air temperature increases above 5 °C, the renewal period of lucerne, grains (winter wheat) and the majority of fruit plants and spring regrowth of pasture grasslands start. This transfer of air temperature is favourable to sow cotton and summer corn. The period with $T \geq 10$ °C lasts 200–220 days per annum. The best conditions to sow heat-loving crops (cotton) begins around 1–15 April when the mean daily air temperature increases above 15 °C; during this time, the soil at a depth of 10 cm also warms up to 16–18 °C (Fig. 10.1).

Climate change, apart from anthropogenic influences, is a significant impact factor on the hydrologic cycle of the river (Stucker et al. 2012). The most hazardous and complex consequences are the increasingly frequent extreme floods and droughts (Dukhovny et al. 2008). In the Syrdarya basin, the frequency of years in which droughts occurred between 1990 and 2007 is less, compared to those which occurred during 1950–1990, while the occurrence of floods (prob. ≤ 25 %) and extreme floods (prob. ≤10 %) increased, respectively, by 1.4 and 2 times (Dukhovny et al. 2008). In drought years, water withdrawal rates increase at the main intake points. Consequently, surface runoff rates from fields are minimised and collector-drainage water (CDW) is used as an additional source instead. Therefore, low water years are characterised by low CDW effluents into the rivers. And, in order to cope with the water shortage, water withdrawal from CDN increases twofold (Dukhovny et al. 2012).

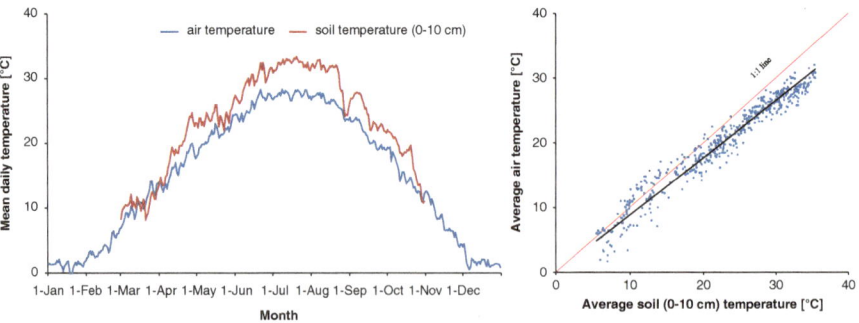

Fig. 10.1 Average daily air and soil temperatures (*left*) and their relationship (*right*) (Source: Fergana agro-meteorological station: averaged from 2001 to 2011)

10.2.3   Demography

The valley is a highly populated area in Central Asia (CA). An average population growth rate of 1.5–2 % per year is common in the provinces of FV (Dukhovny and Stulina 2012). According to different sources, population density in the valley ranged between 25 and 621 inhabitants per km^2 in 2010 (Table 10.1), being as high as 320–620 persons/km^2 in Uzbek provinces and as low as 25 persons/km^2 and 90 persons/km^2 in Kyrgyz and Tajik provinces, respectively. All provinces in the valley are dominated by high rural populations as in Table 10.1 with densities of 200–500 persons/km^2.

Given the importance of agriculture for the whole FV, natural resources, such as land and water, have historically been amongst the most important factors in the development of the region (Qadir et al. 2009). Data from other sources depict that 44–45 % of the irrigated lands of the Syrdarya Basin are located in the FV (Toryanikova and Kenshimov 1999; UNEP et al. 2005). However, the amount of irrigated lands available shared by the three countries is already limited and demand for scarce natural resources will continue to rise with population growth. Hence, the size of the population depending upon these resources is consequently a key factor in political security and environmental issues. High population density also increases the risk of depleting natural resources (Dukhovny and Stulina 2012), and therefore, competition and even conflict for their control would be self-explanatory.

10.2.4   Land Use and Agricultural Production

The agricultural system in Central Asian countries, especially in Uzbekistan, has experienced continuous intensification with increased cultivation of winter wheat in the last two decades. Once the predominant crop in FV, cotton cultivation has declined rapidly since the 1990s while the cropping area under cereals for food security is increasing together with gardens and vineyards (primarily by the introduction of new lands in adyrs) (Fig. 10.2). The decline of lands under forage crops can be explained by the reduction in livestock in all CA states since 1991 (Lioubimtseva and Henebry 2009) and by the increase in winter wheat and consequent cultivation of secondary crops (maize for silage, sorghum, legumes, etc.) after the wheat harvest. Although highly profitable unregulated cash crops (vegetables, grapes and fruits) are extensively grown in the provinces, cotton and wheat (*Triticum aestivum L.*) are still the dominant crops in the valley taking up more than 46 % of irrigated lands. Upland cotton (*Gossypium hirsutum* L.) is cultivated more extensively than pima cotton (*G. barbadense* L.) due to its short growing period and relatively higher yield (WARMAP 1997; Ibragimov et al. 2008).

During the Soviet period, 3:6 or 3:7 crop rotations (3 years alfalfa and 6–7 years cotton) were recommended (Nerozin 2010) and considered as one of the methods to decrease the soil salinity (by reducing soil evaporation and lowering groundwater level (GWL)). After independence, this rotation was radically changed and nowadays includes 2:1 or 2:3 (2 years winter wheat and 1–3 years cotton). The cotton in Uzbekistan is cultivated in the first half of April and harvested in September (last harvest in November); winter wheat is broadcast seeded in late September (most cases in October) and harvested from mid-June till mid-July. In the remaining period, following the winter wheat harvest, secondary crops such as rice, maize, sorghum, sunflower or vegetables are sown from July to November.

According to Dukhovny et al. (2012), agriculture in FV contributes between 20.8 % (Fergana province) to 58 % (Sogd province) to the gross regional product.

The total production of cotton, due to reducing areas sown in the valley, (see Fig. 10.2) decreased from 2116.7×10^3 t in 1980 to 971.8×10^3 t in 2010. In contrast, in the same period, the cereal production increased almost sixfold (Fig. 10.3). The production of cash crops (e.g. vegetables) also increased in line with their increased growing area. These enabled the gross agricultural production to increase

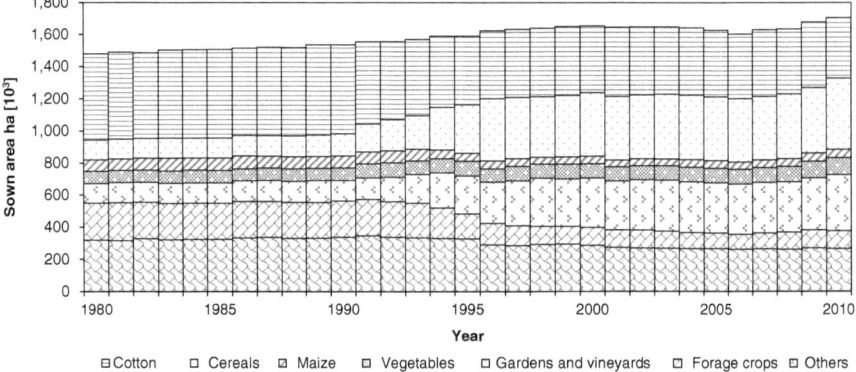

Fig. 10.2 Cropping acreages in Fergana Valley during 1980–2010 (Source: CAWATERinfo 2012)

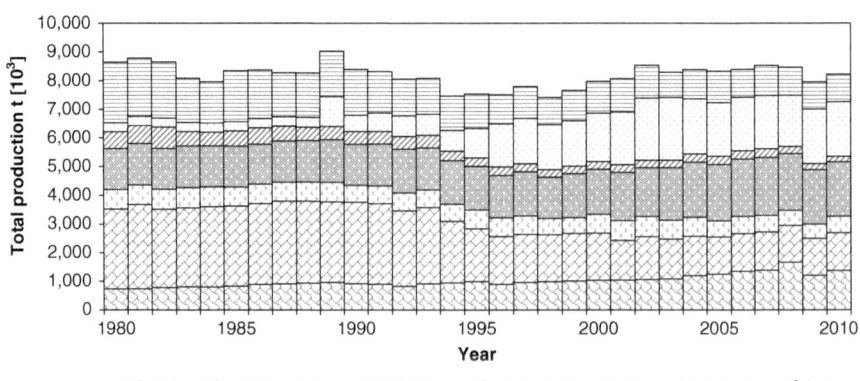

Fig. 10.3 Total production of agricultural crops (10^3 t) in the Fergana Valley during 1980–2010 (Source: CAWATERinfo 2012)

on average from USD 162 per capita in 2001 to USD 240 per capita in 2010 in the valley provinces (Dukhovny et al. 2012).

However, agricultural production in the valley is hampered not only by the unreliable irrigation water supply but also by the unreliable supply of fertilisers and seeds, inappropriate crop cultivation methods, limitations in the allocation of seasonal credits and funding of capital investments, limitations in marketing and processing of produce, control of cropping patterns and poor agricultural development informational services. These issues will need to be addressed through agricultural reforms and measures dealing with the intensification of the agricultural production system.

10.3 Water Resources Management in the Fergana Valley

Water resources in FV are predominantly of a transboundary nature. Most of the region's surface water resources are generated in the mountains of the upstream countries Kyrgyzstan and Tajikistan.

Figure 10.4 shows the main irrigation and drainage systems in the valley and, indeed, provides an insight into the complexity of this unique water management system. Taking into account the dozens of large and hundreds of small towns; thousands of rural settlements with various industrial, utility and other enterprises; thousands of kilometres of irrigation and drainage systems and dozens of reservoirs and pump stations available within the seven provinces of the three riparian countries in this scheme, one can understand that harmonising the management system is not only difficult but also the collection, comparison and assessment of data from various water resources are also key and challenging tasks in themselves.

At present, the main problems in managing the existing irrigation infrastructure in FV are linked to the following shortcomings and constraints:

- The state maintains administrative governance and financial support for the upper layer of water hierarchy without involving the integrated water resources management (IWRM) mechanisms at a national level. Moreover, local stakeholders' engagement in water management within the vertical hierarchy meets resistance from the state.
- The establishment of water user associations (WUAs) and canal management organisations with involvement of the community in the form of a union of canal water users (UCWU) and canal water committees (CWC) has certain effects on the enhancement of uniformity, stability and water supply. However, institutional restructuring is impossible without a strong legal framework. In general, the financial, organisational, legal and technical aspects of these institutions have not been solved (including WUA support by Ministries of Agriculture and Water Management).
- There are no contractual obligations and methodical interrelationships between basin administrative irrigation systems (BAIS), hydrogeologic melioration expeditions (HGME), administrative irrigation systems (AIS) and WUAs.

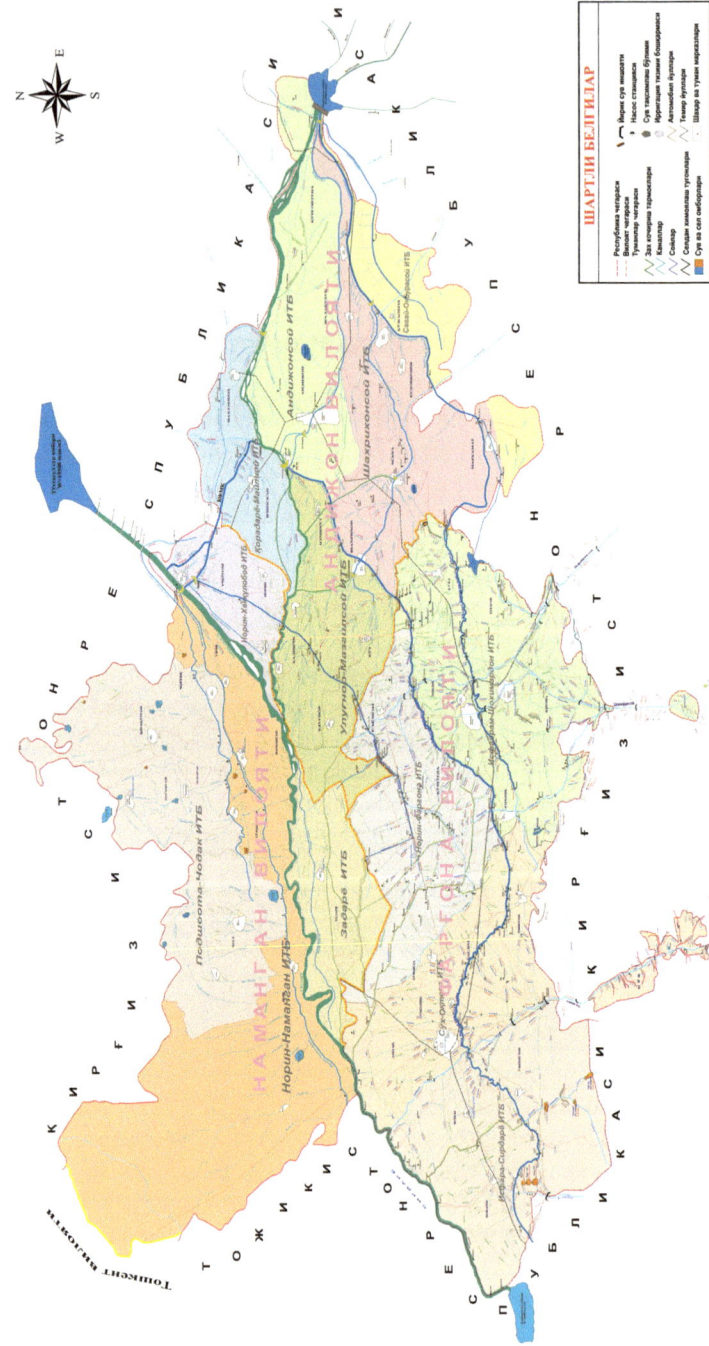

Fig. 10.4 Irrigation and drainage network in the Fergana Valley (Source: author's composition from three BAIS maps Uzbekistan)

- Land degradation and control measures are weakly monitored.
- Consequently, issues of practical support for farmers and end water users in improving land and water productivity have not been solved.
- There is no general programme on water saving taking into account climate change.
- Water accounting and hydrometrics at all water hierarchy levels from transboundary rivers to the ultimate water user should become a strategic objective.
- There are no informational systems and software packages to solve operational water distribution tasks at both canal level and WUA level.
- The majority of the irrigation and drainage infrastructures are outdated and require frequent maintenance and repairs and, consequently, huge investment.

10.3.1   Hydrological Characteristics

The natural annual runoff of the Syrdarya River averages 37.9×10^9 m^3 and ranges from 18.3 to 72.5×10^9 m^3 (CAWATERinfo 2012). Surface water resources of the Naryn and Karadarya tributaries, which are generated in the Kyrgyz region, are estimated to amount to 13.7 and 7.1×10^9 m^3/year accordingly (UNECE 2011). These rivers are strongly regulated by major reservoirs and dams (Kayrakkum, Chardara and Koksarai along the Syrdarya River; Naryn, Krupsai and Uch-Kurgan in the Naryn River and Andijan, Teshiktash, Kujganyar and Bazar-Kurgansky in the Karadarya River).

In addition, a number of small tributaries feed its runoff. The contribution of the transboundary small rivers (TSR) within the valley to the Syrdarya River ranges from 2.9 to 4.1×10^9 m^3/year (Dukhovny et al. 2012). However, due to intensive irrigation, most of the rivers, especially on the left bank, do not reach the Syrdarya River anymore. Moreover, the natural hydrological regime of the Syrdarya River within the valley is disturbed by numerous irrigation withdrawals, water storage and return waters into the river (SIC ICWC 2004).

10.3.2   Irrigation Network

Before the Soviet government in CA, a set of ring irrigation systems existed in the FV, mainly on removal cones of Sokh, Isfara, Isfairam-Shakhimardan, Andijan, downstream part of the Naryn, Akbura and Aravan, while the lands located in the desert and desert-steppe part of the Central Fergana were undeveloped (Khamraev et al. 2011).

During the Soviet administration period, the construction of large dams and water reservoirs in the mountainous areas of upstream countries (Tajikistan and Kyrgyzstan) was initiated for irrigation purposes as a first priority and for hydropower generation as a second priority. In contrary, the lands in downstream countries (Uzbekistan, Kazakhstan and Turkmenistan) were suitable for practising irrigated agriculture and for growing water-intensive agricultural crops (cotton, rice, cereals, etc.).

As a consequence, irrigated lands for agricultural production, particularly in the Uzbek part of the valley, expanded rapidly, mainly for cotton monoculture (Kandiyoti 2005). These expansions resulted in the development of lands in the Central Fergana (Laktaev and Ermenko 1979) and promoted the construction of a number of large main canals and CDN within the command area of the Big Fergana Canal (BFC), South Fergana Canal (SFC) and North Fergana Canal (NFC) as well as Akhunbabayev's Canal, Big Andijan Canal (BAC) and Big Namangan Canal (BNC) (Fig. 10.5). These led to the improvement of irrigation infrastructure and changes to the Syrdarya River flow regime. Due to these constructions, 92 % of the Syrdarya River's flow was regulated (Toryanikova and Kenshimov 1999).

The length of inter-farm canals in the valley is 10,474 km (Table 10.2), of which 5010 km are lined to reduce seepage. The on-farm network has a total length of 53,570 km, and only 18 % of which is lined.

The irrigation infrastructure is in a poor condition and much of it is worn out. Considering the low efficiency of the irrigation system in FV, ranging from 55 % in the Fergana province to 63 % in Namangan province (Ikramov 2007), main and inter-farm canals require reconstruction and anti-seepage measures. Existing regulation structures need to be rehabilitated and new ones constructed; outlet structures and flow measurement stations also need to be constructed. Measures to improve the operation and management system of the irrigation infrastructure are also needed. The on-farm network is very complex, passing through settlements, and is frequently lined with trees, which complicates construction/reconstruction work.

10.3.3 Collector-Drainage Network

The expansion of the irrigated lands in the valley from 675,000 ha in 1932 (Irrigation of Uzbekistan 1975) to 1,572,970 ha in 2010 (CAWATERinfo 2012) has resulted in increasing water withdrawal from main rivers and induced the problems of land degradation and contamination of the Syrdarya River water resources. In their natural state, many of these lands are poorly drained. Therefore, intensive development of irrigated lands in the second half of the twentieth century increased irrigated lands with an artificially drained area by about 685,834 ha. About 70 % of the CDN out of a total length of 31,654 km is on farm. In total, 2861 operating vertical drainage wells, including provisions for running the pumps and surface drains to dispose the excess of water, were built in the valley as a network to lower the groundwater level. These drainage wells are mainly installed in areas with high soil permeability and preferably fresh groundwater that can be reused for irrigation. Consequently, this system is operation and maintenance intensive and requires a continuous electrical power supply. Therefore vertical drainage, a complicated engineering complex, was not used at full capacity after the collapse of the Soviet Union.

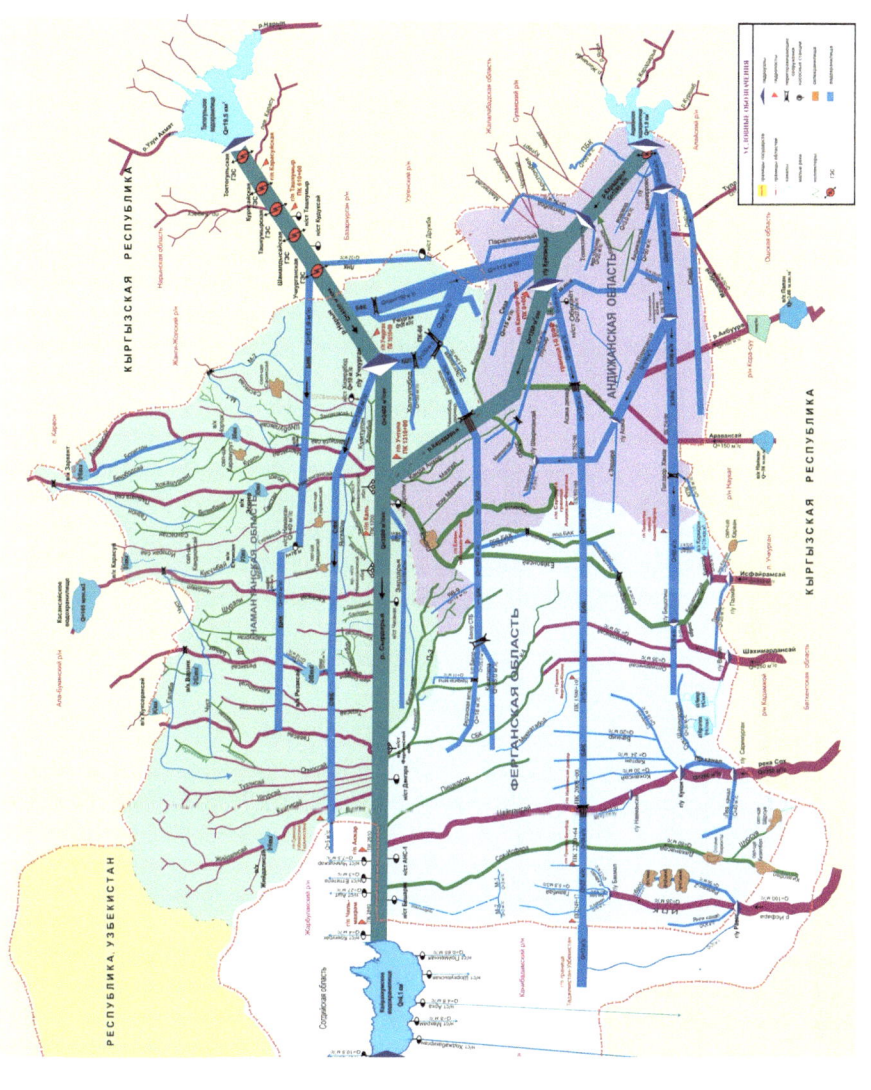

Fig. 10.5 Linear scheme of the irrigation network in the Fergana Valley (Source: author's composition based on UNDP 2007 and BWO Syrdarya maps)

Table 10.2 The type and length of the irrigation network in provinces of the Fergana Valley

Provinces	Gross irrigated lands (10^3 ha)	Inter-farm irrigation network				On-farm irrigation network			
		Total length (km)	% earthen	% concrete	Specific length (m/ha^{-1})	Total length (km)	% earthen	% concrete	Specific length (m/ha^{-1})
Andijan	274	2200	53	47	8.0	12,000	79	21	43.9
Fergana	366	2400	47	53	6.6	18,400	89	11	50.3
Namangan	282	2600	61	39	9.2	8200	81	19	29.0
Osh	184	1212	70	30	6.6	2934	91	9	15.9
Djalalabad	133	656	54	46	4.9	3632	77	23	27.3
Batkent	31	361	N/A	N/A	11.8	N/A	N/A	N/A	N/A
Sogd	304	1045	29	71	3.4	8402	73	27	27.7

Data for 2010
Data source: CAWATERinfo (2012), Dukhovny et al. (2012), NWO EECCA (2010)
Note: N/A – no information available

It also needs to be noted that irrigated lands in Kyrgyzstan provinces have not been strongly affected by the deficit of drainage (Dukhovny et al. 2014). For example, the irrigated lands of Kyrgyzstan within the valley are characterised as in good condition (90–96 % of the irrigated lands) in terms of land reclamation, while irrigated lands in Uzbek and Tajik parts of the valley are relatively poor owing to shallow GWL with higher mineralisation and varying levels of soil salinity.

The design parameters (e.g. cross section, slopes, bottom width) in most of the open drainage systems were altered, and a discharge capacity was reduced due to over siltation and eutrophication. This is mainly connected with ageing of the CDN since it was constructed more than 40 years ago on the one hand and low effectiveness of the drainage systems that have decreased to half capacity due to lack of investments in operation and maintenance on the other. To maintain the system in an effective manner, it is necessary to carry out regular and comprehensive technical maintenance (including observation and minor, major and emergency repairs, etc.). Periodicity of these works was set up during Soviet period and is still carried out. However, to cope with land restructuring and changes in cropping patterns, these regulations need to be reviewed and renewed.

10.3.4 Management Structures

The Fergana province has been chosen to describe the water management structures in the valley due to its hierarchy level being similar to the other provinces (Dukhovny et al. 2009).

The water requirement for irrigation is based on cropping patterns and scheduled by the centrally controlled state water management organisations (WMO) in 10-day intervals. Within the irrigation system, the cropping acreage planned for the forthcoming growing season thus needs to be submitted to water user associations (WUAs) at the end of February with the latest clarification by 1 June for the summer period and 1 December for the winter period (Mirzaev et al. 2009). WUAs as a key organisation within the hierarchy of WMOs (water supplier) and farmers including other water users (water consumer) then estimate planned water use within the territory. As a rule, the water-use plan is carried out using a bottom-up approach starting from the field plots up to the water intake point in main canals and integrates all plans within the hierarchy levels (Dukhovny et al. 2009; Mirzaev et al. 2009). Water distribution is achieved by applying a top-down method (Fig. 10.6) based on water-use limits (quotas). The Ministry of Agriculture and Water Management of the Republic of Uzbekistan (MAWR RUz) establishes the seasonal limits (quotas) based on the forecasted water availability in the sources (Basin Water Management Organisation, BWO Syrdarya and UzHydromet). The basin administrative irrigation system (BAIS) establishes the limit quotas for the districts based on the limit given by MAWR RUz, AIS to WUAs and WUAs to farmers and the other water users within their territory.

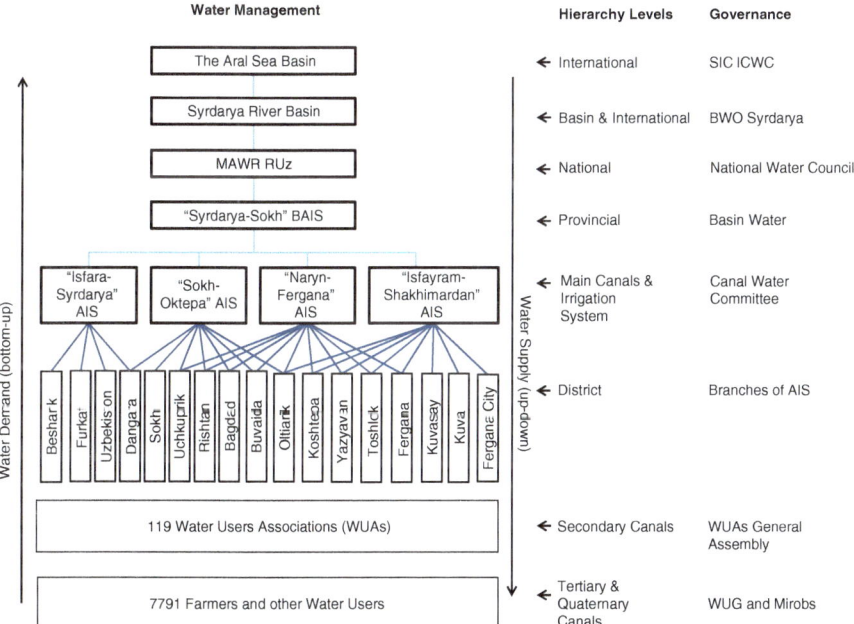

Fig. 10.6 Water management, hierarchy levels and governance linked to Fergana province (Source: Kenjabaev 2014. Note: The number of WUAs and farmers given as for December 01, 2011 according to BAIS "Syrdarya-Sokh" 2012)

10.3.5 Agricultural Water Use

In the valley, agricultural development and the well-being of the rural population depend on the water availability in the transboundary small rivers (TSRs) and on the flow regime of the Syrdarya River.

The agricultural sector constitutes the largest (consumptive) water user in the valley. More than 90 % of water resources actually withdrawn out of $10–14 \times 10^9$ m³/year are allocated to the agricultural sector.

According to Dukhovny et al. (2012), the share of water withdrawal from TSRs and CDN for land irrigation in the valley in the last decade is, respectively, on average 26 and 2.7 %. The water withdrawal from TSRs in the Uzbek part of the valley is as high as 86 %, because 62 % of the irrigated lands of the valley are located in the provinces of Uzbekistan (Jaloobayev 2007). The highest share of water withdrawal from CDN in the valley occurs in provinces located downstream.

Supported by the materials of the Institute of Hydrogeology and Engineering Geology, the Central Asian Scientific Research Institute of Irrigation, SANIIRI (2008) has assessed the annual groundwater resource in the valley to be 8.2×10^9 m³/year, of which about 1.3×10^9 m³/year and 1.9×10^9 m³/year are annually used for drinking and irrigation (including pumping by the vertical drainage wells during non-vegetation period) purposes.

Well balanced in the Soviet era, centrally planned water resource distribution amongst riparian countries changed after gaining the independence and thus required improvements. Consequently, a legal framework for cooperation on the Syrdarya River was put into place in the early 1990s (Dukhovny 2010).

However, looking at the then agreed arrangements on water allocation, these have not been fully implemented to present, or it has proven impossible to agree on water allocation. One limiting factor is that the energy sector (e.g. hydropower) is not addressed by the existing regional organisations engaged in water management cooperation (Rakhmatullaev et al. 2010). Therefore, water supply in FV is not sustainable due to the non-inclusion of the hydropower regime of Naryn and Syrdarya Rivers.

10.4 Ecological Impact of Water Management and Irrigation Practices

Environmental quality issues are significant in relation to irrigated agriculture. Subsurface drainage water quality reflects the groundwater quality and soil water constituents of the soil being drained. Because of the soil salinisation, inefficient irrigation and inadequate drainage, large amounts of water are annually withdrawn from the rivers for land washing (the so-called salt leaching). This activity implies considerable volumes of return water containing salts, agrochemicals and trace elements to rivers and streams associated with the discharge of the CDW.

There is a hydrodynamical connection between groundwater and filtration fluxes from the irrigation canals and drains. The groundwater mainly recharges from the surface water. Hence, shallow groundwater levels, especially in the lower and central parts of the valley, are a major problem. Recent increases in areas under rice in Burgandin Massif in Kadamzhai district of Batken province have been causing waterlogging in most areas in Bagdad, Altyarik and Rishtan districts of the Fergana province (MAWR RUz 2009). Further development of lands in the highlands in Kyrgyzstan (IMF 2012) has aggravated the situation.

Although FV is a relatively water-rich region compared to the lower course regions along the Syrdarya River, water shortage is common in some areas because of mismanagement of water resources, e.g. excessive water use for irrigation and leaching, and water allocation within the systems. As irrigation in an arid climate is needed to create optimal soil moisture for plant growth, it significantly alters natural hydrogeological soil and other land reclamation conditions, causing the salinisation and waterlogging of soils (Dukhovny et al. 1979). The groundwater level rises because of the high water seepage from the canals (about 52 % of main canals and 82 % of field canals are earthen (Table 10.2)); available artificial CDN has deteriorated, while natural drainage is insufficient.

Although shallow GWL enables the reduction of irrigation norms due to the contribution of capillary rise, at the same time, it brings up the existing salts, if they exist, into the adjacent soil layers. To cope with the situation, Dukhovny et al. (1979) proposed to increase the canal system's efficiency and stop water flow in canals during October–March. However, high capital investment and regular maintenance work are required to increase the system's efficiency. Stopping the water flow in canals during October–March is impossible under the current situation due to the following:

Cultivation of cereals, particularly winter wheat (at the expense of the reducing cotton area), has increased since the 1990s (see Fig. 10.2).

Release of winter water from the Toktogul reservoir and the pressure on the Syrdarya River (Abbink et al. 2005). Distribution of excessive winter flows through canals to the fallow lands and temporary storage of water (groundwater banking) in the aquifers of the FV could become a solution until reliable water-energy consensus is found between upstream and downstream countries (Karimov et al. 2010).

The leaching of salt-affected lands, particularly in Central Fergana, is recommended in the winter periods (World Bank et al. 2013).

Improper management of irrigation and drainage systems and altering the hydrological regime of the Syrdarya River have led to the deterioration of water quality from upstream to downstream (SIC ICWC et al. 2011). The water salinity observed in the Syrdarya River has increased in the last 40 years from 300 to 600 mg/l in the upper reaches to 3000 mg/l in the lower reaches within the valley, with a prevailing salt composition of $MgSO_4$, $Ca(HCO_3)_2$, $NaCl$ and $CaSO_4$. Moreover, concentrations of some metals including sulphates and chlorides have increased overall. This has impaired the water resources for drinking purposes in the middle course and sometimes serves as a source of diseases, such as hepatitis, typhoid and gastrointestinal disorders causing morbidity in the local people (SIC ICWC et al. 2011).

10.5 Economical Necessities to Maintain Irrigation Infrastructures

It is important to note that there is a certain discrepancy between the priorities of the government and those of the water users in irrigated farming. The government, in spite of providing the appropriate volume of water resources for irrigated agriculture, is interested in decreasing expenses involved in the transportation and diversion of excess water and allotment of released water between sectors of economy including the demands of environmental supply, e.g. in increasing productivity of water use and water saving, whereas water users are interested in gaining the maximal profit from their agricultural farming in irrigated lands, especially as there is no charge for the volume of water used. Therefore, water supply-related problems should be solved on a foundation of a balancing of interests and by searching for compromises. To this end, applying sound scientific practices, approaching problem solving rationally and neutrally and, whenever necessary, adequately participating the relevant stakeholders are regarded as essential.

Irrigation and drainage were subject to significant investment in the Soviet era, but water management did not occur at a correct level (Ikramov 2007), i.e. the degree of water use was very high, resulting in a rise of GWL and salinity, which in turn led to degradation of the agricultural land. Often, construction of the canals and their maintenance were not thorough, bringing the irrigation and drainage system in dire straits even before the independence of the Central Asian countries in 1991.

After independence, the situation has worsened considerably. The maintenance of the canals has been repeatedly postponed, and consequently, many irrigation and drainage systems have come into disrepair. In many regions, water supply has been undertaken intermittently. Moreover, farmers often do not have enough choice in what to grow and often have limited access to information regarding seed quality, agricultural inputs, services and markets that would enable them to adapt to new conditions. In these circumstances, interruptions in the water supply can lead to catastrophic consequences. Yields per hectare have declined sharply, depressing agricultural incomes for both farms and states even more. With reduced incomes, agricultural workers have fewer funds for maintenance, meaning the infrastructure is continuing to deteriorate, water is becoming even less reliable, and this is a cycle that is ongoing.

Central governmental departments, which once controlled the operation and maintenance of irrigation and drainage facilities in Central Asia during the Soviet era, have weakened considerably. Shrinking budgets mean that wages in water management organisations and their local offices have dropped significantly and qualified technical employees have left (brain drain).

10.5.1 Water Productivity

As Seckler et al. (2003) stated, "current efficiency of water use is so low, especially in irrigation, that most, if not all, of future water needs could be met by increased efficiency alone, without development of additional water supplies". Therefore, substantially increasing water productivity in agriculture is essential to meet the goals of food and environmental security. To achieve it, research that spans the wide range of analysis is required.

Studies of Kenjabaev (2014) in Fergana province showed that irrigation water productivity for cotton and wheat tended to increase while the irrigation water supply decreased. Deficit irrigation could be one of the factors ensuing water productivity variability for cotton and wheat. Ergo, there is a high potential for farmers to save irrigation water through the improved water management and agronomic practices that ensure high water productivity.

10.6   Conclusions

Management of irrigation and drainage infrastructure is very complex in the Fergana Valley. Although temporary water withdrawal from collector-drainage systems is complicating water accounting, irrigation water management within hydrographic units is in accordance with the morphology of the river basin.

Improved water management should provide reliable (with minimal fluctuations) and manageable flow rates to all water consumers within the canal command area. Currently, the process of agricultural restructuring is not complete; therefore, questions about the financial and economic aspects in agriculture and water management are still in the development stage.

Deterioration of irrigation and drainage systems and lack of maintenance are common problems in the region. Specific water consumption is high because of water losses, evaporation and overwatering. Efforts have been made in riparian countries to enhance irrigation systems and their efficiency, particularly through introduction of the integrated water resources management (IWRM) principles (Dukhovny et al. 2013); however, a shortage of financial resources for renovation, maintenance and further upscaling within the entire basin persists. Further efforts are also needed to improve water efficiency; to increase effectiveness of irrigation systems, including repairs and maintenance of the existing infrastructure; to switch to less water demanding crops and to limit the irrigated land area.

Environmental impact assessments of planned transboundary projects should be carried out in a more systematic manner, involving the countries and populations affected. This is particularly relevant for planned hydropower projects in Kyrgyzstan and Tajikistan. Also, cooperation on the management of reservoirs can bring benefits by addressing the needs of different sectors; different reservoirs in a cascade can have complementary operating modes. Developing small-scale hydropower projects, which do not disrupt water flows and are less damaging to the environment, could be considered as an option for energy generation.

Transboundary monitoring needs to be significantly strengthened, especially that of water quality. Research on groundwater, which plays a potentially important role in sustaining ecosystems and limiting land degradation, should also be intensified.

Developing and training employees will lay the foundation for unlocking people potential in the implementation of the best management practices (BMP) and integrated water resources management (IWRM) principles.

References

Abbink K, Moller L, O'Hara S (2005) The Syr Darya River conflict: an experimental case study. Discussion paper no 2005–14 for the Centre of Decision Research and Experimental Economics, Nottingham

Abdullaev I, Kazbekov J, Manthritilake H et al (2010) Water user groups in Central Asia: emerging form of collective action in irrigation water management. Water Resour Manag 24:1029–1043. doi:10.1007/s11269-009-9484-4

Baker N (2011) The Fergana Valley: a soviet legacy faced with climate change. In: ICE Case Studies. http://www1.american.edu/ted/ICE/ferghana.html. Accessed 11 Feb 2015

CAWATERinfo (2012) Regional information system on water and land resources in the Aral Sea Basin (CAWater-IS). http://www.cawater-info.net/bd/index_e.htm. Accessed 11 Feb 2015

Dukhovny V (2010) Water resources management in Central Asia. Achieving the consensus between water and energy sectors. Scientific Information Center of the Interstate Commission for Water Coordination (SIC ICWC), Tashkent

Dukhovny V, Stulina G (2012) Implementation IWRM in Ferghana Valley. Paper presented at the 5th international scientific conference BALWOIS, Balkan Institute for Water and Environment, Ohrid, 28 May–2 June 2012

Dukhovny V, Baklushin M, Tomin E et al (1979) Horizontal drainage in irrigated lands. Kolos, Moscow (in Russian)

Dukhovny V, Sorokin A, Stulina G (2008) Should we think about adaptation to climate change in Central Asia? Adaptation to climate change: regional challenges in light of world experiences. Scientific Information Center of the Interstate Commission for Water Coordination (SIC ICWC), Tashkent

Dukhovny V, Sokolov V, Manthrithilake H (2009) Integrated water resources management: putting good theory into real practice. Central Asian Experience. Scientific Information Center of the Interstate Commission for Water Coordination (SIC ICWC), Tashkent

Dukhovny V, Sokolov V, Horst M et al (2012) The dynamics of present water balance in Fergana Valley. Scientific Information Center of the Interstate Commission for Water Coordination (SIC ICWC), Tashkent

Dukhovny VA, Sokolov VI, Ziganshina DR (2013) Integrated water resources management in Central Asia, as a way of survival in conditions of water scarcity. Quat Int 311:181–188. doi:http://dx.doi.org/10.1016/j.quaint.2013.07.003

Dukhovny V, Yakubov S, Kenjabaev SM (2014) Role of drainage water management in irrigated agriculture of Central Asia. In: Proceedings of the 12th ICID international drainage workshop, "drainage on waterlogged agricultural areas", St. Petersburg, 23–26 June 2014

Ibragimov P, Avtonomov V, Amanturdiev A et al (2008) Uzbek scientific research institute of cotton breeding and seed production: breeding and germplasm resources. J Cotton Sci 12:62–72

Ikramov R (2007) Underground and surface water resources of Central Asia and impact of irrigation on their balance and quality. In: Lal R, Suleimenov M, Stewart B et al (eds) Climate change and terrestrial carbon sequestration in Central Asia. Taylor & Francis, London, pp 97–107

IMF (2012) Medium term development program of the Kyrgyz Republic for 2012–14. International Monetary Fund, Washington, DC

Irrigation of Uzbekistan (1975) Present condition and perspectives of irrigation development in the Syrdarya Basin, vol II. Fan, Tashkent (in Russian)

Jaloobayev A (2007) IWRM financial, economic, and legal aspects: the example of the "IWRM-Ferghana" project. In: Wouters P, Dukhovny V, Allan A (eds) Implementing integrated water resources management in Central Asia. Springer, Dordrecht, pp 45–53

Kandiyoti D (ed) (2005) The cotton sector in Central Asia: economic policy and development challenges. The School of Oriental and African Studies, London

Karimov A, Smakhtin V, Mavlonov A et al (2010) Water "banking" in Fergana Valley aquifers – a solution to water allocation in the Syrdarya River Basin? Agric Water Manage 97:1461–1468. doi:http://dx.doi.org/10.1016/j.agwat.2010.04.011

Kenjabaev SM (2014) Ecohydrology in a changing environment. Dissertation, Justus-Liebig-Universität Gießen

Khamraev SR, Islamov UI, Curtis A et al (eds) (2011) Water resources management in Uzbekistan. Global Water Partnership of Central Asia and Caucasus (GWP CACENA), Tashkent

Laktaev N, Ermenko G (1979) Development of some ameliorative measures to develop lands in Central Fergana. SANIIRI, Tashkent

Lioubimtseva E, Henebry GMM (2009) Climate and environmental change in arid Central Asia: impacts, vulnerability, and adaptations. J Arid Environ 73:963–977. doi:http://dx.doi.org/10.1016/j.jaridenv.2009.04.022

MAWR RUz (2009) Ferghana Valley water resources management project – phase 1: resettlement policy framework and specific resettlement action plan. Ministry of Agriculture and Water Resources of the Republic of Uzbekistan, Tashkent

Mirzaev N, Saidov R, Ergashev I et al (2009) Manual on preparation and organization of plans of water use and water distribution. Scientific Information Center of the Interstate Commission for Water Coordination, Tashkent (in Russian)

Nerozin S (2010) Year-round use of irrigated lands (guidebook for farmers). Water productivity improvement on plot level. Scientific Information Center of the Interstate Commission for Water Coordination (SIC ICWC), Tashkent (in Russian)

NWO EECCA (2010) Charter of the regional network of water (basin) organizations from Eastern Europe, Caucasus and Central Asia. http://www.eecca-water.net/content/view/1150/77/lang,english/. Accessed 11 Feb 2015

Qadir M, Noble A, Qureshi A et al (2009) Salt-induced land and water degradation in the Aral Sea basin: a challenge to sustainable agriculture in Central Asia. Nat Res Forum 33:134–149. doi:10.1111/j.1477-8947.2009.01217.x

Rakhmatullaev S, Huneau F, Le Coustumer P et al (2010) Facts and perspectives of water reservoirs in Central Asia: a special focus on Uzbekistan. Water 2:307–320. doi:10.3390/w2020307

Reddy MJ, Muhammedjanov S, Jumaboev K et al (2012) Analysis of cotton water productivity in Fergana Valley of Central Asia. Agric Sci 3:822–834. doi:10.4236/as.2012.36100

RUz SC (2006) Environmental situation and utilization of natural resources in Uzbekistan: facts and figures 2000–2004. State Committee of the Republic of Uzbekistan on Statistics, Tashkent

SANIIRI (2008) Development of engineering-ameliorative principles to increase water availability in irrigated lands of Uzbekistan: Syrdarya River Basin. Final report on scientific research work. SANIIRI, Tashkent (in Russian)

Seckler D, Molden D, Sakthivadivel R (2003) The concept of efficiency in water resources management and policy. In: Kijne JW, Barker R, Molden D (eds) Water productivity in agriculture: limits and opportunities for improvement. CABI Publishing, Wallingford, pp 37–51

Sic ICWC (2004) Drainage in Aral Sea Basin: towards a strategy of sustainable development. Scientific Information Center of the Interstate Commission for Water Coordination, Tashkent (in Russian)

SIC ICWC, UNECE, CAREC (2011) Water quality in the Amudarya and Syrdarya River Basin. Analytical report. United Nations Economic Commission for Europe, Tashkent

Stucker D, Kazbekov J, Yakubov M et al (2012) Climate change in a small transboundary tributary of the Syr Darya calls for effective cooperation and adaptation. Mt Res Dev 32:275–285. doi:10.1659/MRD-JOURNAL-D-11-00127.1

Syrdarya-Sokh (BAIS) (2012) Data on dynamics of water user's association (WUA) development during 2000–2011. BAIS "Syrdarya-Sokh," Fergana

Toryanikova R, Kenshimov A (1999) Current status of water quality in the Syr Darya River Basin. Environmental Policies and Institutions for Central Asia (EPIC), Almaty

UNDP (2007) Water. Critical resource for Uzbekistan's future. United Nations Development Programme, Tashkent

UNECE (2011) Second assessment of transboundary rivers, lakes and groundwaters. United Nations Economic Commission for Europe, New York/Geneva

UNEP, UNDP, OSCE, NATO (2005) Environment and security. Transforming risks into cooperation. Central Asia: Ferghana/Osh/Khujand area. United Nations Environment Programme, Geneva

WARMAP (1997) The water user and farm management survey (WUFMAS). In: CAWATERinfo. http://www.cawater-info.net/index_e.htm. Accessed 24 Feb 2015

World Bank, SIC ICWC, SPA SANIIRI (2013) Assessment of the results of previous pilot projects on irrigation and drainage in Central Asia. In: Knowledge base on water and land resources use in the Aral Sea Basin. http://www.cawater-info.net/bk/water_land_resources_use/english/english_ver/wb_projects_en.html. Accessed 11 Feb 2015

Chapter 11
Where Water Meets Agriculture: The Ambivalent Role of Water Users Associations

Timothy Moss and Ahmad Hamidov

Abstract This chapter investigates the role of water users associations (WUAs) in managing the Fergana Valley's irrigation system at a local level. WUAs were established in the Uzbek section of the Fergana Valley only from the early 2000s onwards and are generally not regarded as having been effective to date, although individual instances of modestly successful WUAs indicate their future potential as viable entities for collective modes of water management. This chapter begins by explaining the origins, purpose and structure of WUAs in the Fergana Valley as set out in policy guidelines and then contrasts this with a study of how they are working in practice. In the concluding section, the effectiveness of WUAs in the Uzbek section of the Fergana Valley is assessed in terms of criteria derived from the literature. This chapter reveals that Uzbekistan's WUAs lack the funding, water user representation and resources to tackle the major structural problems confronting Fergana Valley's post-socialist irrigation system. Their heavy dependence on powerful institutional regimes for irrigation and for agriculture also severely restricts their action. There exist important exceptions, where WUAs are exploring innovative ways of coping with the enormity of their tasks, in isolation and in collaboration, but these represent only a small minority of WUAs in the region and are, to a large extent, dependent on temporary donor funding.

Keywords Water users associations • Water governance • Institutions • Path dependency • Transformation • Fergana Valley • Integrated water resources management • Participation • Irrigation • Education

T. Moss (✉)
Leibniz Institute for Regional Development and Structural Planning (IRS),
Flakenstraße 28-31, 15537 Erkner, Germany
e-mail: MossT@irs-net.de

A. Hamidov
Department of Agricultural Economics, Humboldt-Universität zu Berlin,
Philippstraße 13, 10099 Berlin, Germany

© The Author(s) 2016
R.F. Hüttl et al. (eds.), *Society - Water - Technology*, Water Resources
Development and Management, DOI 10.1007/978-3-319-18971-0_11

11.1 WUAs in the Global Discourse on Integrated Water Resources Management

Over the past century, the world has witnessed an almost threefold increase in the total area of irrigated agriculture (Ostrom 1992). Representing 20 % of total cropland, irrigated agriculture produces more than 40 % of the world's total agricultural output (Perry 2007, p. 369). Expansion of irrigated agricultural areas has been a key component of nation-building processes (Kreutzmann 2015, in this volume) and, in recent decades, the construction and modernisation of large-scale irrigation facilities has become the target of massive investment programmes of international donor agencies in developing and transition countries. However, towards the late twentieth century, the management and operation of irrigation systems, particularly at the community level, proved an increasing fiscal burden for many governments. As central government funds in many countries were reduced, maintenance standards declined and irrigation infrastructure began to deteriorate at a serious rate (World Bank 2007). One of the main solutions widely voiced from the early 1990s onwards has been the creation of so-called water users associations (WUAs), in which farmers are given more responsibility to manage and maintain local irrigation systems themselves.

WUAs have been defined as "a voluntary, nongovernmental, non-profit entity established and managed by groups of farmers located along one or several watercourse canals" (Winrock International 2007, cited in Gunchinmaa and Yakubov 2010, p. 166). Ideally, WUAs are set up by a group of farmers and other water users along one or more hydrological subsystems or watercourses to collectively manage, operate, maintain and develop a local irrigation and drainage system. Membership is based on contracts and/or agreements between the members and the WUA. In accordance with WUA by-laws, their main responsibilities generally include:

- Ensuring reliable distribution of water amongst water users
- Determining and collecting irrigation service fees
- Resolving disputes on water use and management of the irrigation system in an appropriate, transparent and democratic manner
- Maintaining, refurbishing and improving the irrigation system in the WUA operational area

Within the global debate on water resources management, the wide appeal of WUAs can be attributed not merely to the inherent advantages emanating from the self-organisation of local irrigation infrastructures by the farmers that use them but also to the various ways in which WUAs – on paper at least – play to the dominant discursive paradigm of integrated water resources management (IWRM) (for a definition of IWRM, see GWP 2000, p. 22). Within the global discourse on IWRM, increasing the involvement and responsibility of water users in water management issues has become a central tenet for successful implementation. Decentralising decision-making powers and strengthening the role of local water users is also acknowledged as a core element of IWRM in Central Asia and post-socialist transition countries in general (Dombrowsky et al. 2011; Dukhovny et al. 2013). Globally, WUAs are projected to play a critical role in promoting IWRM reform at the community level.

Whilst the aspirations for WUAs are high, it is now recognised that, to be successful, they are hugely dependent on favourable conditions. On the basis of past experiences worldwide, Merrey (1996, p. 4, also cited in Gunchinmaa and Yakubov 2010, p. 168) cites four principles as preconditions for an effective WUA. These are (1) a supportive institutional environment, (2) the capacity to operate and maintain infrastructure, (3) the benefits of user participation exceeding the costs and (4) effective collective choice arrangements. We will use these four principles to assess the performance of the Fergana Valley WUAs in Sect. 11.4.

Experiences from around the world indicate that some WUAs do live up to the aspirations placed in them and benefit from favourable institutional frameworks, at least in part. Frequently cited instances of largely successful WUAs include those created in Mexico in response to the economic crisis of the late 1980s, when responsibility for local irrigation management and infrastructure was transferred to water users (Rap and Wester 2013). In Turkey, WUAs were set up as part of a national decentralisation policy in response to the inability of the State Hydraulic Works (DSI) to continue funding the operation and maintenance of irrigation infrastructure (Uysal and Atis 2010). In Nepal, experience indicates that farmers with long-term irrigation management traditions can, through the development of autonomous and self-governing WUAs, improve communication, develop their own agreements, regulate compliance and sanction those who do not comply with their own rules (Ostrom 2000, p. 4). Such groups were able to distribute water equitably and keep their irrigation systems better maintained than those systems run by central government agencies (*ibid*). These examples corroborate the expectations placed in WUAs to enable people within a community to pool their resources (e.g. knowledge, expertise and money), to allocate water more effectively amongst members, to keep irrigation and drainage infrastructure in a good condition, to resolve water-related disputes and to impose sanction mechanisms against noncontributors or rule-breaking individuals.

Many other examples from around the world, however, present a very different story of failure and conflict. The difficulties encountered in setting up and operating WUAs are manifold and include financial, political, institutional and administrative constraints. The driving forces for establishing WUAs differ hugely from country to country, depending on political and economic circumstances (World Bank 2007). In most places, the development of WUAs was promoted either through government programmes or by external donor-funded investment projects. These reforms often disregarded local knowledge and experiences. Instead, policies drew on blueprint models with little consideration for the specific sociopolitical context of a country and its institutional capacities. Moreover, implementation mostly followed a top-down bureaucratic approach, allowing little space for the active involvement of civil society (Theesfeld 2005; Yalcin and Mollinga 2007; Abdullaev et al. 2010).

A closer review of the literature on post-socialist countries shows that the strong push of the World Bank and other donor agencies to establish WUAs in communities neither provided for functional local irrigation sector management nor involved local water users in their creation (Theesfeld 2005). As a result, many WUAs (e.g. in Bulgaria) terminated after one irrigation season (*ibid*). Sehring (2009) reports that newly established WUAs in post-socialist Kyrgyzstan and Tajikistan are undermined by informal practices such as patronage and unauthorised water withdrawal. If the WUAs' tasks are to deliver water to its members in due time, keep irrigation and drainage canals maintained, punish those who do not comply with rules and resolve conflicts within the WUA, they were barely effective on the ground (*ibid*). Instead, when the head of the WUA was a local patron, he/she was in most cases able to ensure compliance with water rules using the authority of his position (Sehring 2009). Despite Central Asian governments issuing decrees on the establishment of WUAs and creating thousands of WUAs within a very short period of time, in reality most of them exist on paper only (Wegerich 2009; Abdullaev et al. 2010). Consequently, many commentators see WUAs in Central Asia as inactive and not financially viable (Wegerich 2000; Abdullaev et al. 2010). A low level of user participation and frequent external interventions into WUA internal procedures are additional constraints for WUAs' malfunctioning in Central Asia (Schlüter et al. 2010). In the following section, we take a closer look first at the origin and purpose of WUAs in the Fergana Valley (Sect. 11.2) and then study how they are working in practice (Sect. 11.3) before assessing their effectiveness in terms of the criteria listed above (Sect. 11.4).

11.2 WUAs as Building Blocks of Irrigation Management in the Fergana Valley

The emergence and practice of WUAs in the Fergana Valley cannot be understood without appreciating the history of irrigation management in the region. The chapter by Hermann Kreutzmann, analysing the trajectory of large-scale irrigation from Tsarist times to the present day, provides the necessary insight into the economic, political and sociocultural factors which have shaped this major water engineering project (Kreutzmann 2015, in this volume). In our chapter, we recollect three key factors of Fergana's irrigation system: its strong path dependency in both material and social terms, the powerful interdependence of water management and agricultural production and the conflictual nature of transboundary cooperation between today's independent post-socialist republics.

11.2.1 Early Origins of WUAs

In the Fergana Valley, WUAs were first established in the mid-1990s in Kyrgyzstan (Sehring 2009) but only in the early 2000s in Tajikistan (*ibid*) and in Uzbekistan (Hamidov 2007; Abdullaev et al. 2009). In all instances, the driving forces behind these reforms included the deterioration of on-farm water infrastructure, unequal distribution of water, frequent conflicts amongst water users, inefficient irrigation methods and significant reduction of the state budget for local irrigation administration. In the case of Uzbekistan, for example, during 1991–2000, there was a serious deterioration of the secondary and tertiary canal systems because former collective and state farms could not operate and maintain irrigation canals due to inadequate funding. This resulted in low yields and, subsequently, low incomes for farmers. Meanwhile, the distribution of irrigation water became severely unequal, especially for downstream farmers. Disputes amongst farmers over water increased. It became important to introduce farmer-based organisations in place of the poorly functioning collective farms. In 1996, the Ministry of Agriculture and Water Resources (MAWR) of Uzbekistan contracted the Irrigation and Water Problems Research Institute (former SANIIRI) to study the experience of different countries with WUAs and to set up a framework for the establishment of WUAs. In 1999, SANIIRI completed the study and presented its recommendations to MAWR. As a result, the first WUA was created in Uzbekistan in February 2000 in the Khorezm Region (Hamidov 2007). The first WUAs in the Uzbek section of the Fergana Valley were founded in January 2002 (MAWR RUz 2014).

Critical behind the emergence of WUAs was the IWRM Fergana project funded by the Swiss Agency for Development and Cooperation from 2001 onwards. This transboundary project helped institutionalise IWRM as the guiding policy paradigm for the region (Gunchinmaa and Yakubov 2010, p. 166; Dukhovny et al. 2013). A key element of this IWRM-inspired reform process was to transfer operation and maintenance responsibilities for on-farm irrigation to water users in the form of WUAs. This new governance structure at the lower level was, in theory at least, intended to encourage local farmers to act collectively in managing and maintaining their irrigation systems. Additional institutional arrangements, such as the irrigation service fee, were also introduced by this reform.

11.2.2   Organisational Structure of WUAs

Unlike in neighbouring Kyrgyzstan and Tajikistan, there is no specific law on WUAs in Uzbekistan (Gunchinmaa and Yakubov 2010, p. 173). The legal basis for their status and organisation is provided primarily by the Cabinet of Ministers' Decree No.8, approved in 2002, which determines WUAs as the entity responsible for irrigation management at on-farm level in Uzbekistan. Subsequently, the reformed national Water and Water Use Act of 29 December 2009 clarified the role of WUAs, specifying for instance that WUAs are required to deal with internal conflicts relating to water distribution amongst water users (Art. 2.1) and that they should cooperate with state agencies in protecting and conserving water (Art. 10). National policy guidelines require WUAs to be organised around two units: a decision-making body and a management body (ADB 2006). Figure 11.1 provides an example of the typical structure of a WUA in Uzbekistan. The upper

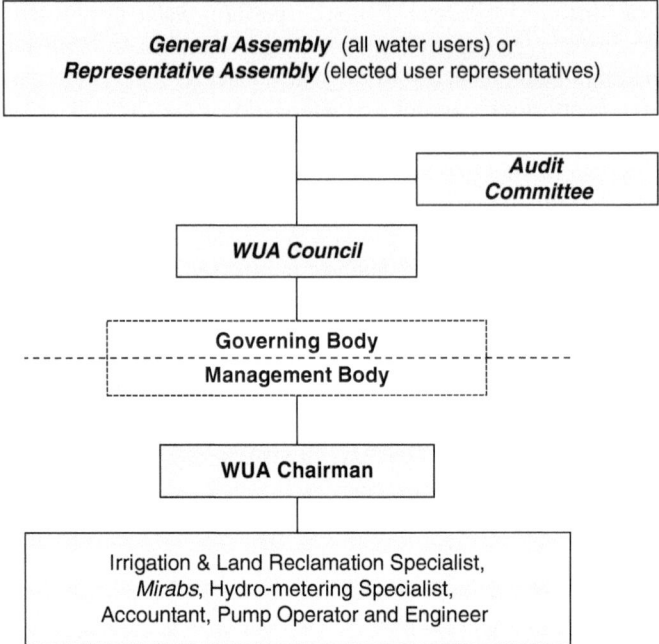

Fig. 11.1 Internal structure of a typical WUA in Uzbekistan (Source: ADB 2006)

part of the hierarchy comprises the general assembly, the WUA council and the audit committee; the lower part comprises the management team of chairman and specialist staff.

The supreme governing body of the WUA is the *general assembly*. The policy guideline specifies that the assembly should meet twice a year, when all members of the WUA are invited and encouraged to participate. When a member is unable to attend, he/she assigns a representative (usually a neighbouring farmer in the same WUA) to report on key messages emerging from the meeting. The main topics of discussion include electing a chairman, determining irrigation service fees (ISFs), evaluating the performance of the WUA management team and identifying the dates for maintaining on-farm irrigation canals through *khashars*, voluntary action of groups of people living in the community. In her study, Zavgorodnyay (2006) points to the fact that, in practice, the general assembly does not play a major role in the decision-making process. Instead, the chairman closely collaborates with the director of local machinery tractor parks (which provide agricultural machinery services to WUA members) and the head of the village citizens' assembly to reach decisions (*ibid*).

The primary task of a *WUA council* is to supervise and monitor the financial and technical activities of the WUA. Members of the WUA council work without salaries and are required to meet at least once a month. Typically, a WUA council has about five members elected by the assembly. The task of the *audit committee* is to conduct an annual review of all financial records and bank transfers of the WUA. The committee members are elected during the general assembly meeting and they report directly to the assembly. A WUA accountant assists the committee during the assessment.

Additionally, there is a *WUA management team* to manage a WUA's activity on a daily basis. This team includes the WUA manager, elected by the general assembly, an accountant and technical staff (e.g. *mirab* – or "water master" – and engineers), with its activities paid for through ISFs. The manager's responsibility is to handle irrigation-related activities in the WUA territory, identify potential irrigation canals to be maintained and initiate *khashars*, prepare the annual budget of the WUA and calculate irrigation service fees (ISFs) for both members (farmers) and non-members (e.g. local households). The WUA accountant is responsible for the overall financial activities of the association. He/she is in charge of collecting ISFs.

11.2.3   Spatial Reform of WUAs

A major reform to WUAs in Uzbekistan subsequent to their creation in the early 2000s was their spatial reorganisation around hydrographic, rather than administrative, units in accordance with IWRM principles (Dukhovny et al. 2013, p. 184; see Chap. 8, Moss and Dobner 2015, in this volume). With Decree No. 320 of the Cabinet of Ministers of July 2003 on "improving the organisation of water resources management", the Uzbek state started reforming its water sector based on a policy of decentralisation. A key element of this reform was to create a basin irrigation systems authority (BISA) in place of the former regional water resources management department (*OblVodKhoz*) and to establish irrigation systems authorities (ISAs) as well as main canal authorities (MCA), subordinated to BISA, in place of district water resources management departments (*RayVodKhoz*). As a result, the number of local water organisations was reduced. For example, 12 administrative organisations at provincial level were replaced by 10 BISAs in 2003. In the Fergana Valley within Uzbekistan, which includes three provinces (Andijan, Fergana and Namangan), three BISAs were created: Naryn-Karadarya, Naryn-Syrdarya and Syrdarya-Soh. Furthermore, 13 ISAs and 5 CMOs were also established as a result of the 2003 decree.

At the lower level of the water management hierarchy, this decree called for reorganising WUAs spatially around the local canal network, rather than administrative territories. Wegerich (2009) points out that at the WUA level the shift did not incorporate the drainage system but focused solely on the infrastructure delivery system. Since this reform, equity in the allocation of water supply per hectare has increased within the Khorezm province, the lower part of the Amudarya basin (*ibid*). Dukhovny et al. (2013) also report that the transition to the hydrographic principle in the Fergana Valley of Uzbekistan brought positive outcomes in terms of water distribution equity between up- and downstream communities. However, they also acknowledge that the transition to hydrographic units of management is not yet complete in the Fergana Valley as a whole (*ibid*).

Table 11.1 Distribution of the number of WUAs in the Uzbek section of the Fergana Valley

Provinces	No. of WUAs	No. of members	Total irrigated area (ha)
Andijan	109	6390	246,278
Fergana	124	6098	322,167
Namangan	147	5793	273,104

Source: MAWR RUz (2014)

11.2.4   Current Status of WUAs

By the end of 2013, 1510 WUAs had been established in Uzbekistan, serving nearly 57,000 individual farmers and covering about 4.2 million ha (MAWR RUz 2014). Within the Fergana Valley, comprising three provinces of Uzbekistan, the total number of WUAs is 380. On average, each WUA covers about 2200 ha, varying significantly between different provinces (see Table 11.1). As a result of the state's land consolidation policies of 2008, 2009 and 2010 (Moss and Dobner 2015, in this volume), the number of members representing each WUA in the valley has sharply declined and, on average, each WUA has now about 50 members. According to data from MAWR RUz (2014), despite the fact that WUAs were initially created from the beginning of the 2000s in the Fergana Valley, most associations were re-established based on hydrographic principles and reregistered at the Ministry of Justice as a non-governmental and non-profit organisation only in the late 2000s.

11.3   Fergana Valley's WUAs in Practice

Created by central government decree yet without a clear regulatory mandate or financial support, the WUAs in Uzbekistan have faced from the beginning an uphill struggle. On paper, they are an integral piece in the hierarchical jigsaw of Uzbek water management, filling the institutional void for on-farm irrigation systems left by the collapse of collective farms. Expectations in their performance were – and still are – high, although for a variety of reasons. State water authorities want WUAs to keep the existing irrigation system going, donor organisations hope WUAs can generate more participatory forms of integrated water resources management, and farmers want WUAs to provide them with the water they need for an increasing variety of crops. Caught between these multiple claims in a context of severe socio-economic transformation and deep-seated authoritarian rule, WUAs face an unenviable task. How, then, are they working in practice? To what extent are they fulfilling the expectations placed in them? How far do they meet the principles of success as defined in the international literature on WUAs?

Taking the Uzbek section of the Fergana Valley as our case study area (see Fig. 11.2), we investigate in this section past and current experiences of WUAs working at the interface of water management and agricultural production on the

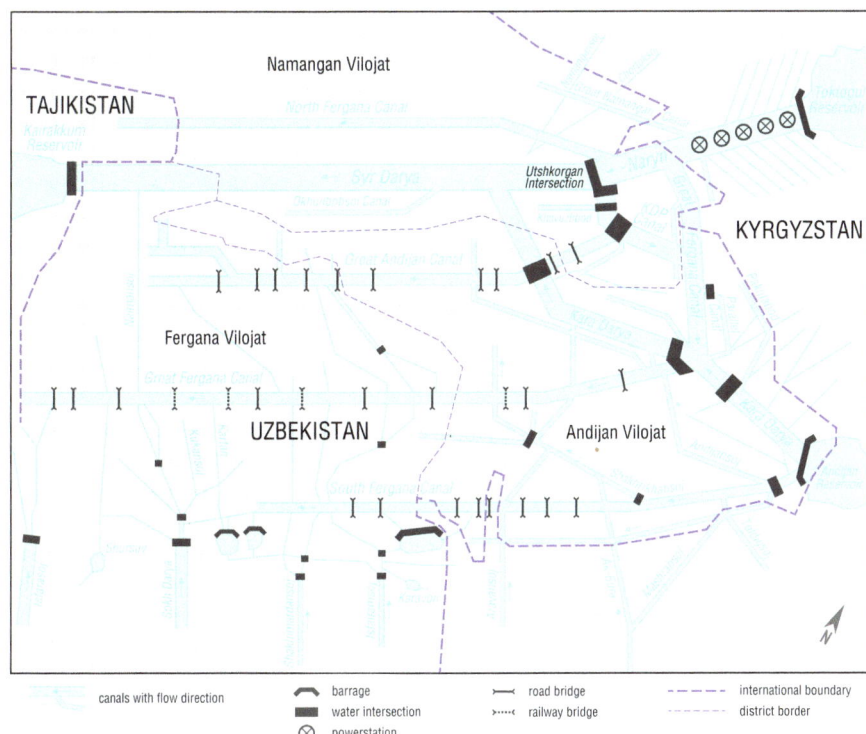

Fig. 11.2 Stylised map of the Main Canals in the Uzbek section of the Fergana Valley (Source: H. Kreutzmann 2015 in this volume pp. 113–127)

farm level. We are interested in revealing how water policy and regulations get translated into irrigation management practices on the ground and what informal "rules in use" (Ostrom 1990) are emerging in response. To this end, we have analysed the literature on WUAs in the Fergana Valley – in particular studies emerging from the IWRM Fergana Valley project funded by the Swiss Agency for Development and Cooperation – and field notes taken during a study trip visiting various WUAs and irrigation authorities in the region in May 2014. The section explores four practices pertinent to the role of WUAs: practices of representation, leadership, regulation and education.

11.3.1   Practices of Representation

The potential value of WUAs – following the global discourse on IWRM – lies in them providing an institutional basis for water users in an irrigation area to manage available water resources more effectively and equitably through collective action.

This depends on the active participation of the farmers using irrigated water in decision-making processes and management practices of WUAs. The ways in which water users are represented in WUAs are, thus, of critical importance. In Uzbekistan, the top-down, technocratic nature by which WUAs were created in the early 2000s set the pattern for the hierarchical and largely unrepresentative design of WUAs in the Fergana Valley and elsewhere. Although a prime task of WUAs is to resolve disputes over water allocation, requiring close collaboration with farmers, water users themselves were neither consulted nor informed about the creation of the WUAs and the reorganisation of water management this implied. "Therefore the water users considered the WUAs as another water administration imposed on them, and not a way of introducing collective action water management" (Abdullaev et al. 2010, p. 1035).

Representation of water users within WUAs is generally very weak (Dukhovny et al. 2008). Council members of WUAs are elected from amongst the water users of the irrigation area served, but they are not active in the decision-making process. Most WUA councils are ineffective, leaving authority in the hands of the management body of paid officials and, in particular, the WUA chairman. The influence of individual farmers is limited, as a rule, to the annual general assembly of the WUA, a platform for the WUA management to inform members rather than a forum for open debate (WUA Tomchikul). The old Soviet-style command system tends to stifle participatory forms of governance (Gunchinmaa and Yakubov 2010). Noting also the effects of limited funding, Schlüter et al. conclude from their research that:

> WUAs in Uzbekistan have in reality a poor decision-making mandate, are strongly influenced by patronage networks and interventions of the khokims (local governors) into their internal processes and suffer from a lack of financial means which all contributes to their malfunctioning. (Schlüter et al. 2010, p. 629)

In the Uzbek context, then, the term "water users association" is something of a misnomer. WUAs are de facto bodies set up by the government to secure state production targets for cotton and wheat by maintaining the existing irrigation system and ensuring the required allocation of water to the farmers. In short, WUAs in Uzbekistan may be acting on the behalf of water users but not at their behest.

One indictment of the ineffectiveness and unrepresentative nature of WUAs is the recent emergence in the Fergana Valley of water user groups (WUGs) as self-help initiatives organised by water users themselves (Dukhovny et al. 2008; Abdullaev et al. 2010). Here, farmers "have taken water management into their own hands", developing their own modes of collective action at the local level to manage water as a common pool resource (Abdullaev et al. 2010, p. 1031). Created by farmers themselves, the WUGs in the Fergana Valley have received financial support and advice from the IWRM project funded by the SDC and implemented by SIC-ICWC and IWMI. The number of WUGs in the South Fergana Canal area alone rose rapidly from 23 in 2006 to 160 in 2008 (Abdullaev et al. 2010, p. 1039). Some have emerged out of former collective farm brigades, some out of an extended family and others in response to a local water distribution conflict. WUGs not only own and maintain pumps and clean the smaller irrigation canals but they also introduce their own rules for water distribution and monitor water allocations, generating new levels of transparency to irrigation services (Abdullaev et al. 2010). They have no legal status but are proving increasingly influential as a voice for water users rights, drawing on the authority of informal institutions, social norms and local leaders (on informal institutions of irrigation in post-Soviet Central Asia, see Sehring 2009). The inclusion of WUG leaders in WUA councils is regarded by some commentators as a major step towards transforming WUAs into more participative, responsive organisations acting in water user interests (Abdullaev et al. 2010). These WUG leaders are raising the concerns of local farmers in the WUAs, demanding action and offering options for cooperative ventures. Their degree of influence depends heavily, however, on the willingness of the WUA chairman to share power.

A second attempt to improve representation in the South Fergana Canal area is the emergence of a Canal Water Users Union of South Fergana Canal in 2005, representing the first and only umbrella organisation for WUAs in any region in Uzbekistan. It was created with funding from the IWRM Fergana Valley project with the task of advising new WUAs and resolving conflicts within and between them over water allocation (Abdullaev et al. 2009). All 43 WUAs in the top two reaches of the South Fergana Canal are members of the union. It works primarily by inviting WUA representatives to its central office to discuss differences and is unusual in that its activities focus not on technical issues but on people and their interaction. Even though the project – together with its funding – has been terminated, the federation continues to operate, paying its seven staff members with fees collected from its member organisations. This, in itself, is an indication that the federation is regarded as providing a service of value to water users.

Issues of representation are by no means restricted to structures and procedures within and between WUAs but relate to higher levels of decision-making, where WUA interests need articulating. Referring to the representation of WUA leaders in councils and committees at subbasin and basin levels, Dukhovny et al. reflect:

> Our experience shows that the management of WUAs and Canal Water User Committees do not participate enough in the processes of water resources planning, allocation, and management, as well as in decision making relating to maintaining and rehabilitating of water infrastructure and seeking funding sources. (Dukhovny et al. 2008, p. 28)

Within the IWRM pilot areas, there are positive signs that WUAs are being given a voice in the water councils of basins and subbasins and in the water committees of irrigations systems, but few water officials are prepared to view such forms of representation as beneficial to their own position (Dukhovny et al. 2008, p. 29).

11.3.2 Practices of Leadership

How WUAs are run is, in practice, highly dependent on the people in charge and their modes of leadership. The allocation of water via irrigation systems is traditionally the responsibility of the *mirab* or "water master". This practice has not changed with the emergence of WUAs. The *mirab* is the sole operator of the weirs, sluices and settings regulating the amount of water distributed to each farmer. He/she is the one who sanctions farmers that manipulate weirs to their advantage, imposing fines or a reduction in the water allocation. A WUA is heavily dependent on its chairman, who is occasionally the only person in full-time employment. This dependency on persons of authority is a product not only of a WUA's internal structure and staffing but also of a strong cultural reliance on hierarchical forms of rule in a community. In many WUAs, old elites remain figures of considerable influence. This raises the risk of patronage, clientelism and expected codes of conduct (Schlüter et al. 2010). Here lies the ambivalence of leadership in the Fergana Valley. Strong leadership of a WUA or WUG is needed to gain the respect and support of water users internally and water officials externally. The lack of leadership skills is widely regarded as a serious impediment to the performance of WUAs. However, strong leadership can be a vehicle for sustaining predominant power relations in a community at the expense of inclusive, collaborative modes of governance.

11.3.3 Practices of Regulation

On paper, the tasks of WUAs in regulating water flows via irrigation systems are straightforward. WUAs are expected to draw up a water use plan for their area, distribute water to their members according to this plan, monitor flow rates on irrigation and drainage canals, keep records of water use, repair and maintain irrigation and drainage systems and conduct land reclamation and drainage (Dukhovny et al. 2008). In practice, their ability to regulate water flows is severely constrained not only by limited resources but also by the stringent quotas set by higher authorities. The water use plan developed by the WUAs in a bottom-up process of data collection from the farmers on the basis of their water requirements for planned crops is frequently rendered redundant by the top-down practice of setting water use quotas by state bodies. These quotas emerge from the national Ministry of Agriculture and Water Resources and are then passed down the water management hierarchy, being translated into quotas for the Basin Irrigation Systems Authority, then for the Irrigation Systems Authority at district level and finally for the WUAs and the farmers they serve (Kenjabaev 2014, pp. 16, 26–27; see Fig. 11.3). Even this top-down allocation regime does not work as intended in practice. Inadequate monitoring and lack of staff make it difficult for WUAs to ensure adherence to the quotas for irrigated water at the field level. More fundamentally, there is a lack of incentive for most actors at the local level to adhere to the water-use limits. As Kenjabaev discovered in his research, "[…] neither WUAs nor farmers and other water users have an incentive to know the applied and delivered amount of water as no price is set for water in Uzbekistan" (Kenjabaev 2014, p. 27, footnote 6). The result is extensive non-compliance of the quotas set by the water authorities.

The problem is further complicated by the different modes of water regulation applied to cash crops on the one hand and household crops on the other. Whereas

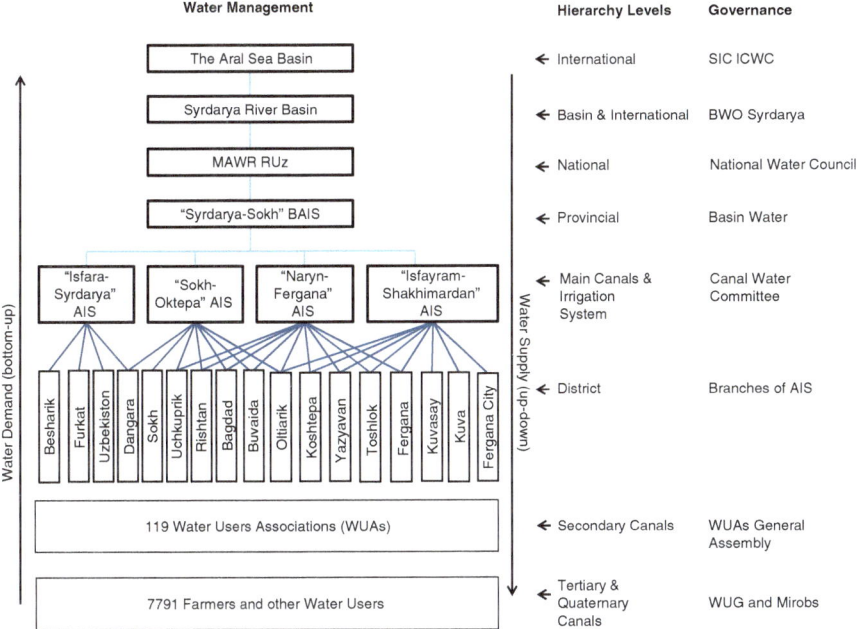

Fig. 11.3 Structure of water management organisations in Uzbekistan (Source: Kenjabaev 2014, p. 27)

water allocated for the compulsory crops of cotton and wheat is tightly regulated by the WUAs, enabling them to collect fees in return for services, for voluntary crops (such as fruit and vegetables) grown by individual households or small cooperatives of local residents, water is poorly regulated and rarely paid for. Here, again, there is emerging evidence of water users developing their own forms of collective action in response to an institutional void. In the WUA Kadyrjon-Azamjon, an unusual system of water management has developed whereby a community citizen council organises water allocation and cost collection itself.

Practices of regulation are, indeed, closely bound to practices of payment for irrigation services rendered. WUAs do not receive any state funding as a rule, although temporary subsidies can be made available from water management organisations for specific measures, such as the reduction of water losses. Also, in Uzbekistan, water fees are not levied, in line with the principle that water is a public good. Instead of charging for water used, WUAs are dependent on fees for the irrigation services they provide. These irrigation service fees were introduced in Uzbekistan only in 2001. They can vary according to the size of the farm and to the type of crop. For example, farmers in the WUA Tomchikul pay 45,000 soum[1] per ha for (private) orchards but only 20,000 soum per ha for cotton fields (excursion in WUA Tomchikul). The incentives for farmers to conserve water are, under this system,

---

[1] Soum is the Uzbek currency. 1 Euro ≈ 3100 UZS.

minimal. Furthermore, the non-payment of fees for irrigation services is a major problem for almost all WUAs in the region, seriously limiting their scope for action and, thereby, undermining their effectiveness. An attempt in the mid-2000s in the Bukhara province to price water on a consumption, rather than a per-unit-of-area, basis failed for political reasons out of fear of protests by the WUA's members (interview with the Swiss Agency for Development and Cooperation). Experiences such as this reinforce a technical-managerial approach to the work of WUAs, in which engineering solutions are applied to what are essentially social problems. Under such circumstances, it is perhaps not surprising that ecological issues, relating either to local watercourses and water-based landscapes or the state of the entire river basin, are not addressed by WUAs at all.

11.3.4   Practices of Education

Several WUAs that are active – in particular those selected as pilots for IWRM projects – are offering training for their members on how to optimise irrigation practices. Faced with a large number of new farmers, following land reforms, with little or no experience of managing crops themselves, many WUAs pursue an educational mission. The WUAs provide information and expertise to their members in the form of events, brochures, experimental and demonstration sites and advice by agricultural advisors (see Fig. 11.4). Beyond technical knowledge, this educational programme is intended to encourage greater self-responsibility amongst farmers accustomed to relying on the state but also to promote greater willingness to pay the WUAs for the services they provide.

This educational mission does not stop at the boundaries of the WUAs but is a central feature of irrigation management in post-Soviet Uzbekistan in general and of the IWRM project in the Fergana Valley in particular. Substantive training programmes for irrigation managers are provided by the Central ICWC Training Centre in Tashkent, with branches in the provinces of Andijan, Fergana and Khodjent (see Fig. 11.5). Training schemes funded by the IWRM Fergana Valley project provide extension services at the level of BISAs, ISAs and WUAs, with the active support of the Swiss Agency for Development and Cooperation. They also target the creation of more WUGs in the Fergana Valley. The emphasis placed on scientific knowledge and its application in the field resonates powerfully with Soviet practices of training irrigation specialists and Soviet imagery of the heroic water engineer making barren lands fertile in the name of national modernisation.

Nevertheless, these educational schemes have a number of drawbacks. Firstly, their reliance on project-based funding makes their long-term sustainability questionable. Secondly, they focus largely on technical and managerial issues, offering little help to practitioners when it comes to dealing with disputes over water allocations, developing an appreciation of environmental impacts and constraints or exploring economic incentives for water conservation. What is lacking is a multidisciplinary, cross-sectoral and basin-oriented approach to irrigation training. In this

Fig. 11.4 Information and training material on irrigation practices at the WUA Kadyrjon-Azamjon (Photo: T. Moss)

Fig. 11.5 A model irrigation scheme at the Marhamat Vocational College in Andijan (Photo: T. Moss)

Fig. 11.6 The research station of SANIIRI: a shadow of its former self (Photo: T. Moss)

sense, the educational programmes on offer are struggling to find their way between a Soviet tradition reliant on huge amounts of technical data generated by large teams of specialised scientists on the one hand and modern requirements for different kinds of knowledge provided under the auspices of a much reduced budget for irrigation research on the other (see Fig. 11.6).

11.4 WUAs as Models for Emulation?

The experience of WUAs in the Uzbek section of the Fergana Valley – and not only there – is a sobering one. Whilst it is perhaps inevitable that the WUAs could never meet all the expectations made of them at their inception, their general lack of impact on irrigation services and water governance is indicative of deeper, structural weaknesses in water and land management. WUAs in Uzbekistan are, in essence, not really water users associations at all, in that they were created by central government decree, permit only minimal representation of water users and operate to satisfy state targets for cash crops in accordance with state quotas for water allocation. Thus they fail to meet the definition of WUAs given in Sect. 11.1. They also fail to fulfil any of Merrey's four principles for successful WUAs, also cited above. The institutional environment within which they operate is not supportive, providing for little funding whilst demanding a high degree of conformity (1). Their capacity to operate and maintain irrigation infrastructure is severely limited not only by a lack of money but also by an inability to mobilise support from water users (2). Many WUAs in the region are clearly not convinced that the benefits of water user partici-pation outweigh the transaction costs involved (3). As a result, WUAs generally have proven unable to develop effective collective choice arrangements in manag-ing water and water infrastructures (4). In terms of the first core challenge set out in Chap. 8, (Moss and Dobner 2015, in this volume) – relating to the water-land nexus – they are caught between two powerful institutional regimes, for irrigation and for agriculture, respectively, with each placing unrealistic and unsustainable demands on WUAs. Prospects for integrated rural development, taking a cross-sectoral approach to the water-land-food nexus, are currently poor. The story of WUAs in the Fergana Valley is also a rich portrayal of the tensions emerging from parallel trends of path dependency and transformation – the second core challenge addressed in this part of the book. The legacy of the Soviet era lives on in many informal institutions and formal structures of irrigation management. At the same time, the WUAs are having to cope with enormous changes in the wake of post-Soviet land reforms, transboundary relations and emergent modes of governance.

There are important exceptions to this negative picture, situations where WUAs are exploring innovative ways of coping with the enormity of their tasks, but they represent only a small minority of WUAs in the region and are, to a large extent, dependent on temporary donor funding. The rapid growth of WUGs is perhaps the most promising recent development in irrigation governance in the Fergana Valley, as it represents an emergent desire of farmers to make their own collective arrange-ments for water management. It is, however, too early to reach any judgement on how they are likely to work in the future and what impact they might have on the practices of WUAs. What current experiences of WUGs indicate is that strengthen-ing the decision-making powers of water users can generate amongst them a greater sense of collective responsibility for managing water and land in a more sustainable way and that this can, in favourable circumstances, stimulate learning processes at the WUA level too. In terms of the future, this recent development would seem to resonate with what Lankford and Hepworth call the "bazaar" model of water man-agement (2010). Rather than a "cathedral" model of hierarchical, monocentric rule, the authors suggest that a polycentric model (the "bazaar"), in which various organ-isational forms coexist, is especially suited to situations characterised by little reli-able data, fluctual water supply and demand and under-resourced regulatory agencies – all factors prevalent in the Fergana Valley.

Field Trips and Discussions

1. Central Office of Fergana Irrigation District, Fergana, 4 May 2014
2. Office of South Fergana Canal Management, 5 May 2014
3. WUA Kadyrjon-Azamjon, 5 May 2014
4. Federation of WUAs of South Fergana subbasin, 5 May 2014
5. SANIIRI Research Station, 6 May 2014

6. Automated irrigation management system, Uchkurgan, 7 May 2014
7. WUA Tomchikul, 7 May 2014
8. Marhamat Vocational College, Andijan, 7 May 2014
9. Swiss Agency for Development and Cooperation, 8 May 2014

References

Abdullaev I, Kazbekov J, Manthritilake H et al (2009) Participatory water management at the main canal: a case from South Ferghana canal in Uzbekistan. Agric Water Manag 96:317–329. doi:10.1016/j.agwat.2008.08.013

Abdullaev I, Kazbekov J, Manthritilake H et al (2010) Water user groups in Central Asia: emerging form of collective action in irrigation water management. Water Resour Manag 24:1029–1043. doi:10.1007/s11269-009-9484-4

ADB (2006) Republic of Uzbekistan. Guidebooks for water users' associations in Uzbekistan. Technical assistance consultant's report. Asian Development Bank, Tashkent

Dombrowsky I, Horlemann L, Hagemann N (2011) Integrated water resources management in post-socalist countries – a comparison of Mongolia and Ukraine. Paper presented at the international conference on integrated water resources management, Dresden, 12–13 Oct

Dukhovny V, Sorokin A, Stulina G (2008) Should we think about adaptation to climate change in Central Asia? Adaptation to climate change: regional challenges in light of world experiences. Scientific Information Center of the Interstate Commission for Water Coordination (SIC ICWC), Tashkent

Dukhovny VA, Sokolov VI, Ziganshina DR (2013) Integrated water resources management in Central Asia, as a way of survival in conditions of water scarcity. Quat Int 311:181–188. doi:http://dx.doi.org/10.1016/j.quaint.2013.07.003

Gunchinmaa T, Yakubov M (2010) Institutions and transition: does a better institutional environment make water users associations more effective in Central Asia? Water Policy 12:165–185. doi:10.2166/wp.2009.047

GWP (2000) Integrated water resources management. Technical advisory committee, background paper no. 4. Global Water Partnership, Stockholm

Hamidov A (2007) Water user associations in Uzbekistan: review of conditions for sustainable development. Working Paper for the World Bank Institute (WBI). Washington, DC

Kenjabaev SM (2014) Ecohydrology in a changing environment. Dissertation, Justus-Liebig-Universität Gießen

Kreutzmann H (2015) From upscaling to rescaling – the Fergana Basin's transformation from Tsarist irrigation to water management for an independent Uzbekistan. In: Huettl RF, Bens O, Bismuth C, Hoechstetter S (eds) Society – water – technology: a critical appraisal of major water engineering projects. Springer, Dordrecht, pp 113–127

Lankford B, Hepworth N (2010) The cathedral and the bazaar: monocentric and polycentric river basin management. Water Altern 3:82–101

Merrey DJ (1996) Institutional design principles for accountability in large irrigation systems. International Irrigation Management Institute (IIMI), Colombo

Moss T, Dobner P (2015) Between multiple transformations and systemic path dependencies. In: Huettl RF, Bens O, Bismuth C, Hoechstetter S (eds) Society – water – technology: a critical appraisal of major water engineering projects. Springer, Dordrecht, pp 101–111

Ostrom E (1990) Governing the commons: the evolution of institutions for collective action. Cambridge University Press, Cambridge

Ostrom E (1992) The rudiments of a theory of the origins, survival, and performance of common-property institutions. In: Bromley D (ed) Making the commons work: theory, practice, and policy. ICS Press, San Francisco, pp 293–318

Ostrom E (2000) Reformulating the commons. Swiss Polit Sci Rev 6:29–52. doi:10.1002/j.1662-6370.2000.tb00285.x

Perry C (2007) Efficient irrigation; inefficient communication; flawed recommendations. Irrig Drain 56:367–378. doi:10.1002/ird.323

Rap E, Wester P (2013) The practices and politics of making policy: irrigation management transfer in Mexico. Water Altern 6:506–531

MAWR RUz (2014) Statistical report. Department of Water Resources Balance and Development of Water Saving Technologies. Ministry of Agriculture and Water Resources of the Republic of Uzbekistan, Tashkent, Uzbekistan

Schlüter M, Hirsch D, Pahl-Wostl C (2010) Coping with change: responses of the Uzbek water management regime to socio-economic transition and global change. Environ Sci Pol 13:620–636. doi:10.1016/j.envsci.2010.09.001

Sehring J (2009) Path dependencies and institutional bricolage in post-soviet water governance. Water Altern 2:61–81

Theesfeld I (2005) A common pool resource in transition. Determinants of institutional change for Bulgaria's postsocialist irrigation sector. Shaker, Aachen

Uysal ÖK, Atis E (2010) Assessing the performance of participatory irrigation management over time: a case study from Turkey. Agric Water Manag 97:1017–1025

Wegerich K (2000) Water user associations in Uzbekistan and Kyrgyzstan: study on conditions for sustainable development, Occasional paper no. 32. School for Oriental and African Studies, London

Wegerich K (2009) Shifting to hydrological boundaries – the politics of implementation in the lower Amu Darya Basin. Phys Chem Earth 34:279–288. doi:http://dx.doi.org/10.1016/j.pce.2008.06.003

Winrock International (2007) What is WUA? A guideline for Central Asia, Little Rock, AR

World Bank (2007) Global development of farmer water user associations (WUA): lessons from South-East Asia. World Bank Institute (WBI), Washington, DC

Yalcin R, Mollinga PP (2007) Water users associations in Uzbekistan: the introduction of a new institutional arrangement for local water management. Amu Darya case study – Uzbekistan. Center of Development Research, Bonn

Zavgorodnyay D (2006) Water user associations in the Republic of Uzbekistan. Theory and practice. Dissertation, Center of Development Research (ZEF), University of Bonn

Chapter 12
Theory, the Market and the State: Agricultural Reforms in Post Socialist Uzbekistan Between Economic Incentives and Institutional Obstacles

Bernd Hansjürgens

Abstract Water pricing is seen as an important element in efficient water resource management. By providing information about water resource scarcity, water prices can make explicit the value of water and can set adequate incentives for water users to use water more sustainably. However, designing efficient water resource pricing schemes is dependent on many prerequisites that are hard to fulfil. In this chapter, we contrast the prerequisites of water pricing with real-world contexts in the Fergana Valley. We show that many prerequisites for water pricing are not met in this area, so that water pricing reforms are unable to perform the functions usually associated with water prices. Nevertheless, it is possible to articulate a number of steps toward a reform of the agricultural sector which may at least point the way towards a more sustainable use of water resources.

Keywords Water scarcity • Fergana Valley • Water resources • Efficient water pricing • Water services • Interdependence of orders • Design elements • Institutional obstacles • Agricultural sector • Paths for reform • Water rights • Water users associations

12.1 Introduction: The Need to Price Water Resources in the Fergana Valley

Water scarcity refers to a situation in which water demand exceeds water supply. While many people assume that water scarcity is a problem of absolute scarcity (i.e. water not being available at all), the economic approach addresses water scarcity as a relative problem, meaning that the needs of those who wish to use water for certain

B. Hansjürgens (✉)
Department of Economics, Helmholtz Centre for Environmental
Research – UFZ, Permoserstraße 13, 04318 Leipzig, Germany
e-mail: bernd.hansjuergens@ufz.de

© The Author(s) 2016
R.F. Hüttl et al. (eds.), *Society - Water - Technology*, Water Resources
Development and Management, DOI 10.1007/978-3-319-18971-0_12

purposes (irrigation, drinking water, industrial production, ecological functions and services, etc.) cannot be met with available water resources. As with most resources on Earth, water scarcity is a far-reaching and ubiquitous problem.

The Fergana Valley in Uzbekistan is characterised by severe water scarcity. Although the annual amount of water in the region is regarded as sufficient – due to high precipitation rates in the upstream river areas – there are seasonal fluctuations: There is too much water available in the wintertime, leading to flooding in the Valley, whereas particularly during the summer months, the needs of water users often cannot be met. As a consequence, water resources have to be managed, through allocation to users according to certain decision-making rules or specific allocation mechanisms (e.g. quotas).

In the literature, several factors are mentioned as being the cause of water scarcity in the Fergana Valley (Giese and Sehring 2007; Dukhovny and de Schutter 2011). While these causes are diverse in nature, they can nevertheless be grouped into two major categories:

- One set of sources refers to the transboundary character of water resources in Central Asia (e.g. Libert et al. 2008; Dukhovny and de Schutter 2011, p. 280; Eschment 2011; Wegerich et al. 2012). After the breakdown of the Soviet Union, the then-Soviet provinces became independent states. During the Soviet era, the interests of upstream users and downstream users were (more or less) balanced via state planning: The downstream water users Uzbekistan, Turkmenistan and Kazakhstan had to deliver energy in the winter to the upstream users Kyrgyzstan and Tajikistan, while the upstream users had to provide sufficient water to the downstream users in the summertime. This situation changed completely in the early 1990s. The upstream users increasingly used the water resources for hydropower production in winter to safeguard their energy. The result was (and still is) that too much water was delivered to downstream users in winter, while water for downstream users (mainly for the irrigation of cotton plants) was insufficient in the summer. Thus, the Central Asia water crisis is generally regarded as being the result of a lack of (international) cooperation and coordination (Dukhovny and de Schutter 2011, p. 281).

- A second group of authors refers to the inefficiency of water resource management in the Fergana Valley (e.g. Abdullaev et al. 2007; Schlüter et al. 2010, p. 621; Abdullaev and Atabaeva 2012; Kenjabaev and Frede 2015, in this volume). These authors focus on the argument that despite the existence of large infrastructure facilities for water irrigation in the Fergana Valley (i.e. the Fergana Canal and a comprehensive canal system with far-reaching grids), there are high water losses due to the outmoded infrastructure of the canal system and an inefficient use of water (Abdullaev et al. 2007; Libert et al. 2008, p. 18; Dukhovny et al. 2009). Water losses in the Fergana Valley are estimated to be in the order of nearly 50 % of total water availability (Giese and Sehring 2007, p. 12). A decline in financial resources is seen as a key reason for this. While in the past the water sector was among the strongest sectors in Uzbekistan in terms of financial state flows (Abdullaev and Atabaeva 2012, p. 106), the amount of money devoted to sustaining the technical systems and the management of the irrigation system has decreased in both absolute and relative terms (*ibid*, p. 10). Today there are clear signs of problems in sustaining the infrastructure all over Uzbekistan.

Although the problems of water management are multicausal and mutually interconnected, in this contribution, we focus on the second line of argument, namely, the situation within the Fergana Valley. We want to focus on those factors of water management that can be influenced independently by the Uzbekistan government. Many proposals have been put forward with the aim of improving the water management system in the Fergana Valley. The most important policy recommendations and measures include, inter alia, the following elements (Dukhovny and de Schutter 2011, p. 319):

- Management information systems
- A training and qualification programme
- IWRM
- Development of a monitoring system
- Evaluating water demand and adjusting irrigation
- Introduction of water prices

Although there is no doubt that the decline of the water sector in the Fergana Valley can only be overcome by a comprehensive approach covering (at least) several of the above-mentioned elements, in the following we refer to just one aspect that is part of nearly every proposal: the role of water resources pricing. We want to shed some light on the potential of pricing mechanisms and on how these mechanisms can serve as incentives for more efficient resource use in the agricultural sector. At the same time, we want to address the institutional prerequisites for introducing water prices and to highlight the obstacles that prevent water prices from performing their functions. By explicitly focusing on the design elements and prerequisites of water pricing mechanisms on the one hand and on the institutional obstacles involved in the case of the Fergana Valley on the other, we want to illustrate how difficult it is to apply theory-based recommendations to the specific context-dependent situations found in real-world societies, with their historical path dependencies and institutional restrictions. By focusing on these aspects, we see our contribution as taking a similar line of reasoning as other chapters in this volume that point to the need to combine technical, economic and societal perspectives in order to develop well-functioning major water engineering projects (MWEPs).

The following section will outline some "economics of water pricing". In elaborating the preconditions for introducing an effective and efficient water pricing scheme, we want to elucidate the notion that any successful instrument has to be based on "good" design. We also want to raise awareness that designing such instruments is an "art" that can rarely be achieved in real-world policy processes, even in Western countries with their – compared to Central Asian states – much more effective institutions and governance structures. In the following section, we will focus on the institutional settings that shape the conditions in the agricultural sector in Uzbekistan. Our use of the term "institutional settings" refers to the formal and informal rules, norms, customs and habits in Uzbekistan's agricultural sector that shape farmers' behaviour. We will show that efficient pricing (in the sense that economic incentives are set for water users) cannot easily be achieved under such conditions. This leads to the conclusion that proposals for institutional reforms in relation to major water infrastructure projects cannot be based on a "nirvana approach" but have instead to be examined carefully with respect to the specific political, administrative and cultural conditions at hand. Nevertheless, in the final section we point out some possible pathways for step-by-step reforms that at least point in the direction of water-related reforms.

12.2   Some Economics of Water Pricing

Setting prices for environmental goods such as water resources serves several functions (Rogers et al. 2002; Young 2005):

1. *Financing function*: By introducing water levies (be it a fee, charge, tax, special contribution, or some other form of public financing), public revenues are generated that can be used for investments in the water sector. So one goal of introducing water prices is to generate financial resources in order to finance water-related infrastructures.
2. *Information function*: It is assumed that water prices, if properly calculated, reveal the "true value" of water resources. Demonstrating the true value of water by pricing water resources may serve to reflect the value a society accords to water resources. This is a prerequisite if water users are to obtain the necessary information about water scarcity and to carry the "full costs" of water when making decisions about water resource use.
3. *Incentives function*: By providing price signals that have to be integrated into decision-making, water users are incentivised to calculate the cost of water (including private and social costs) as part of their decisions regarding their use of water resources. Thus, private and public decisions will be changed, and water saving can be achieved if water users are confronted with the full costs of water. Water pricing can thus be seen as a management principle that leads to more sustainable water use. For many water experts, then, setting adequate incentives is a fundamental prerequisite for achieving efficiency in water resource use.

Interestingly (but not surprisingly), current discussions about water pricing in the Fergana Valley focus solely on the financing function. There are several reasons for this:

- First, since the building of the Fergana Canal and the Fergana Valley irrigation grid system in the late 1930s, water management has been considered mainly as a technical enterprise: Building, sustaining and managing the technical infrastructure have primarily been seen as the main task of water management engineering. All technical, managerial and financial resources have been devoted to building and

maintaining the technical systems. Water problems have predominantly been perceived as problems of infrastructure supply and only to a lesser degree (if at all) as
problems of demand management. "Water management strategies are still strongly
"Soviet" in approach, regarded by state actors as purely "technical", because other
dimensions – economic, social and political – are "fixed" through the state regulation" (Abdullaev and Mollinga 2010, p. 85).

- Second, as a consequence, the reasons for the decline in infrastructure were (and
  still are) associated mainly with a lack of financial resources in the water sector
  (in addition to the lack of international cooperation). This has been aggravated
  by the fact that the influence of the water management authorities has decreased
  since the end of the Soviet era.
- Third, the organisations and public authorities responsible for water management have not had any experience in the use of water pricing mechanisms as a
  means of steering users' behaviour in a certain direction. They emerged and
  developed within a system of state-driven supply management and control
  (Eschment 2011, p. 20) and are therefore not familiar with the functioning of
  market-based environmental instruments such as prices. Not surprisingly, the
  other two functions of water pricing – providing information and setting incentives – have been widely ignored in decision-making.
- Fourth and finally, the design of adequate water prices becomes more difficult
  the more the state actors' interests are oriented towards protecting the agricultural sector. As will be shown in detail below, the planting of cotton as a major
  state crop is still given top priority in Uzbekistan. There is little to no political
  will to make the agricultural sector pay for the water services it receives.

From an economic perspective, however, the information and incentive functions
are crucial elements of water pricing policy. The idea behind economic thinking is
that users of water resources should bear the full costs of their resource use. This
leads to two problems that have to be solved (Hansjürgens 1997; Gawel 2001):

1. What are the full costs of water and its services?
2. Who should bear the full costs of water services?

12.2.1   What Are the Full Costs of Water and Its Services?

What costs should form the basis for pricing water resources? Or, more precisely,
what cost elements should be included? Figure 12.1 illustrates possible answers
(Rogers et al. 2002; Young 2005). From an economic perspective, the cost of water
includes all the resources that are required to deliver a unit of water to the end user.
These costs comprise not only the operation and maintenance (O & M) costs of
daily running water-related supply systems (e.g. purchased raw water, electricity for
pumping, labour, repair materials, input costs for managing and operating storage,
distribution and treatment plants) but also the costs that have to be spent to build up
and consolidate the necessary infrastructures (depreciation charges, interest

Fig. 12.1 Costs of water – full cost of a single use (Adapted from Rogers et al. 2002, p. 7)

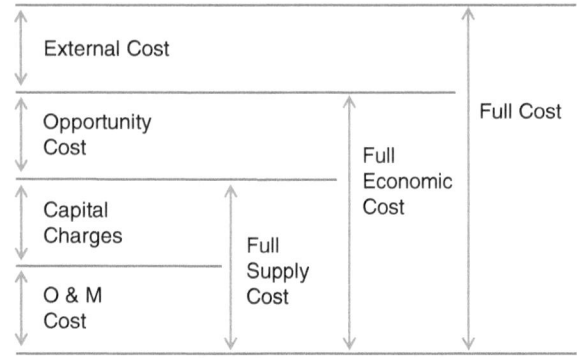

charges). Thus, the capital costs of financing water-related infrastructures have to be included. In addition, there are two further cost categories that have to be taken into account from an economic perspective. "Opportunity costs" refer to the fact that consuming water for one purpose deprives other users of water (if water is used, say, for agricultural irrigation, it is no longer available to private households or ecosystems). Using the money for water services means that the foregone net benefits from alternative uses is a cost. Opportunity costs are zero if there are no alternative uses, i.e. if there is no shortage of water. Finally, water use may be associated with external costs, i.e. costs imposed on other actors. These may be the costs associated with polluting or extracting water, i.e. environmental costs. If upstream users use water for energy production, this may lead to flooding; from the perspective of the upstream water user, this is an "external" cost, but it is nevertheless a cost borne by society. Similarly, the costs of irrigating land may lead to changing groundwater tables or to water losses for downstream users; this is a cost for uninvolved third-party groups or society as a whole. These costs are "external" from the perspective of the agricultural users.

Of course, calculating these costs is not an easy task. Even in industrialised countries with the highest technical standards and properly functioning governance structures, it is extremely difficult. The reason is that environmental costs in particular are hard to define, as it is almost impossible to calculate external effects (Baumol and Oates 1988). The point here is that from an economic perspective, all these costs should nevertheless – at least in principle – be included in water pricing in order to fulfil the prerequisite that water prices fully reflect their information and incentive function (Rogers et al. 2002, p. 9).

Following these definitions of costs of water resource use, the pricing mechanism can follow three perspectives (Hansjürgens 1997):

1. A "refinancing perspective" suggests that only past and present costs of water supply systems are financed via water prices; this includes investment costs, variable O & M costs and foreign capital costs.
2. A "company perspective", aimed at preserving the value of the infrastructure, additionally includes calculatory depreciation (based on current costs) and cal-

culatory capital costs (for using internal financial resources for investment purposes).
3. The "economic perspective" of water pricing includes not only private costs, as is the case in the refinancing and company perspective, but also so-called social costs, which refers to additionally including environmental and resource costs. The latter refers to societally assessed external effects where the calculation of these social costs is a major (albeit largely unresolved) challenge.

12.2.2   Who Should Bear the Full Costs of Water Services?

Even if the task of defining the (full) costs has been tackled successfully, the second question is how (according to which rules) costs should be distributed among water users. In order to steer human behaviour, the costs of water have to be transmitted directly to the water users following their use of the resource. The reason is that, in the economic model, it should be the water user who decides upon strategies and means to change his or her behaviour. Two arguments can be put forward for this, based on fairness and allocation, respectively:

1. The water user is the person who benefits directly from using water resources. Accordingly, following the polluter pays principle, this person should be the addressee of pricing in order to bear the costs of resource use. It is generally considered fair that the user of a resource should pay for that use.
2. Water users are normally those who know best how to change their water-related behaviour, because they have the most comprehensive knowledge of ways to avoid water use (Hansjürgens 2001): They can avoid water use by reducing consumption, they can use water more effectively, or they can reduce pollutants so that water quality is safeguarded. Thus, the polluter pays principle is not only a principle of fairness but also a principle of allocative efficiency.

Having explained the two basic steps of defining water prices – defining the *full* costs of water and distributing these costs among water users in a way that confronts users with the true costs of the water resources they use – we now turn to the design issues of water pricing. Our aim in highlighting these issues is to make two points clear: (1) Water pricing is an "art" that involves resolving some very difficult methodological and design-specific "technical" problems. (2) Water pricing is a political challenge that involves mitigating the influence of those who lose power in the process of setting price incentives.

- *Who pays the water price?*
  The water price should be paid by the entity (individual, private household, company, farmer, etc.) that uses the water resources. But who is the user of the resources? Answering this question is more difficult than it seems at first sight. Is it the water company that draws water from groundwater tables? Or is it the

water users association (WUA) that receives the water from the water company and allocates it to the farmer? Or is it the farmer himself or herself?

Different water pricing schemes can be developed depending on how the water user is defined. If it is only the water provisioning company that has to pay water prices, a water saving incentive is only set (if at all) for the provisioning company. The question arises whether and how the water company will pass on the price to the end users. Will this happen explicitly by levying fees or charges or implicitly by raising the water price? How can it be ensured that the end users are confronted with the true costs of their resource use? Are the end users those entities that can best reduce water resource use or avoid the pollution and degradation of water resources? In addition, many types of water user can be subject to pricing: the user of potable water, the discharger of wastewater, the owner of a house (if the house is rented, the owner is not necessarily the same as the user), the individual farmer, etc.

• *What is the price for?*
 If the user of water resources is identified and addressed by the water price, a second question is "what does the payer pay for?" Is it water consumption or wastewater discharge (or both)? Or is it the grid and delivery system that is chosen as a basis for pricing water? If the farmer is chosen as the addressee of water pricing, is the price related to his or her real consumption or is it related to the farm size? Is it a fixed price derived from a certain formula, or is the price set according to "actual" (real) use (requiring the monitoring of use along with measuring devices)?

• *How much does the payer have to pay?*
 The payment is usually referred to as the price of water (in the narrow sense). Bearing in mind the comments above regarding the definition of "full costs" of water, however, several issues have to be resolved: Should the water price be based on average costs or on marginal costs? An average cost-based water price would lead to identical costs for all water users, while the marginal cost-based water price would burden the users according to their specific (marginal) resource use (the amount of water they use). From an economic perspective, marginal cost pricing is superior to average cost pricing because end users are directly confronted with their resource use. However, for marginal cost pricing, measuring and monitoring issues have to be resolved. For both types of pricing, it is necessary to ask which cost components are included: full costs including external costs or full supply costs or O & M costs? Do the costs refer to short-term costs or long-term costs (including new investments)? If current costs (instead of historical costs of investment) are chosen, what is the adequate discount rate?

• *What is the tariff?*
 The water tariff determines how the price is imposed on different consumers. Here, several options are possible: The tariff could be based on a fixed rate, i.e. irrespective of consumption (this is called a flat rate), or on a variable rate where the price per unit of assessment base is linear, progressive or degressive. A block tariff can be chosen, or the price can be single part (with only a fixed or a variable component) or two part (with a fixed and a variable component). It could be

uniform (the same for all users) or else differentiated (by user groups). All these features can also be found in different variants, e.g. different tariffs can be combined.

These design elements, which can be taken from textbooks on water management or taxation (public finance), have been presented here to demonstrate that water pricing is a rather complicated issue, where a broad range of "technical" design questions have to be addressed. Apart from these design questions, which are difficult to resolve (due to a lack of institutional and managerial capacity and expertise), there is another factor that might impede the introduction of water prices: If water prices are introduced, there will be individuals or groups in society that have to face higher cost burdens. There are losers compared to the status quo where no water prices exist. In many cases, these "losers" are powerful interest groups that undertake huge political efforts to prevent the introduction of water prices or – if their introduction as such cannot be avoided – seek to influence the design of the water prices in a way that ensures they themselves are not burdened. The agricultural sector can be seen as such a pressure group in most countries, regardless of the societal system (decentralised market-based economy or more centralised planned economy).

Having explained the key elements of "the economics of water pricing", we now turn to the situation in the Fergana Valley. We want to illustrate that the conditions in the Valley do not provide the necessary basis for introducing any type of efficient water pricing.

12.3 Political and Institutional Obstacles

12.3.1 Protection of the Agricultural Sector

"Cotton has been a major crop in Uzbekistan at least from the time of the Russian empire" (Abdullaev et al. 2007, p. 112). During Soviet times, Uzbekistan became one of the main producers and the second largest exporter of cotton in the world. Central Asia became the "cotton basket" of the Soviet Union (see also Weinthal 2006; Abdullaev and Atabaeva 2012, p. 107). The factors that made this development possible were, first, the growing proportion of land under irrigation, achieved by expanding the Fergana Valley irrigation system and securing sufficient amounts of water during summer periods,[1] and, second, the state order of the Soviet Union that prescribed the planting of nearly 100 % cotton in the area (Abdullaev et al. 2007, p. 113; Abdullaev and Atabaeva 2012, p. 107).

After the breakdown of the Soviet Union in 1991, the former provinces became independent states. Although the existing rules governing water allocation between

---

[1] "The hydraulic mission has started its journey in Central Asia" (Abdullaev and Atabaeva 2012, p. 104). "Hydraulic mission" means here that in Soviet times, "water professionals started to develop their way of working, translating and adapting many unrealistic, idealistic ideas of communist party into 'real-life' situations" (*ibid*, p. 107). At the same time "Central Asia became part of 'heroic' history of irrigation engineering" (*ibid*, p. 105).

the Central Asian states were safeguarded by means of the 1992 "Agreement on Cooperation in Joint Management, Use and Protection of Interstate Sources of Water Resources" (Libert et al. 2008, p. 11), the water situation changed dramatically because the Central Asian states acted independently, following their own interests (Libert et al. 2008; Sehring 2009). The upstream users did not allocate sufficient water to the downstream users in the summertime, while the energy provision from downstream to upstream users in the wintertime was no longer guaranteed. The two principles contained in International Water Law that there should be "equitable and reasonable utilisation of transboundary water resources [and that] the obligation of one State should not cause significant harm to another State through its use of shared water resources" were violated (Libert et al. 2008, p. 12).

However, this did not mean that Uzbekistan's agricultural policy "really" changed (Abdullaev and Atabaeva 2012, p. 105). Its focus was and still is on cotton production. The state response to the new political situation was to seek independence in the provisioning of food by import substitution. This had become a necessity since the import of wheat from other Soviet states ceased in the early 1990s. Thus, Uzbekistan's politicians were forced to focus their primary policy objective on safeguarding people's nutrition. As a consequence, wheat became the second important crop that was imposed by the state. In addition to structural changes, which focused on more individual freedom for farmers and reduced farm sizes (Abdullaev et al. 2007), "the Uzbekistan government's decision to promote wheat self-sufficiency is largely responsible for the recent decline in cotton production" (Weinthal 2006, p. 20).

Nevertheless, cotton production is still a key element of Uzbekistan's agricultural policy (see Fig. 12.2). A large proportion of Uzbekistan's export revenues derive from cotton production, while the sector provides a high percentage of employment (Weinthal 2006, p. 19; Abdullaev et al. 2007, p. 113). As the export of cotton is organised predominantly by state authorities, state revenues are highly dependent on this source. For this reason, the regulations stipulating which crops farmers are to plant have not really changed. Instead of imposing a single "cash crop", as in the Soviet era, farmers are now required to plant two cash crops. Today, nearly 85 % of state production is reserved for these two "cash crops".[2] Nevertheless, the production of cotton has decreased over time (see Fig. 12.2).

With respect to the recommendation to introduce water prices, the developments described here have an important implication: In Uzbekistan, there has hardly been any political will to introduce (efficient) water prices in the agricultural sector. Burdening the agricultural sector with additional state-imposed prices was – and still is – a taboo. Political decision makers have an interest in protecting the agricultural sector from the imposition of water-related prices. This has been fully embedded in the tradition of Uzbekistan's political decision makers, where all past decisions were oriented towards prioritising – and not burdening – the agricultural

---

[2] One should be aware that in Uzbekistan "cash crop" refers to certain crops that are exported and therefore generate revenues; thus, primarily, cotton can be described as a "cash crop" in this sense. Wheat is primarily required for meeting the population's food needs – and not for acquiring "cash".

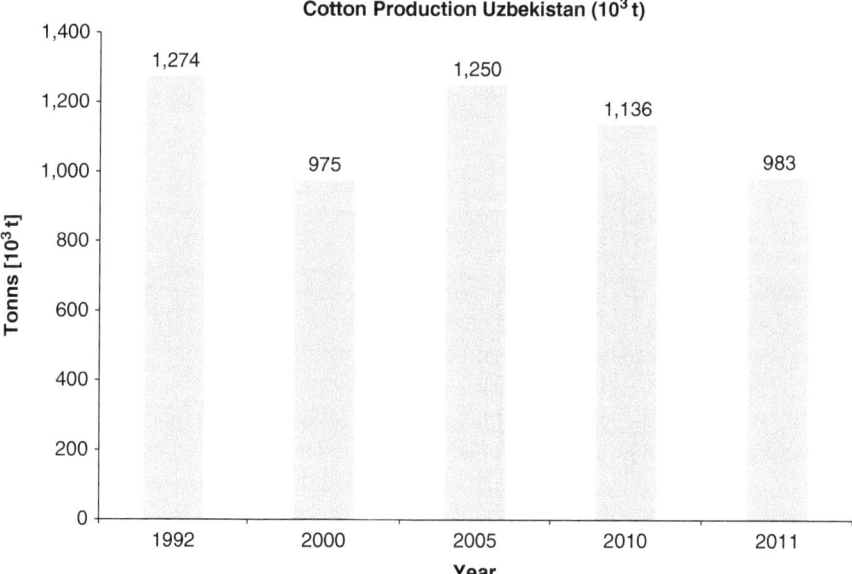

Fig. 12.2 Cotton production in Uzbekistan (10³ t) (Source: Modified from Shamukhitdinova and Adelt 2013, p. 7). It should be noted that the official statistics might underestimate the amount of cotton produced, as some of the cotton produced might not be sold to "official state authorities" but to neighbouring countries through informal ("black") markets

sector. Instead of paying for water, farmers receive significant subsidies for water irrigation. "The obtainment of free inputs is an implicit side-payment from the state to the farmers that allows them to use water for other purposes [...] [F]armers have few incentives to conserve water" (Weinthal 2006, p. 22). Although water scarcity has increased dramatically in the region over the last 25 years, this tendency to favour agriculture has not been overcome to this day.[3]

A further element that aggravates these tendencies even further can be seen in the fact that water management issues in Central Asia have been focused on supply-side management measures to date. The history of water management in Russian times and later in the Soviet era of Central Asia is part of the "heroic history of irrigation engineering" (Abdullaev and Atabaeva 2012, p. 105). All water-related problems were considered as technical problems to be solved with the assistance of large-scale technical solutions. Thus, the notion of steering demand by changing

---

[3] It should be noted here that this is similar to many other countries where a felt need to protect the agricultural sector can be observed. See, for example, the reference to Israel in this volume where the authors state that the concern to protect the agricultural sector is "deeply rooted in the Zionist movement" (Bismuth et al. 2015, in this volume, pp. 253–275). Similarly, in Germany, the water extraction charge *(Wasserpfennig* or *Wasserentnahmeentgelt)*, which is levied in some German states, must be paid by water users – and not (!) by the agricultural sector. Instead, the revenues are used to compensate farmers for income losses due to reduced use of nitrogen and fertilisers in water protection areas. The polluter pays principle is thus reversed.

water users' consumption patterns does not even seem to exist. In fact, when such ideas were presented by the author at a conference on water-related problems in Central Asia in May 2014, for most conference participants they lay outside the scope of their perceptions of water management practices for the region.

12.3.2   No Freedom for Individual Choice

A second obstacle to efficient water pricing can be seen in the fact that the address-ees of water prices do not have a chance to respond economically. While on paper there have been attempts to transform the former Soviet Union central planning system into a market economy, many changes in the economic and political system in Uzbekistan have remained gradual. The economic and political system in Uzbekistan was (and still is) shaped by the modes and characteristics of central planning regimes. Abdullaev and Atabaeva (2012) therefore speak of a "slow trans-formation" process. "But the voluntary change of a state apparatus in the case of Central Asian countries only happened with minor changes that remain to be sur-face changes" (Abdullaev and Atabaeva 2012, p. 105). "All states of the [Central Asian] region have returned in different forms back to control of agriculture in dif-ferent forms" (*ibid*, p. 108). Moss and Hamidov (2015, in this volume) speak of a "gradual approach to market reform".

From an economic perspective, the decisive point is that private property in the agricultural sector is not fully realised; farmers do not have full powers of decision-making regarding their inputs and outputs. To understand this issue, farmers' deci-sions have to be seen in their specific institutional contexts: Farmers are obligated to plant the two "cash crops", cotton and wheat, on their land. They have to buy all the necessary inputs (seed, fertilisers, etc.) on credit. They are even told what to plant, when to plant and where to plant. In addition, they cannot decide by themselves when and how to irrigate their fields. They are also told when to harvest the crops (Weinthal 2006, p. 20). There is hardly any autonomy with respect to the resources they input or the quantity and quality of their output. The farmers receive a fixed price for their harvest which allows them to pay back their loans. However, the price is far lower than the world price for cotton so that private profits cannot be made. "Because the government has always granted [the farmers] debt relief, they have little incentive to become more efficient by reducing their inputs and managing their level of debt like individual private farmers. [...] As a result, [they] still operate without any hard-budget constraints" (Weinthal 2006, p. 21).

Certainly, there have been some changes in the system since independence. The growing number of smaller farms (compared to the huge kolkhozes) has led to a decrease of the previously centralised decision-making procedures in collective farming that were dominated by the heads of agricultural units. Instead, de-collectivisation has led former members of the collective farms as well as citizens with no agricultural experience to become individual farmers (Weinthal 2006, pp. 19–22; Abdullaev and Atabaeva 2012, p. 109). These farmers have gained partial access to production inputs, but water access has still been dominated by the former collective decision makers and/or the newly emerging water users associations (see below). The individual farmers apply different means (power, money, influence, good connections, etc.) to gain access to water to irrigate their crops (for a case study in the Fergana Valley, see Abdullaev and Mollinga 2010, p. 93). Thus, in times of increasing scarcity, water allocation became subject to bargaining and "good relationships". It was not the water price, however, that controlled the distribution of water among users but rather other factors that determined the outcome of these socio-economic processes. "The influence and control of the water management processes have been considerably reduced due to individualisation of the agricultural processes. The water professional's role has been replaced by set of water control strategies of newly emerging water users: large, small, subsistence farmers, fisheries and constructors, etc., who started to compete for water for their purposes. The water management became more socio-technical process not only technical" (Abdullaev and Atabaeva 2012, p. 109).

Choices about land use – which crops to plant on given areas of land – are defined by state quotas. "Even if farmers fulfil their cotton production quota, they can still be penalised if the area they plant with cotton is less than the requirement" (Abdullaev et al. 2007, p. 116). Farmers' decisions are therefore mainly characterised by restricted property rights or the absence of property rights for resource use. The term "property rights" refers to the rights to use a resource or a good independently, i.e. to buy, rent or sell a resource or a good (Alchian 1965; Eggertsson 1990). Defined property rights are a prerequisite for using goods and resources efficiently. They are the basis for independent and efficient decisions in a market economy. It is only when individual decision makers face the full consequences of their behaviour that they will take all the consequences of their decision into account. Furthermore – and this is crucial in the present context – property rights are also a prerequisite for the proper functioning of pricing mechanisms.

Market prices only make sense if there are clearly defined property rights. Prices deliver information that forms the basis for individuals' choices. They set incentives to react economically to changing supply and demand conditions and thus to changing conditions of scarcity. Following this line of argument, water prices set by politicians make sense only if there is a clear definition of such property rights and if property rights are allocated to individual decision makers. Farmers can only follow price incentives if they have the freedom to choose among several alternatives. Therefore, if water prices are to incentivise resource users, this can only be accomplished if the resource users have freedom of choice. Clearly defined property rights where decisions about goods and resources are connected with private property are indispensable for the use of water prices as a policy instrument.

These considerations can be traced back to the early proponents of the idea of a market economy such as Walter Eucken (1890–1950). Eucken stated that systemic changes in a societal system (from centralised planning to a market economy) can only be achieved if *all* its elements (market economy for the economic system, democratic decision-making for the political system, separation of powers, legal provisions for protecting individuals' rights) are changed simultaneously. There is an "interdependence of orders" (Eucken 1952/1990). If only certain elements of system transformation are taken into account, transformations will fail.

There is a second closely related argument that highlights the relevance of individual responsibilities: the polluter pays principle. This principle states that the polluter or user of a resource has to pay for the costs associated with resource use. In general, this principle is primarily seen as a distributional principle, contributing to the fairness of environmental burden sharing between resource users. It should be noted, however, that it also follows an underlying efficiency principle: The resource user knows best how to use a resource efficiently (Hansjürgens 2001). This efficiency argument also emphasises that water pricing schemes should address the user of resources – seen here as the individual who knows best how to change his or her behaviour.

As long as farmers' property rights in the Fergana Valley are not clearly defined, it is likely that water prices will not work properly. Water prices may succeed in generating public revenues that may in turn help to finance infrastructure investments. However, they will fail as an information and steering mechanism. Thus, the general theoretical claim that water prices lead to more efficient water resource use may fail unless real-world obstacles in the agricultural sector are taken into account.

12.3.3   Lack of Clearly Defined Water Rights and Unclear Role of Water Users Associations

Closely related to the aforementioned argument regarding property rights (which referred to farmers' discretionary powers to use their land) is the role of property rights with regard to water – water rights. Setting price incentives also requires property rights for water uses. However, farmers in the Fergana Valley lack not only property rights over the use of inputs and outputs, they particularly lack property rights over water resources (an input in agricultural production); water management is subordinate to agricultural needs (Abdullaev and Mollinga 2010, p. 93).

Historically, the allocation of water was placed in the hands of water users associations (WUAs). WUAs not only own and maintain pumps and clean the smaller irrigation canals, they also introduce their own rules for water distribution and monitor water allocations (Abdullaev et al. 2010). These organisations were implemented in the early 2000s (Moss and Hamidov 2015, in this volume) as a reaction to the water allocation problems in Uzbekistan at that time. The increasing individualism of the agricultural sector after decentralisation, involving a transition from the former Soviet type of large-scale farming to a more diverse structure with smaller fields and a considerably higher number of farmers, led to a mismatch between agricultural and water-related institutional structures. The latter still relied on the rules for allocating water to the cash crop-producing state farms, while the need to allocate water between different uses (cash crops of cotton and wheat, but also privately planted crops like rice and vegetables) and different levels of decision-making increased. Abdullaev and Mollinga (2010, pp. 93–98) provide an institutional analysis of water-related decision-making, pointing to the fact that, in addition to the newly created water users associations, agricultural quotas for cotton and wheat production, ancient rules about "water turns", neighbourhood relationships and clans also had a major impact on de facto decision-making. As a result, water users associations are not seen primarily as an organisation that represents the needs of water users but rather as another water administration body imposed on farmers in a top-down manner (Abdullaev et al. 2010, p. 1035). As one reaction, in addition to water users associations, water user groups emerged, as "true" representatives of farmers and water users (Abdullaev and Mollinga 2010), where the "farmers have taken water management into their own hands" (Moss and Hamidov 2015, in this volume, pp. 149–167).

It is not necessary to go into the details of institutional change here. Rather, the key argument here is that water users associations, as the decisive organisational units for solving water irrigation issues, were (and still are) relatively weak so that there has been no opportunity to introduce pricing at this level. As the water users associations are not financed by the state, they depend heavily on the fees and charges they impose on farmers. However, these fees and charges rarely reflect the situation of water scarcity, as they are usually based on farm size and the type of crop (Moss and Hamidov 2015, in this volume, pp. 149–167). Consequently, the incentives for using water more effectively are relatively weak. The payment is seen more as a membership contribution (rather like a tax) than as a fee or charge that is imposed in exchange for services received. There is no connection between the water use and the revenues raised, as it is not water itself but the area of land that is subject to taxation; the tax base and the tariff also lack incentives for efficient water use. Needless to say that the prerequisites for monitoring – which requires metering not only at the level of the main canals but also at the level of secondary and tertiary canals and grids – are not met (Weinthal 2006, p. 19). Also, compliance is not achieved as many members of WUAs just do not pay the charges imposed on them. The result is, as Kenjabaev (2014, p. 27) notes, "neither WUAs nor farmers and other water users have an incentive to know the applied and delivered amount of water as no price is set for water in Uzbekistan."

It is not surprising that water allocation in many cases does not reflect the rules of the WUAs, but is embedded in historical decision-making, which means that former elites and power constellations are more important than written rules. Against this background, the pricing of water resources has no chance of being introduced and implemented. Recalling once more Walter Eucken's (1952/1990) observation that there is an "interdependence of orders", pricing as one element of dealing with water scarcity makes little sense if other issues such as well-defined property rights in the agricultural and the water sector, effective governance structures and a clear division of responsibilities are not resolved.

12.4 Concluding Remarks

Water pricing is considered by many economists and water experts to be a key element in water management. It is seen as a crucial step towards coping with the adverse effects of increasing water scarcity. This paper has contrasted the theoretical idea of water pricing with the practical obstacles evident in Uzbekistan. Uzbekistan is a post-Soviet era country where the process of transformation into a market economy with decentralised decision-making has been rather slow. Although the water-related irrigation system in the area is among the oldest and most extensive in the world, there is almost no integrative management of water resources that incorporates the possibility of incentivising water users.

It has become clear that water-related technologies are closely connected with socio-economic developments. Using water for agricultural purposes is a practice deeply rooted in Uzbek society so that irrigation is given high priority over other water uses. These institutional (cultural, legal and economic) roots constitute a powerful obstacle to introducing water prices in the area. Path and context dependencies that have emerged over decades cannot be overcome in a short period of time.

It can thus be concluded that in this geographical area, the challenge of increasing water scarcity cannot (yet) be met by introducing water prices. Without a comprehensive reform of property rights (including land and water-related property rights), water prices will not work properly. These findings contradict not only economists' recommendations that are based on a theoretical "nirvana approach" but also the opinion of water experts and practitioners in Central Asia that focus exclusively on financial factors when calling for water prices as a key element in solving the water scarcity problems in the area.

What, then, are the alternatives? Indeed, are there any? The answer to these questions would go beyond the scope of this paper and so cannot be given here. But there are certain steps that could be taken which potentially point in the right direction. One could be to reduce the amount of cash crops to be grown, thus giving farmers more freedom to plant crops in accordance with their own interests. Another step would be to strengthen the role of the WUA in the sense that they might actually become the kind of organisations they already are on paper, namely, ones that represent farmers' interests. It remains an open question whether Uzbekistan will go in such a direction.

References

Abdullaev I, Atabaeva S (2012) Water sector in Central Asia: slow transformation and potential for cooperation. Int J Sustain Soc 4:103–112. doi:10.1504/IJSSOC.2012.044668

Abdullaev I, Mollinga PP (2010) The socio-technical aspects of water management: emerging trends at grass roots level in Uzbekistan. Water 2:85–100. doi:10.3390/w2010085

Abdullaev I, Giordano M, Rasulov A (2007) Cotton in Uzbekistan: water and welfare. In: Kandiyoti D (ed) The cotton sector in Central Asia economic policy and development challenges. The School of Oriental and African Studies. University of London, London, pp 112–128

Abdullaev I, Kazbekov J, Manthritilake H et al (2010) Water user groups in Central Asia: emerging form of collective action in irrigation water management. Water Resour Manag 24:1029–1043. doi:10.1007/s11269-009-9484-4

Alchian A (1965) Some economics of property rights. Il Politico 30:816–829

Baumol WJ, Oates WE (1988) The theory of environmental policy, 2nd edn. Cambridge University Press, Cambridge

Bismuth C, Hansjürgens B, Yaari I (2015) Technologies, incentives and cost recovery: is there an Israeli role model? In: Huettl RF, Bens O, Bismuth C, Hoechstetter S (eds) Society – water – technology: a critical appraisal of major water engineering projects. Springer, Dordrecht, pp 253–275

Dukhovny VA, de Schutter J (2011) Water in Central Asia – past, present, future. Taylor & Francis, London

Dukhovny V, Sokolov V, Manthrithilake H (2009) Integrated water resources management: putting good theory into real practice. Central Asian Experience. Scientific Information Center of the Interstate Commission for Water Coordination (SIC ICWC), Tashkent

Eggertsson T (1990) Economic behavior and institutions. Cambridge University Press, Cambridge

Eschment B (2011) Wasserverteilung in Zentralasien. Ein unlösbares Problem? Friedrich Ebert Stiftung, Berlin

Eucken W (1952/1990) Grundsätze der Wirtschaftspolitik, 6th edn. Mohr, Tübingen

Gawel E (2001) Effizienz im Umweltrecht. Grundsatzfragen wirtschaftlicher Umweltnutzung aus rechts-, wirtschafts- und sozialwissenschaftlicher Sicht. Nomos, Baden-Baden

Giese E, Sehring J (2007) Regionalexpertise – Destabilisierungs- und Konfliktpotenzial prognostizierter Umweltveränderungen in der Region Zentralasien bis 2020/2050. Wissenschaftlicher Beirat der Bundesregierung Globale Umweltveränderungen (WBGU), Berlin

Hansjürgens B (1997) Gebührenkalkulation auf Basis volkswirtschaftlicher Kosten? Anwendungsprobleme und Lösungsmöglichkeite. Archiv für Kommunalwissenschaften 36:233–252

Hansjürgens B (2001) Das Verursacherprinzip als Effizienzregel. In: Gawel E (ed) Effizienz im Umweltrecht. Grundsatzfragen wirtschaftlicher Umweltnutzung aus rechts-, wirtschafts- und sozialwissenschaftlicher Sicht. Nomos, Baden-Baden, pp 381–396

Kenjabaev SM (2014) Ecohydrology in a changing environment. Dissertation, Justus-Liebig-Universität Gießen

Kenjabaev SM, Frede H-G (2015) Irrigation infrastructure in Fergana today: ecological implications – economic necessities. In: Huettl RF, Bens O, Bismuth C, Hoechstetter S (eds) Society – water – technology: a critical appraisal of major water engineering projects. Springer, Dordrecht, pp 129–148

Libert B, Orolbaev E, Steklov Y (2008) Water and energy crisis in Central Asia. China Eurasia Forum Quart 6:9–20

Moss T, Hamidov A (2015) Where water meets agriculture: the ambivalent role of the water users associations (WUAs). In: Huettl RF, Bens O, Bismuth C, Hoechstetter S (eds) Society – water – technology: a critical appraisal of major water engineering projects. Springer, Dordrecht, pp 149–167

Rogers P, de Silva R, Bhatia R (2002) Water is an economic good: how to use prices to promote equity, efficiency, and sustainability. Water Policy 4:1–17. doi:10.1016/S1366-7017(02)00004-1

Schlüter M, Hirsch D, Pahl-Wostl C (2010) Coping with change: responses of the Uzbek water management regime to socio-economic transition and global change. Environ Sci Pol 13:620–636. doi:10.1016/j.envsci.2010.09.001

Sehring J (2009) The politics of water institutional reform in neo-patrimonial states: a comparative analysis of Kyrgyzstan and Tajikistan. VS Verlag für Sozialwissenschaften, Wiesbaden

Shamukhitdinova L, Adelt S (2013) Die Wiederbelebung zentralasiatischer textiler Handwerkstechniken im Prozess der Nationsbildung in Usbekistan. Zentralasien-Analysen 72:2–8

Wegerich K, Kazbekov J, Mukhamedova N et al (2012) Is it possible to shift to hydrological boundaries? The Ferghana Valley meshed system. Int J Water Res Dev 28:545–564. doi:10.1080/07900627.2012.684316

Weinthal E (2006) Water conflict and cooperation in Central Asia. Prepared as background paper for the UN human development report 2006. http://hdr.undp.org/sites/default/files/weinthal_erika.pdf. Accessed 9 Apr 2015

Young R (2005) Determining the economic value of water: concepts and methods. RFF Press, Washington, DC

Part IV
The Lower Jordan Valley – The Red Sea-Dead Sea Conveyance Project and Its Complex History

Chapter 13
Water Resources, Cooperation and Power Asymmetries in the Water Management of the Lower Jordan Valley: The Situation Today and the Path that Has Led There

Christine Bismuth

Abstract This chapter aims at providing an overview of the uses and the state of the water resources in the Lower Jordan Basin. The years 2007/2008 serve as a base line to describe the specific situation, marked by a succession of drought years during the first decade of the twenty-first century. An overview of the major treaties and agreements between the riparians and implications on the water management situation is presented. The nature of the relations between the different parties is analysed.

Keywords Water resources • Lower Jordan Basin • Yarmuk • Dead Sea • Red Sea • RSDS Conveyance Project • Water uses • Water balance • Water conflict • Israel • Jordan • Palestinian Territories

13.1 Water Resources of the Lower Jordan Basin

The Red Sea–Dead Sea (RSDS) Conveyance Project is the latest attempt to widen the range of available solutions to overcome the water stress in the Lower Jordan Basin and to stabilise the level of the Dead Sea. One of the major arguments for the RSDS Conveyance Project from its supporters is that no other options for long-lasting solutions are available. This argument is contested by most of the environmental protection groups, who fear the associated risks of the project for the Arava valley, the Dead Sea and the Red Sea itself. Those groups push for more water conserving solutions.

C. Bismuth (✉)
Interdisciplinary Research Group Society - Water - Technology,
Berlin-Brandenburg Academy of Sciences and Humanities,
Jägerstraße 22/23, 10117 Berlin, Germany

Helmholtz Centre Potsdam - GFZ German Research Centre for Geosciences,
Telegrafenberg, 14473 Potsdam, Germany
e-mail: bismuth@gfz-potsdam.de

R.F. Hüttl et al. (eds.), *Society - Water - Technology*, Water Resources
Development and Management, DOI 10.1007/978-3-319-18971-0_13

In order to understand the present situation of the Lower Jordan Basin and the impact of the planned project, a comprehensive view on the development and use of the water resources in the region is compulsory. Not only decisions with regard to water management but also institutional and societal settings had influence on the status of the water resources, the choices taken and the remaining options.

During the last 50 years, the Lower Jordan Basin was subject to important changes, resulting in distinct consequences for both the region's water resources and the Dead Sea as the final recipient of the Jordan River.

It is quite a challenge to provide a comprehensive overview of the region's water balances for the following reasons:

- Statistical data and methods vary between the different main riparians Israel, Palestine and Jordan, a common methodology is not applied.
- Due to the different conveyance systems (Israel's National Water Carrier, "Disi" conveyance system, desalinated seawater conveyance), it is not possible to restrict the observations to the Lower Jordan Basin. Instead, the water databases from Israel, Jordan and the West Bank have to be taken into account.
- Water data for the Kingdom of Jordan are partly based on projections as in the case of developed surface water sources. Climate change scenarios, which suggest a significant decrease in precipitation rates for the region, are not fully included in the projections.
- Even though groundwater abstraction and groundwater levels are regularly measured by the Israeli Water Authority (IWA) and its Hydrological Service, the latest groundwater data published in 2013 by the Israel Central Bureau of Statistics (ICBS) date from 2008. Since that period, water management in Israel has changed, and the most recent data on use and availability of water resources are in fact from 2013 (Israel Water Authority 2014). In its reports, the IWA does not distinguish between groundwater and surface water sources, but differentiates between various qualities of the consumed water (recycled, brackish, saline or potable).
- As the Israeli water distribution system stretches over the West Bank, groundwater resources are partially used jointly, whereas Palestinians have no direct access to Jordan. From an outside perspective, it is most difficult to identify Palestinian and Israeli water sources and abstractions.
- Another point of imprecise and controversial information is the quantity of water, which the Kingdom of Jordan receives from Syria via the Yarmuk River. The Jordanian–Syrian agreement on use of the water of the Yarmuk has fixed 200×10^6 m³/year for Jordan. In fact, Jordan does not receive more than 50×10^6 m³/year according to Prof. Salameh, member of the Jordan Royal Water Committee (Shami 2013).
- The figures which we present in Figs. 13.2 and 13.3 are in fact not values for the Lower Jordan Basin but values for Israel, the West Bank and Jordan. Because of the water transfers into the basin, the unknown exact locations of the abstractions, especially from the West Bank, we cannot draw a picture for the Lower Jordan Basin, but we look on the entity of the three countries. For simplistic reasons, we call that entity "Lower Jordan Basin". Another source of imprecise information is the division between the West Bank and Gaza strip. Under correct

circumstances, the figures for water uses of the Gaza strip should not have been included, and the water inflow from Israel into Gaza should have been taken into consideration. But as we cannot make a point with concern to exact and verified information concerning the West Bank and Gaza, we consider the resulting differences as negligible as they would not change the overall picture.

Keeping in mind those limitations, an overview of the water balance of the Dead Sea is presented in Fig. 13.1 and for the Lower Jordan Basin in Fig. 13.3.

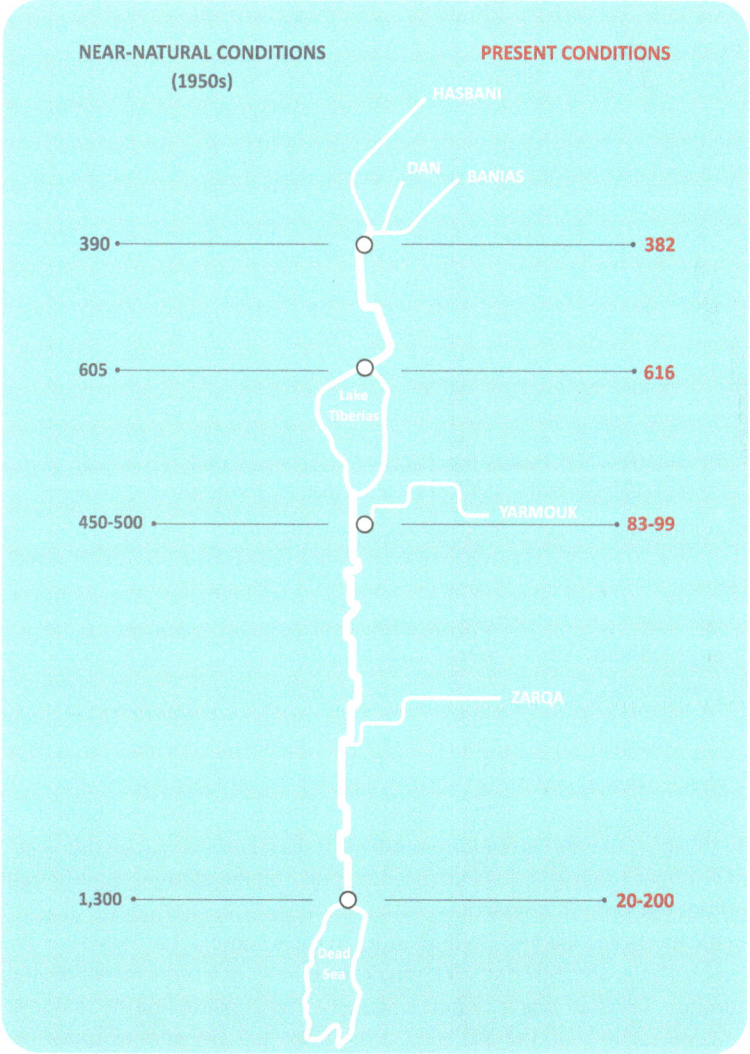

Fig. 13.1 Natural and present water balance of the Jordan River and the Dead Sea (Source: UN-ESCWA and BGR 2013. Source: Compiled by ESCWA-BGR based on Courcier et al. 2005; GRDC 2011, HSI 1944–2008)

While the inflow to Lake Tiberias did not show important variations since the 1950s, the outflow of the lake into Jordan was considerably reduced by the diversion of the lake's waters into Israel's National Water Carrier. The National Water Carrier transports the water from the northern part of Israel to the southern parts. The carrier provides both drinking water and water for irrigation to Israel. In order to enhance the quality of the water in the carrier, saline sources of the lake have been diverted away from the lake into Jordan, raising its salinity levels.

Water abstraction from Jordan itself and the Yarmuk, as one of its major tributaries, has increased significantly over the years, resulting in a minor inflow of $20–200 \times 10^6$ m³/year into the Dead Sea, compared to an inflow of about $1,3 \times 10^9$ m³/year 60 years ago.

A further $200–300 \times 10^6$ m³/year is abstracted from the Dead Sea itself by the Jordanian and Israeli Dead Sea potash companies. According to the survey of TAHAL and Geological Survey of Israel (GSI) (2011, p. 5), a minimum inflow of 700×10^6 m³/year would be indispensable to stabilise the current sea level.

Addition of all abstractions leads to the known consequence for the Dead Sea: Its level is declining with a rate of about 1 m annually (Bismuth et al. 2015a, in this volume, pp. 89–98).

13.1.1 Water Uses and Water Abstractions

The different uses for Israel, the Palestine territories and Jordan are presented in Fig. 13.2. Agriculture is the most important water user in the region, even though its contribution to the gross domestic product (GDP) is rather low with 3.3 % for Jordan (World Bank 2013a, p. 23) and 1.4 % for Israel (2012 GDP at current prices, Israel Central Bureau of Statistics (2013). In 2008, the GDP for agriculture in the Palestinian Territories was 4.8 % (Palestine Economic Policy Research Institute 2010).

Figure 13.3 shows the water abstraction for Israel, Jordan and the West Bank by type of source. The data do not present abstractions from non-approved wells in the West Bank and in Jordan. The Palestinian Water Authority does reveal information neither on the use of desalinated water nor on wastewater. According to the same sources, existing storm water harvesting structures (dams, cisterns and agricultural ponds in the West Bank) have a bulk potential of around 5.45×10^6 m³ (Palestinian Water Authority 2012). But as neither the year nor the actual volumes of abstractions are pointed out, and the amount is quite negligible, we did not consider this value.

We have selected the year 2007 in order to use a comparable data basis for Jordan and Israel. The years 2007 and 2008 mark the climax of a sequence of drought years and turning points in water policies, namely, the construction of major seawater desalination facilities at the Mediterranean shore and key administrative and legal reforms. Therefore, data from those years appear to provide a valid basis to evaluate progress and failures. Changes in the use and abstraction of water in Israel from 2007 until now will be discussed in Bismuth et al. (2015b, in this volume, pp. 253–275).

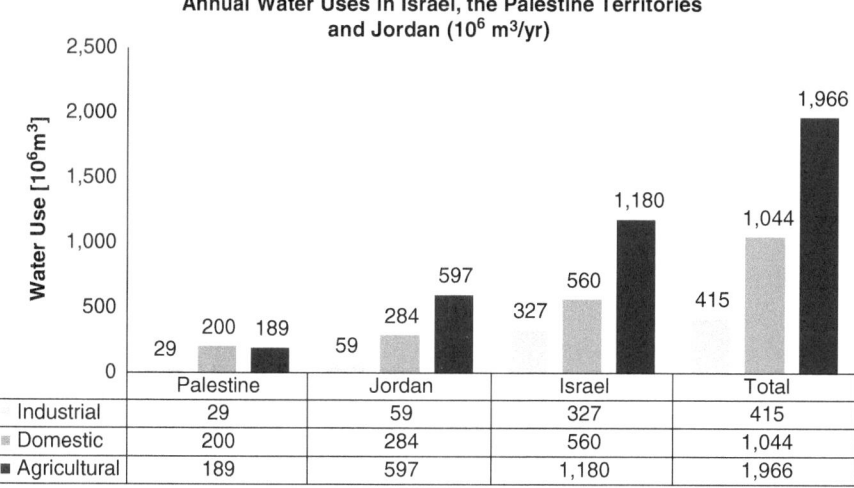

Fig. 13.2 Water uses in Israel, the Palestine Territories and Jordan (Source: 2007 data for Israel are derived from Israel Central Bureau of Statistics (2012), data for Palestine are based on FAO Aquastat (2014) and data for Jordan are taken from Jordan Ministry for Water and Irrigation (2009))

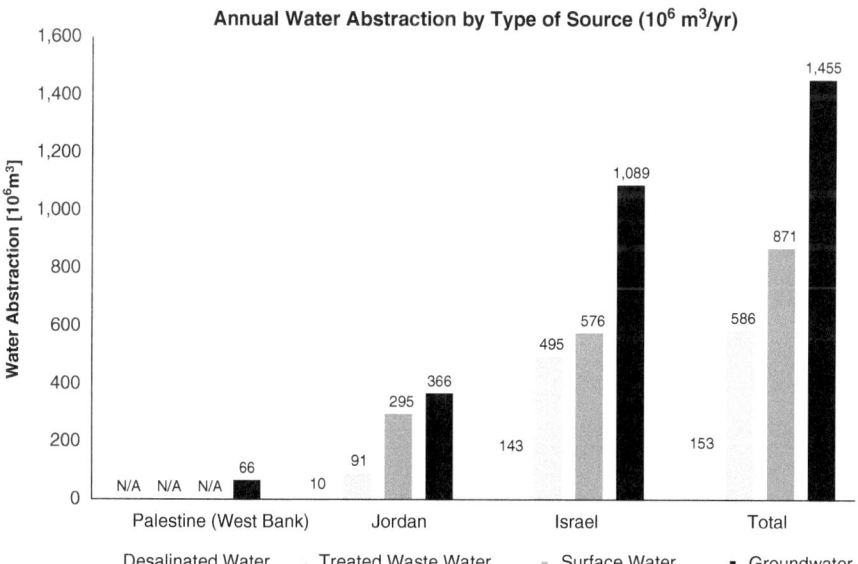

Fig. 13.3 Water abstraction by type of source (Source: 2007 data for Israel are from the Israel Central Bureau of Statistics (2012, p. 19), 2007 data for Jordan are from Jordan Ministry for Water and Irrigation (2009, p. 1–6) and 2011 data for the West Bank are from the Palestinian Water Authority (2012, p. 26))

In 2007, groundwater has been the most important source of water uses in the Lower Jordan Basin. The total water abstraction for 2007 summed up to about 3131×10^6 m³. Israel released around 234×10^6 m³ after use to the subsurface of which 88×10^6 m³ is lost due to leakages. Jordan released around 55×10^6 m³ to the subsurface with the share of losses probably being much higher as the unaccounted water in the municipals, which is around 50 % (Jordan Ministry for Water and Irrigation 2009) and 30 % for the Palestine territories (Palestinian Water Authority 2012).

13.1.2 Water Balance

The overall water balance for the region is presented in Fig. 13.4. We calculated a deficit of around 280×10^6 m³ in 2007 for the whole region.

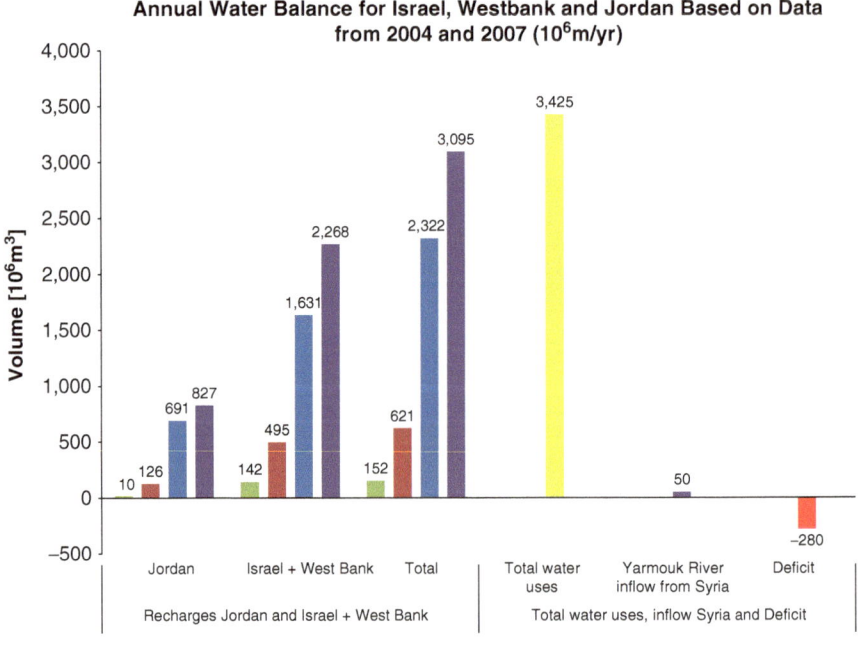

Fig. 13.4 Water balance for Israel, the West Bank and Jordan based on 2004 and 2007 data. Source: According to Weinberger et al. (2012, Table 4), the total average annual recharge for Israel and the West Bank 1631×10^6 m³ was calculated on a period from 1993 to 2009. 2007 data for desalinated water and for effluents were derived from Israel Central Bureau of Statistics (2012, Table 7 and p. 20). For Jordan, we calculated the average annual recharge, the effluents and the amount of desalinated water on a very conservative basis, based on the data from the Jordan National Master Plan from 2004 (Jordan Ministry for Water and Irrigation and German Technical Cooperation 2004, p. 48, Table 3.1 and p. 54) and from Water for Life (Jordan Ministry for Water and Irrigation 2009, executive summary, p. 7)

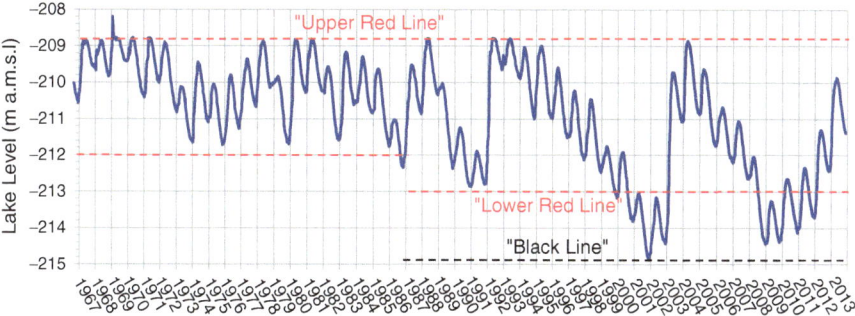

Fig. 13.5 Lake Tiberias lake levels (Source: Markel et al. 2014)

To compensate for a part of the deficit, Jordan used 91×10^6 m³ from nonrenewable groundwater sources in 2007. The most prominent aquifer is the "Disi" aquifer, a transboundary fossil groundwater resource shared with Saudi Arabia.

13.1.3   Environmental Consequences of Current Water Uses

Salt water intrusion to the coastal aquifers, significant decreases in the groundwater levels and resulting rising salinity levels were some of the hidden consequences of the unsustainable water uses. The crossing of the "Lower Red Line" in Lake Tiberias (Markel et al. 2014) marks the threshold where negative ecological consequences occur and where obvious and negative consequences for the National Water Carrier are most probable (see Fig. 13.5). The "Red Line" is a precautionary line defined by the Israeli Water Authority. A continuously lowered water level will lead to rapid salinisation due to penetration of saline water from underground sources into the aquifer (Haran et al. 2008).

For the Lower Jordan River itself, only 5 % of its natural flow is left. By diverting the saline springs of Lake Tiberias, the salinity in the river raised. The remaining water originates from agricultural runoffs, poorly treated effluents and drainage waters (Bamya et al. 2012).

13.1.4   Climate and Demography

Precipitation rates in the region show a high seasonal and annual variability. Comparisons between the average recharge from rainfall estimates for the period 1973 to 1992 with those for the period from 1993 to 2009 show a decline of 11 % for Israel and the Palestinian Territories. For the Lake Tiberias basin as the most important basin to provide the region with freshwater, the decline is particularly high with more than 13 % (Weinberger et al. 2012). Regional climate change impact will most probably lead to decreasing precipitation rates and increase drought events (IPCC 2013).

Expected population growth will aggravate the water scarcity problems. For all concerned stakeholders such as the water authorities and "the Friends of the Earth Middle East", it is evident that traditional water management offers no solution and that demand and supply management has to be altered.

13.1.5 Proposed Strategies

Increasing environmental concerns about the state of the water resources and the environment are calling for the rehabilitation of the Lower Jordan flows. The partial restoration of the Lower Jordan River is an economically viable option as additional income for the riparian populations could be generated from the benefits (Becker et al. 2014).

Israel reacted to the water crisis with an enforced building of seawater desalination facilities and wastewater treatment plants for the reuse of water and with extensive structural and price reforms. The challenges and benefits of those reforms are discussed in Bismuth et al. (2015b, in this volume, pp. 253–275).

Jordan equally started to implement measures like building the "Disi" Aquifer Conveyance Project, aiming to use the fossil waters of this groundwater resource. The recently inaugurated "Disi" pipeline will provide Jordan with an additional annual supply in the order of 100×10^6 m^3 for the next 20–30 years depending on abstraction rates. Consequences of existing use practices and specific challenges for Jordan are addressed in Chap. 15 (Yorke 2015, in this volume, pp. 227–251).

Setting up a seawater desalination plant at Aqaba combined with a conveyance of the remaining brine to the Dead Sea is part of the planned RSDS Conveyance Project. This project and its alternatives will be presented in Malkawi and Tsur (2015, in this volume, pp. 205–225).

13.2 History of Water Conflicts, Cooperation and Treaties

The existing water accords play an important role for the definition of rules for the management of the common water resources. They also have an influence on the abstraction rates and the exchange of water between the riparians. A trilateral project like the RSDS Conveyance Project further demands some level of cooperation. Therefore, we will shortly present the main and relevant agreements in the context of the Lower Jordan Basin, and we will analyse them in regard to the RSDS Project.

13.2.1 The Johnston Plan

In 1953, the United States sent a special envoy, Eric Johnston, to the region in order to mediate an agreement on the Jordan River allocations, later called the Johnston Plan. Though the parties never formally ratified the plan, they have initially adhered to it. According to the plan, 400×10^6 m³ per year were allocated to Israel, 720×10^6 m³ to Jordan and 132×10^6 m³ to Syria (Phillips et al. 2007). But in the 1960s, the parties began to develop projects in excess of the Johnston allocations. This could be considered as one of the reasons for the 1967 war, which gave Israel control over two of the three Jordan headwaters: the entire Lower Jordan River and the Mountain Aquifers in the West Bank (Wolf 2000). The latter are of strategic importance for Israel's provision with groundwater, as major springs in Israel are alimented by those aquifers (Baumgarten 2010). Even though the Johnston Plan is frequently cited as a basis for cooperation agreements between the concerned parties, it falls short in view of sustainable groundwater uses, environmental needs and the impacts of population growth and climate change on the availability of water resources (Mager 2015).

13.2.2 The Agreement Concerning the Utilisation of the Yarmuk Waters

Jordan signed an agreement with Syria concerning the utilisation of the waters of the Yarmuk River (Syrian Arab Rebublic and Jordan 1987). The agreement foresaw the establishment of a joint Syria–Jordan Commission for the implementation of the dam-building works at Maqarin. The dam at the Yarmuk River was finally realised in 2011 with a storage capacity of 110×10^6 m³. So far the dam's reservoir remained unfilled, as droughts and increased consumption in Syria have considerably reduced the annual flow of the Yarmuk River. Since the agreement was signed in 1987, more than 30 dams and more than 300 wells have been erected (UN-ESCWA and BGR 2013).

13.2.3 The Peace Treaty Between Israel and Jordan

Israel and Jordan signed a peace treaty in 1994 (Jordan-Israel Peace Treaty 1994). The peace treaty refers implicitly to the three main principles of international customary water law (rule of equitable and reasonable utilisation, the no-harm rule and the duty to cooperate), but adapts them to the special political situation. The water issues between the two states are settled in Article 6 and in Annex II of the treaty. The parties agree on the allocations of the shared water resources from Yarmuk and Jordan: In the summer season, Israel receives 12×10^6 m³ from the Yarmuk River, and Jordan receives the remaining waters, while Israel is allowed to obtain 13×10^6 m³ during the winter period. The parties agreed also on a storage system, which permits the storage of 20×10^6 m³ allocation during the summer period. In Article 7 of Annex II, the establishment of a Joint Water Committee comprised of three members from each country is fixed. The cooperation of water issues and the exchange of relevant data on water resources are synchronised by the Joint Water Committee. The Joint Water Committee shall survey existing uses for documentation and prevention of appreciable harm. The treaty foresees also the joint establishment of monitoring stations and bans the disposal of wastewater in the rivers without treatment to standards allowing the unrestricted agricultural use.

In fact the Jordan–Israel Peace Treaty is the legal foundation for the development of the Red Sea–Dead Sea (RSDS) Conveyance Project. Israel and Jordan admitted the fact that the natural water resources are not sufficient to meet their needs. The parties agreed to cooperate in the development of new water resources among others, and Israel agreed to transfer desalinated water to Jordan.

The Jordan–Israel Peace Treaty of 1994 did not address any of the other riparian rights or any other aspect of the Jordan River basin except those of the Yarmuk and Jordan River. Any peace treaty between Israel and the Palestinians with concern to water management will therefore interfere with the agreements settled in the Jordan–Israel Peace Treaty (Mager 2015).

13.2.4   The Oslo II Agreement

The most important water issues of concern between Israel and Palestine are the use of the aquifers located in the West Bank, the Eastern, Western and North-Eastern Mountain Aquifers, and the sharing of their resources. The share and distribution of the water resources and the establishment of a Palestinian Water Administration Authority were settled in Annex III, Article 40 of the Oslo II Agreement (Israeli–Palestinian Interim Agreement (Oslo II) 28 September 1995, pp. 318 ff.). In essence the Oslo II Agreement gives the Palestinians the right to establish a Water Administration Authority and acknowledges for the first time in principle Palestinian water rights. The future water demands for the Palestinians have been mutually agreed to be around $70–80 \times 10^6$ m³ per year. The exact allocation is postponed to the Permanent Status Negotiations and Agreement.

The two parties agreed to establish a Joint Water Committee (JWC) for the interim period until a peace treaty between the Palestinians and the Israelis will be settled. Even though the Joint Water Committee has far reaching administrative responsibilities concerning the management of the water resources in reality, the JWC led to the formalisation of discriminatory management practices (Baumgarten 2010, p. 189). All development projects are under the condition of prior approval by the JWC, but as all decisions of the JWC should be reached by consensus, Israel got a de facto veto right. Furthermore, projects outside the areas under administration of the Palestinian Authority (A and B) need the approval of the civil administration, which represents a branch of the Israeli Defence Ministry. This required approval delays necessary projects as it is a long bureaucratic process (The Knesset 2011). As a consequence, the Palestinians are not able to develop their water resources or projects in the desired way.

But also the Palestinians denied approval to some of the proposed projects, as they would serve some of the interests of the Israeli settlements in the West Bank. The Palestinian side categorically turns down any cooperation with the Israeli settlements.[1]

In fact, the JWC in its present form and under the present political circumstances is not an effective instrument to derive solutions and settle conflicts for the most important water management problems of the West Bank, which are the old and insufficient potable water distribution network, the problem of not approved water abstractions, the lacking sewage collection and treatment and the pollution of the streams, wadis and groundwater sources. Fragmented institutions, limited gover-nance due to the occupation and the split between the different Palestinian fractions resulted in the lack of environmental planning instruments, capacities, legislation and enforcement in Palestine. The political lock-in situation between Israel and Palestine concerning the peace process and the power asymmetries between the two parties has also led to an obstruction of the management of the shared water resources. Israel can be criticised for discriminatory water practices (Kislev 2008). According to Kislev, Palestinians receive only 60 % of the water share that Jewish settlements receive, and if water losses are taken into account, the figure might even be less.

13.2.5 The Red Sea–Dead Sea Water Conveyance Study Programme

Even though the Red Sea–Dead Sea Water Conveyance Study Programme under the World Bank represents neither a treaty nor an accord among the three riparians of the Lower Jordan, it represents a first common action within the institution of the World Bank between Israel, Jordan and the Palestine territories. We can draw some important lessons from the study programme both with concern to the relations between the riparians but also with concern to the planning process of the RSDS Conveyance Project as an example of a future MWEP.

As part of the Jordan–Israel Peace Treaty negotiations, the two parties conceived a concept to convey water from the Red Sea to the Dead Sea. As the scope of the project requires international funding, the concept had to be agreed by the Palestinian Authorities. This was done by the submission of a jointly signed letter to the World Bank dated 9 May 2005 requesting to coordinate donor financing and the manage-

---

[1] Interview on 2 January 2014 with A. M. Hindi (Palestinian National Authority, Palestinian Water Authority, Director General National Water Council's Unit) and R. A. El Sheikh (Palestinian Water Authority, Deputy Chairman)

ment of the implementation of the study programme (World Bank 2013b). The three parties announced their agreement at the World Economic Forum at the Dead Sea in May 2005. To finance the estimated costs of the study programme of USD 16 million, the World Bank established a multi-donor trust for finance. It took 5 years to establish the trust in 2010. The donors were France, Greece, Italy, Japan, South Korea, the Netherlands, Sweden and the United States.

Initially, the study programme did not include the Study of Alternatives but only a feasibility study and the environment and social assessment study. It was due to the pressure of the environmental nongovernmental organisations that a study of alternatives was conducted as the last study within a series of different studies.

The terms of reference never considered investigating on the feasibility to generate energy from saline water, but the financial benefices were considered in the study of alternatives Malkawi and Tsur (2015, in this volume, pp. 205–225). This fact is crucial for the implementation of the polluters pay principle (Bismuth et al. 2015b, in this volume, pp. 253–275) and also for the overall costs of the project.

From the beginning of the study programme, the role of the Palestinian Authorities had been quite ambiguous: On the one side, they saw in the project an opportunity to achieve results for their creation of a Palestinian state, and, on the other side, the Palestinians in their majority opposed the project, which resulted in minor active participation in the Study of Alternatives. All three beneficiary parties proposed to the World Bank a list of experts to conduct the study, but on the Palestinian list, only non-Palestinians appeared. It is not that the Palestinians lack qualified expertise among their scientists, but finally a British citizen (Tony Allen) was chosen as the expert to represent the Palestinian's interests. The stakeholder discussions on 20 and 21 February 2013 reflect this ambiguous position of the Palestinians between their needs for more water, their rights on land and resources, their opposition to Israel and the acknowledgements of the Jordanian water needs (www.worldbank.org/rds). The majority of the participants in the stakeholder forum would have preferred to settle water questions in the peace negotiation process, and they feared that with an agreement they would lose their rights on water and land. Some of the participants demanded the rights of the Palestinians to develop their own Potassium companies at the Dead Sea and to construct their own hotel sector at the sea shore, but without outlining where the additional water should come from. The stakeholder discussions in Israel on 18 and 19 February 2013 reflected more on the environmental concerns but also on Israel's concern to support Jordan in its quest for new water sources, while the Jordanian meetings on 14 and 17 February 2013 were centred around the questions of affordable water prices and the economic consequences of the project but also on security and safety aspects.

Only few participants raised the questions on the possible management and controlling structures with regard to the complicated relational setting in the region. This aspect was not adequately addressed neither in the terms of references nor in the presented reports.

13.2.6   The Water Swap Memorandum of Understanding

The Memorandum of Understanding (MoU) (see Bismuth et al. 2015a, in this volume, pp. 89–98) between Israel, Jordan and the Palestinian Territories concerning the Red Sea–Dead Sea Water Conveyance Project is the latest and most concrete agreement between the three parties. The negotiation process already for the agreement to launch the feasibility study supported and conducted by the World Bank has turned out to be rather fastidious and time consuming as the Palestinians saw an opportunity to gain more influence on the shared water resources between Israel and the Palestinian Territories and to use the project as a bargaining chip in the peace treaty negotiation process as well as an instrument to realise national sovereignty (Fischhendler et al. 2013).

While the MoU constitutes an intelligent cost saving solution for Jordan and Israel, based on mutual cooperation, the MoU does not provide any substantial solution for the manifold water management problems between Israelis and Palestinians. The approach as foreseen in the MoU reduces the problems between the two parties merely to quantitative aspects.

Detailed regulations will be fixed in bilateral accords between the parties. This facilitates the realisation of the Israeli–Jordanian agreements, as the approval to build the most needed seawater desalination plant at Aqaba will no longer depend on the Palestinians. But under the present political circumstances between the Palestinians and the Israelis, it is more than unpredictable what this means for the realisation of the conveyance and any other actions undertaken to halt the further decline of the Dead Sea water level. Any common action for the safeguard of the Dead Sea appears to need a more comprehensive approach.

13.3   Conclusions

The data on water uses, on abstractions and on available sources indicate clearly that already in 2007 the three riparians have been in a deficiency situation with important impacts on groundwater resources, river and lake ecosystems and the Dead Sea itself. The discussions within the RSDS Conveyance Study Programme conducted by the World Bank revealed that a technical solution is only one of the several assets required to halt the further decline of the Dead Sea. The results of the RSDS Study Programme do not provide proposals for adequate management, control structures and the financial conditions for such a major project, but also cost calculations specifically for the part of the energy generation remain unsettled.

The existing bases for international cooperation are the different agreements between Israel and Jordan, Israel and the Palestinian Authorities and Jordan and Syria. All three accords (the Jordanian–Syrian agreements, the Jordan–Israel Peace Treaty and the Oslo II Agreement) have in common that they are ambiguous, vague and voluntary and leave room for interpretation for each party (Mager 2015). Rules on compliance and control and mechanisms for mediation in case of a conflict between the parties are not established or not in operation, especially after the Al-Aqsa Intifada in September 2000 (Dombrowsky 2003). The accords might be sufficient for short-term planning and communication, but for a longer perspective and the establishment of a common coherent management strategy, they do not provide adequate structures. Furthermore, the agreements are inflexible with regard to the challenges of regional climate change impacts or to newly arising issues, as they consolidate existing uses.

The agreements recognise only states as legitimate actors, which leads to centralisation and nationalisation of the water management. Both in Palestine and in Israel, the discourse on water is focused on the development of resources seen as a part of the nation-building effort (Trottier and Brooks 2013).

The nature of the relations between Israelis and Palestinians defines the way how water management problems are addressed. The large power asymmetries between the two parties do not only impede the development of intelligent solutions, but they also prevent the Palestinians from developing their own objectives, based on necessities but also on the availability of natural water resources and the principles of sustainability.

Shared water resources mean shared responsibilities and duties. With more water from Israeli desalination plants, quantitative problems might be eased, but nothing is gained to overcome existing power asymmetries or insufficient conflict mitigation instruments. Any measure which builds trust and mutual understanding is most needed in that region. What is furthermore needed is a frank discussion about carrying capacities in the light of climate change and population growth and on the role of agriculture.

Shared water resources mean furthermore that data on the water resources have to be shared in a transparent and reliable way. That could be a first step towards trust building not only between the different countries but also in view of the citizens.

References

Bamya S, Becker N, Saaf EJ et al (2012) Towards a living Jordan River: a regional economic benefits study on the rehabilitation of the Lower Jordan River. Friends of the Earth Middle East, Amman/Bethlehem/Tel Aviv

Baumgarten P (2010) Israel's transboundary water disputes. J Land Res Environ Law 30:179–197

Becker N, Helgeson J, Katz DL (2014) Once there was a river: a benefit-cost analysis of rehabilitation of the Jordan River. Reg Environ Chang 14:1303–1314. doi:10.1007/s10113-013-0578-4

Bismuth C, Hoechstetter S, Bens O (2015a) Research in two cases studies: (1) Irrigation and land use in the Fergana Valley and (2) Water management in the Lower Jordan Valley. In: Huettl RF, Bens O, Bismuth C, Hoechstetter S (eds) Society water technology: a critical appraisal of major water engineering projects. Springer, Dordrecht, pp 89–98

Bismuth C, Hansjürgens B, Yaari I (2015b) Technologies, incentives and cost recovery: is there an Israeli role model? In: Huettl RF, Bens O, Bismuth C, Hoechstetter S (eds) Society – water – technology: a critical appraisal of major water engineering projects. Springer, Dordrecht, pp 253–275

Courcier R, Venot JP, Molle F (2005) Historical transformations of the lower Jordan river basin (in Jordan): changes in water use and projections (1950-2025). Compr Assess Water Manag Agric 9

Dombrowsky I (2003) Water accords in the Middle East peace process. In: Brauch HG, Liotta PH, Marquina A et al (eds) Security and environment in the Mediterranean – conceptualising security and environmental conflict. Springer, Berlin/Heidelberg/New York, pp 729–744

FAO Aquastat (2014) Country fact sheet occupied Palestinian Territory. In: Aquastat. www.fao. org/nr/aquastat. Accessed 27 Jan 2015

Fischhendler I, Wolf AT, Eckstein G (2013) The role of creative language in addressing political asymmetries: the Israeli-Arab Water Agreements. In: Megdal SB, Varady RG, Eden S (eds) Shared borders, shared waters. CRC Press, Leiden, pp 53–74

GRDC (The Global Runoff Data Center) (2011). Available at http://www.bafg.de/GRDC/EN/ home/homepagenode.html. Accessed 6 Jan 2012

Haran M, Samuels R, Gabbay S et al (2008) Quality indicators of the state of chemical pollution in Israel. Israel J Chem 42:119–132. doi:10.1560/QU9Q-XGF9-HUM6-DA61

IPCC (2013) Climate change 2013: the physical science basis. Working group I contribution to the fifth Assessment Report of the Intergovernmental Panel on Climate Change. Cambridge University Press, Cambridge/New York

Israel Central Bureau of Statistics (2012) Satellite account of water in Israel 2007–2008. Israel Central Bureau of Statistics (CBS), Jerusalem

Israel Central Bureau of Statistics (2013) Input, output and domestic product in agriculture, Table 19.14. In: CBS, Statistical abstract of Israel 2013. http://www.cbs.gov.il/reader/shnaton/ templ_shnaton_e.html?num_tab=st19_14&CYear=2013. Accessed 11 Feb 2015

Israel Water Authority (2014) Water authority report 2013. Israel Water Authority, Jerusalem (in Hebrew)

Israeli Government and PLO (1995) Israeli-Palestinian Interim Agreement on the West Bank and the Gaza Strip. Signed in Washington, DC. 28 September 1995 (Oslo II). http://mfa. gov.il/MFA/ForeignPolicy/Peace/Guide/Pages/THE ISRAELI-PALESTINIAN INTERIM AGREEMENT.aspx. Accessed 4 Sept 2015

Jordan-Israel Peace Treaty (1994) Treaty of peace between the state of Israel and the Hashemite Kingdom of Jordan (including texts of Annex II – water and Annex IV – environment), 26 October 1994. http://foeme.org/www/?module=regional_data&record_id=3. Accessed 3 Feb 2015

Kislev Y (2008) Water in the Palestinian localities. In: Israeli water magazine. http://departments. agri.huji.ac.il/economics/teachers/kislev_yoav/yoav-black.pdf. Accessed 11 Feb 2015

Mager U (2015) International water law: global developments and regional examples, Miscellane. Jedermann-Verlag Heidelberg, Heidelberg

Malkawi A, Tsur Y (2015) Reclaiming the Dead Sea: alternatives for action. In: Huettl RF, Bens O, Bismuth C, Hoechstetter S (eds) Society – water – technology: a critical appraisal of major water engineering projects. Springer, Dordrecht, pp 253–275

Markel D, Shamir U, Green P (2014) Operational management of Lake Kinneret and its watershed. In: Zohary T, Sukenik A, Nishry A (eds) Lake Kinneret. Ecology and management. Springer, Dordrecht, pp 541–560

MWI (2009) Water for life: Jordan's water strategy 2008–2022. http://www.irinnews.org/pdf/jordan_national_water_strategy.pdf. Accessed 14 Jan 2015

MWI GTZ (2004) The national water master plan. Ministry of Water and Irrigation, Amman

Palestine Economic Policy Research Institute (2010) Overview of the Palestinian economy. Palestine Economy Policy Research Institute, Ramallah

Palestinian Water Authority (2012) Annual status report on water resources, water supply, and wastewater in the occupied State of Palestine 2011. Palestinian Water Authority, Ramallah

Phillips DJH, Attili S, Mccaffrey S et al (2007) The Jordan River Basin: 1. clarification of the allocations in the Johnston Plan. Water Int 32:16–38. doi:10.1080/02508060708691963

Shami S (2013) Syria further deepens Jordan's water crisis. "Disi" water project provides "temporary relief." In: Thomson Reuters Foundation. http://governance.arij.net/en/?p=51. Accessed 7 Aug 2014

Syrian Arab Rebublic and Jordan (1987) Agreement concerning the utilization of the Yarmuk waters (with annex). Signed at Amman on 3 September 1987. http://www.internationalwaterlaw.org/documents/regionaldocs/Jordan-Syria-1987.pdf. Accessed 4 Mar 2015

TAHAL, GSI (2011) Dead Sea study (final report). http://siteresources.worldbank.org/INTREDSEADEADSEA/Resources/Dead_Sea_Study_Final_August_2011.pdf. Accessed 14 Jan 2015

The Knesset (2011) Israeli-Palestinian cooperation on water issues: presented to the Internal Affairs and Environment Committee. Center for Research and Information, Jerusalem

Trottier J, Brooks DB (2013) Academic tribes and transboundary water management: water in the Israeli-Palestinian peace process. In: Science & diplomacy. http://www.sciencediplomacy.org/files/academic_tribes_and_transboundary_water_management_science__diplomacy.pdf. Accessed 17 Feb 2015

UN-ESCWA, BGR (2013) Inventory of shared water resources in Western Asia: Chapter 6 Jordan River Basin. United Nations Economic and Social Commission for Western Asia; Federal Institute for Geosciences and Natural Resources, Beirut

Weinberger G, Livshitz Y, Givati A et al (2012) The natural water resources between the Mediterranean Sea and the Jordan River. Israel Hydrological Service, Jerusalem

Wolf AT (2000) "Hydrostrategic" territory in the Jordan Basin: water, war, and Arab-Israeli peace negotiations. In: Amery HA, Wolf AT (eds) Water in the Middle East. A geography of peace. University of Texas Press, Austin, pp 63–120

World Bank (2013a) Jordan economic monitor. Moderate economic activity with significant downside risk. World Bank, Washington, DC

World Bank (2013b) Red Sea – Dead Sea Water Conveyance Study Program overview – updated January 2013. http://siteresources.worldbank.org/EXTREDSEADEADSEA/Resources/Overview_RDS_Jan_2013.pdf?&&resourceurlname=Overview_RDS_Jan_2013.pdf. Accessed 4 Mar 2015

Yorke V (2015) Jordan's shadow state and water management: prospects for water security will depend on politics and regional cooperation. In: Huettl RF, Bens O, Bismuth C, Hoechstetter S (eds) Society – water – technology: a critical appraisal of major water engineering projects. Springer, Dordrecht, pp 227–251

Chapter 14
Reclaiming the Dead Sea: Alternatives for Action

Abdallah I. Husein Malkawi and Yacov Tsur

Abstract The sustainable supply of natural water available in the water basin feeding the Dead Sea (comprising of Israel, Jordan and the Palestinian Authority) will soon drop below 100 cubic metres (m^3) per person per year. This has resulted from upstream diversions that over time have deprived the Dead Sea of more than 90 % of its historical inflow and led to a progressive decline of its water level with detrimental effects on the surrounding environment and infrastructure. We examine four alternatives to stabilise or restore the Dead Sea and evaluate the costs associated with each alternative. We also offer a mechanism to pay for the reclamation alternatives based on a surcharge levied on all upstream diversions (including water consumed by the potash industries). The surcharge rates associated with the four alternatives range between zero and USD 0.10 per m^3.

Keywords Dead Sea reclamation • Water scarcity • Environmental amenities • Recycling • Desalination • Study of alternatives • World Bank • Jordan River • Yarmuk River • RSDS conveyance project

14.1 Introduction

Upstream diversions have diminished water flow into the Dead Sea by over $1,500 \times 10^6$ m^3/year during the last 50 years. The most significant diversions have been from the upper Jordan River (mostly by Israel) and the Yarmuk River (mostly by Syria), while the remaining diversions are mostly from side wadis in the Dead Sea eastern escarpment (Salameh and El-Naser 2000; TAHAL and GSI 2011). The Dead Sea water balance is further exacerbated by the additional

A.I.H. Malkawi (✉)
Jordan University of Science and Technology, P.O. Box 3030, Irbid 22110, Jordan
e-mail: mhusein@just.edu.jo

Y. Tsur
The Hebrew University of Jerusalem, P.O. Box 12, Rehovot 76100, Israel
e-mail: tsur@agri.huji.ac.il

© The Author(s) 2016
R.F. Hüttl et al. (eds.), *Society - Water - Technology*, Water Resources Development and Management, DOI 10.1007/978-3-319-18971-0_14

water loss of about 262×10^6 m³/year due to the potash industries of Israel and Jordan (Zbranek 2013). The Dead Sea water level is currently at about 428 m below sea level (mbSL), some 30 m lower than its 1960 level, and continues to decline by more than a metre annually on average (Rawashdeh et al. 2013). The progressive decline in the water level and the ensuing retreat of the shore line have given rise to sinkholes, mud flats and landslides with serious damage to infrastructure and irreversible damage to habitat of unique species. The estimated direct costs range between USD 73 million per year and USD 227 million per year (Becker and Katz 2009).[1]

Stabilising the Dead Sea at its current level requires an additional water inflow of 700–800×10^6 m³/year, while fully restoring the Dead Sea to its historical level (of 395–400 mbSL) would mean increasing the inflow by more than $1,100 \times 10^6$ m³/year (Malkawi et al. 2010; TAHAL and GSI 2011). Reclaiming the Dead Sea, thus, necessitates substantial additional inflows, which raise three interrelated questions: from where (i.e. what are the possible sources of the additional inflow), at what cost and how to pay for the reclamation? The first and second questions have been analysed in a number of studies, and the main findings of these studies will serve as a benchmark and a point of departure for this effort.[2] The third question, regarding who should pay for the Dead Sea reclamation, has not been properly addressed and will receive special attention in this chapter.

To put the Dead Sea reclamation problem in context, we begin in the next section with a brief summary of the current water situation in the Jordan River Basin comprised of Israel, Jordan and the Palestinian Authority – the three riparian parties to the Dead Sea. Section 14.3 discusses the Dead Sea reclamation alternatives involving seawater conveyance from the Red Sea or from the Mediterranean, drawing mainly on the study of alternatives to the Red Sea–Dead Sea Water Conveyance Project (Allan et al. 2014), giving special attention to the costs of the different alternatives. Section 14.4 elaborates on an alternative, first offered in the abovementioned study of alternatives, which takes a long-term perspective (3–4 decades) by combining measures that will be implemented incrementally over time. Section 14.5 discusses the issue of how to cover the costs of Dead Sea reclamation and Sect. 14.6 concludes.

---

[1] These estimates are based on the local population's willingness to pay to prevent further decline of the Dead Sea level. However, the unique characteristics of the Dead Sea imply that the benefit of its preservation extends beyond the region and includes the international community as a whole. The total benefit of preventing the declining of the Dead Sea is therefore likely to be larger than the above range.

[2] See Vardi (1990) and Beyth (2006, 2007) for overviews of past proposals and the recent ensemble of studies coordinated by the World Bank available at http://web.worldbank.org/WBSITE/EXTERNAL/COUNTRIES/MENAEXT/EXTREDSEADEADSEA/0,,contentMDK:21827416~p agePK:64168427~piPK:64168435~theSitePK:5174617,00.html.

14.2 Water Scarcity in the Jordan River Basin

We consider the part of the Jordan River Basin that includes the three Dead Sea riparian parties, namely, Israel, Jordan and the Palestinian Authority (see Fig. 14.1).[3] A useful (albeit rough) index of regional water scarcity is the quantity of renewable (natural) water available per person in a sustainable fashion, obtained by dividing the average annual supply of renewable natural water by the existing population and measured in units of cubic metre (m^3) per year and person (m^3/year per person). Regions whose renewable water supplies fall below 1,000 m^3/year per person or 500 m^3/year per person are said to experience water scarcity or absolute scarcity, respectively (Falkenmark et al. 1989). The 100 m^3/year per person threshold is often mentioned as the supply required to satisfy basic human needs (Gleick 1996). While the supply of natural renewable water is on average constant (with possible trends over the long run, due, e.g. to climate change), the population is expanding quite rapidly in this region, implying that the m^3/year per person index will decline over time and therefore aggravating water scarcity.

Table 14.1 presents the average supply of renewable water in the Jordan River Basin. It shows that the renewable (natural) water supplies (available on a sustainable fashion, i.e. without drawing down stocks) in the region comprised of Jordan, Israel and the Palestinian Authority are on average $2,428 \times 10^6$ m^3/year, and this quantity includes 232×10^6 m^3/year of brackish water (i.e. water with chloride concentration above 400 mg/l, which is unsuitable for drinking and irrigation of many crops without mixing).[4] The total supply of good quality natural water is therefore $2,196 \times 10^6$ m^3/year ($=2,428 - 232$) on average.

Figure 14.2 presents actual (as of 2011) and projected populations for Israel, Jordan and the Palestinian Authority from 1950 to 2050. The m^3/year per person scarcity index is obtained by dividing the average annual water supply ($2,196 \times 10^6$ m^3/year or $2,428 \times 10^6$ m^3/year) by the population. The results are shown in Table 14.2.

As the table reveals, the region as a whole is already far below the absolute scarcity mark of 500 m^3/year per person and will soon enter subsistence scarcity below 100 m^3/year per person. Such an acute scarcity implies that increasing the supply of potable water for domestic uses receives the highest priority. This observation virtually implies that natural (potable) water cannot on its own achieve the goal of Dead Sea reclamation (which, as noted above, requires 700×10^6 m^3/year to 800×10^6 m^3/year just for stabilising the current level) and other sources must be found for that purpose. These other sources are seawater or recycled water. Indeed, most proposals for reclaiming the Dead Sea, either by stopping its decline or restoring its level to its

---

[3] The Jordan River basin contains also parts of southern Lebanon and of southwest Syria. Due to lack of data on these regions, they will not be included in this study.

[4] A detailed account of natural, renewable water supplies (including inter-temporal fluctuations) can be found in Tsur (2014).

Fig. 14.1 The Jordan River Basin. The Upper Jordan River extends between its headwater (at the confluence of the Dan, Banias and Hatzbani) and the Lake Tiberias. The Lower Jordan River is the southern stretch of the river between Lake Tiberias and the Dead Sea (Source: UNEP/DEWA/GRID-Geneva(http://en.wikipedia.org/wiki/Jordan_River#mediaviewer/File:JordanRiver_en.svg))

Table 14.1 Renewable water resources in the Jordan River Basin

	10^6 m³/year	Source
Israel and Palestinian Authority	1,451 (1,683)[a]	Weinberger et al. (2012)
Jordan	745	Ministry of Water and Irrigation (2009; executive summary, p. 7).
Total	2,196 (2,428)	

[a]Average over the period 1993–2009 without the 232×10^6 m³/year of brackish water (*in parenthesis*: total supply with brackish water)

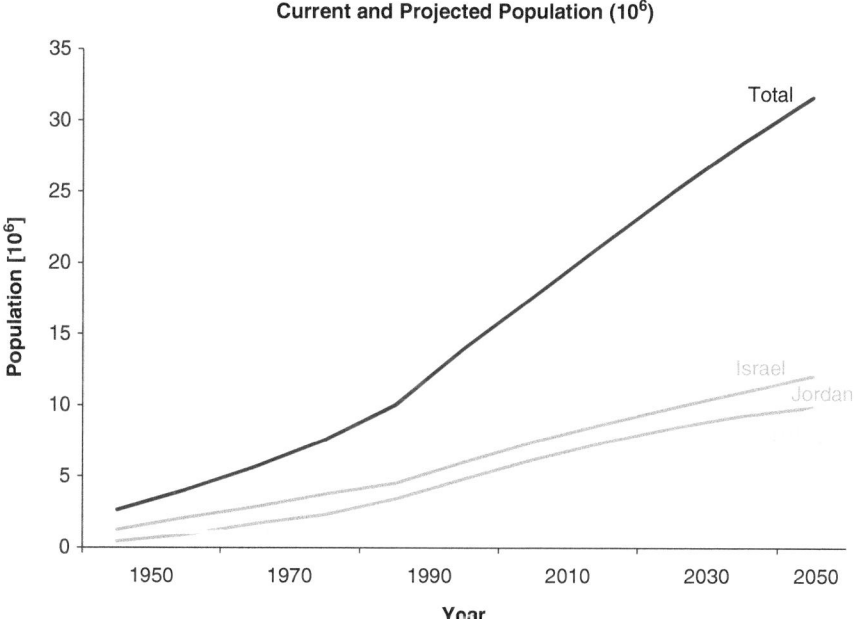

Fig. 14.2 Current (until 2011) and projected population (million) (Source: United Nations (2011))

Table 14.2 Population and annual per person supplies of natural (potable) water

Year	Population (million)	m³/year per person based on $2,428 \times 10^6$ m³/year	m³/year per person. based on $2,196 \times 10^6$ m³/year
2013	18.8	129	117
2030	25	97	88
2050	31.6	77	69

pre-diversion state, involve conveyance of large quantities of seawater from the Mediterranean or from the Red Sea (see Vardi 1990; Beyth 2007 for overviews on past proposals and www.worldbank.org/rds for the ensemble of studies associated with recent "Red Sea–Dead Sea Conveyance Study Program"). These sea-to-sea conveyance alternatives involve large-scale infrastructure projects and require large upfront investment, raising doubts about their feasibility. In the next section, we briefly summarise one Red Sea–Dead Sea Project and two Mediterranean Sea–Dead Sea Projects considered in the abovementioned World Bank studies.

14.3 Water Conveyance from the Red Sea and the Mediterranean Sea

Our cost calculations are based on the most recent data available from Coyne et Bellier's (2014) feasibility study. This feasibility study considers a comprehensive project with the dual goal of reclaiming the Dead Sea and increasing the supply of potable water in the region: upon completion, the project will convey $2,000 \times 10^6$ m^3/year from the Red Sea to the Dead Sea, desalinate 850×10^6 m^3/year that will be delivered mostly to Amman and discharge $1,150 \times 10^6$ m^3/year of brine in the Dead Sea. As we focus on the Dead Sea reclamation, the cost of seawater–brine discharge in the Dead Sea reported here pertains to the cost of a project, the sole purpose of which is to stabilise the Dead Sea water level. This involves the conveyance of up to $1,150 \times 10^6$ m^3/year seawater from the Red Sea or the Mediterranean to the Dead Sea exploiting the elevation difference to generate hydropower.[5] We discuss water conveyance from the Red Sea and from the Mediterranean Sea in turn.

14.3.1 Red Sea–Dead Sea Water Conveyance

The feasibility study of the Red Sea–Dead Sea alternative (Coyne et Bellier 2014) considered two basic alignments that vary according to the method of water conveyance: surface (buried) pipelines or tunnelling. The advantage of the pipeline approach is that it can be implemented in phases over time (by adding pipelines as needed); the disadvantage is that it requires lifting the water to an altitude of 220 m before letting it flow downward to the Dead Sea (at 390–400 mbSL), and this (pumping) operation adds on to the running costs. The tunnel option, on the other hand, does away with the need to lift the conveyed water, but requires complete investment of the entire infrastructure upfront. Due to environmental risks (associated with possible stratification, gypsum crystallisation and algae bloom), it is strongly recommended that the quantities of seawater (or brine) discharge in the

---

[5] A detailed explanation of how the cost of Dead Sea reclamation is calculated, based on Coyne et Bellier's (2014) data, can be found in Allan et al (2014).

Dead Sea will be increased gradually over a period of time (TAHAL and GSI 2011). We therefore focus on the surface pipeline option.[6]

Figures 14.3 and 14.4 present the costs of seawater (or brine) discharged in the Dead Sea from the Red Sea–Pipeline Project under two electricity tariff regimes: under regime A, electricity is bought from and sold to the Jordanian grid at the Jordanian tariffs (which prevailed in 2012); under electricity tariff B, electricity is obtained from the Jordanian grid at the Jordanian tariff and the electricity generated is sold to the Israeli grid at the (higher) Israeli tariffs. The costs are in 2012 prices.

The costs in Figs. 14.3 and 14.4 are calculated as follows: First, the fixed investment (infrastructure construction) cost is calculated using Coyne et Bellier's (2014) data. This cost is then annualised based on the interest rate and the depreciation rate. To this cost, one adds the annual variable cost (O&M, energy), reported in Coyne et Bellier's (2014), to obtain the gross annual cost. The net cost is obtained by subtracting the annual profit of the hydropower plant. The costs in Fig. 14.4 are obtained by dividing the annual costs of Fig. 14.3 by $1,150 \times 10^6$ m³/year which is the quan-

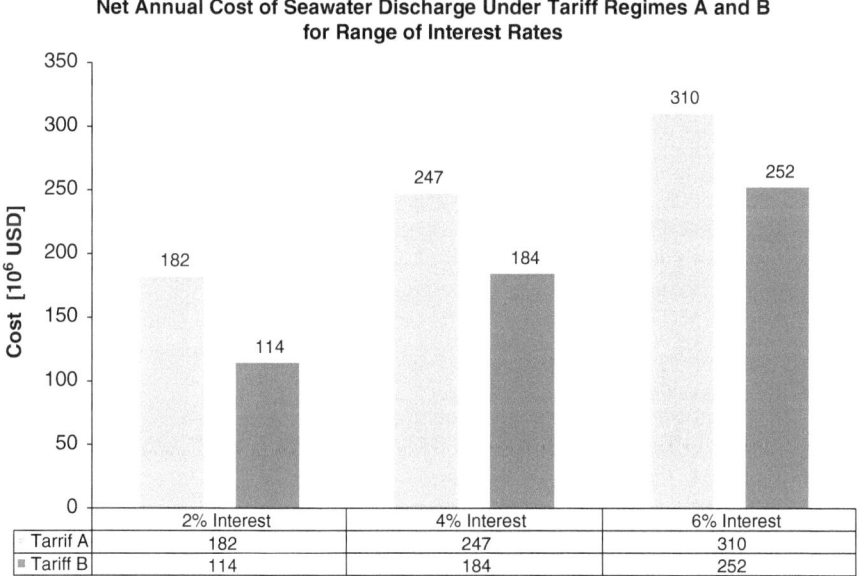

Fig. 14.3 Net annual cost (*million USD*) of seawater discharge in the Dead Sea of the Red Sea Pipeline Project under electricity tariff regimes A and B for a range of interest rates

---

[6] While the tunnel should be constructed in full, it is always possible to increase the flow of water gradually. If the added cost of the upfront investment in building the tunnel, compared to the cost associated with the gradual pipeline project, exceeds the saving of the energy cost required by the former, then the pipeline option is more cost effective. It turns out that the costs of the two alignments are similar, with a slight advantage to the tunnel option at low interest rates (capital cost) and to the pipelines option at higher interest rates (see details as well as a map of the area in Allan et al. (2014)).

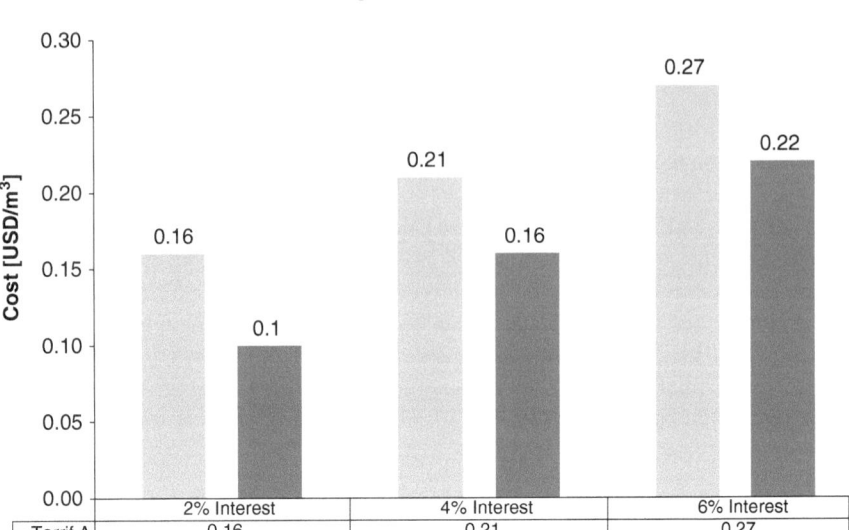

Fig. 14.4 Cost per cubic metre (*USD/m³*) of seawater–brine discharge in the Dead Sea of the Red Sea Pipeline Project under electricity tariff regimes A and B for a range of interest rates (the annual costs of Fig. 14.3 divided by the annual discharge of $1,150 \times 10^6$ m³/year)

tity of seawater–brine to be discharged in the Dead Sea after project completion. Expressing the cost in USD/m³ of discharged seawater will facilitate comparisons with other alternatives. Because the hydropower profits under electricity tariff regime B (hydroelectricity is sold to the Israeli grid) are higher than under electricity regime A (hydroelectricity is sold to the Jordanian grid), the net costs are lower under regime B than under regime A. At a 2 % capital cost (interest rate), for example, the annual cost of a Pipeline Project that discharges $1,150 \times 10^6$ m³/year of Red Sea water (or brine) in the Dead Sea is USD 182 million (USD 0.16/m³) or USD 114 million (USD 0.1/m³) under tariff regime A or B, respectively. All costs and (hydropower) profits are calculated using Coyne et Bellier's (2014) data (see details in Allan et al. 2014, Appendix 2).

For the Dead Sea reclamation to pass a cost-benefit test, the benefit generated by the project must exceed its cost. As common when measuring economic values of environmental amenities, such as in the present case, the benefit of reclaiming the Dead Sea is measured by the willingness to pay to stabilise (or restore) its water level. One study estimates this value between USD 73 million a year and USD 227 million a year (Becker and Katz 2009). Consider, for the sake of illustration, the midpoint of USD 150 million per year. Then, observing Fig. 14.3, the project passes the cost-benefit criterion at 2 % capital cost (interest rate) and electricity tariff regime B. As mentioned above, the study of Becker and Katz (2009) estimates the local population's willingness to pay. The unique historical, cultural and environmental characteristics of the Dead Sea imply that the benefit of its preservation extends beyond the local region and the overall willingness to pay to reclaim the Dead Sea is credibly wider ranging than the above.

The above discussion illuminates the crucial role of the discount rate: with an annual benefit of USD 150 million, the project passes the cost-benefit criterion at 2 % (and tariff regime B) but not at 4 % or above. This raises the question regarding the more appropriate discount rate to use. One approach is to use the discount rate used to evaluate public projects. This discount rate varies across countries as well as across types and durations of the public projects and ranges between 1 and 10 % (see Gollier 2013, p. 8–9, for an overview). A different approach, which accounts for the unique characteristic of the project under consideration, is to use an ecological discount rate, estimated by Gollier (2010) at about 1.5 %.

To sum up, preliminary estimates of the benefit to the local population associated with stabilising (or restoring) the Dead Sea range between USD 73 million a year to USD 227 million a year. Taking the midpoint of USD 150 million a year as a point estimate, a Red Sea–Dead Sea (phased) Pipeline Project passes the cost-benefit criterion (i.e. is justified on economic ground) at an interest rate (capital costs) of about 2 % and electricity tariffs regime B (where the electricity consumed is purchased from the Jordanian grid and the electricity generated is sold to the Israeli grid). The unique characteristics of the Dead Sea imply that the benefit of its reclamation extends beyond the local population and is therefore likely to be larger than Becker and Katz's (2009) estimates. Because the project's benefit extends into the distant future and is environmental (ecological) in nature, a low discount rate, around 1.5 %, is justified.

14.3.2 Mediterranean Sea–Dead Sea Water Conveyance

The main advantage of using the Mediterranean rather than the Red Sea as a source of water conveyance is the shorter distance (see Fig. 14.5). For this reason, most past proposals of water conveyance to the Dead Sea considered the Mediterranean as the preferred source (see Vardi 1990). Following the study of alternatives to the Red Sea–Dead Sea Project (Allan et al. 2014), we consider two Mediterranean Sea–Dead Sea routes: a southern route from Ashkelon to Qumran and a northern route from Atlit (south of Haifa) to Naharayim–Bakura (where the Yarmuk joins the Jordan River) and to the Dead Sea along the lower Jordan River route. We discuss each in turn.

Fig. 14.5 Conveyance routes from the Red Sea and the Mediterranean to the Dead Sea (Adopted from Wikipedia)

14.3.3   Southern Route (Ashkelon → Qumran)

As in the previous case, water can be conveyed via surface pipelines or a tunnel. The topography of the area implies that the former option requires lifting the water more than 800 m before letting it flow downward to the Dead Sea. The cost of pumping the water to such an altitude renders the pipeline option too expensive, and we concentrate only on the tunnel option.[7]

---

[7] As in the previous case, the disadvantage of the tunnel option is that it has to be constructed in full upfront and cannot be phased out over time. However, the shorter distance implies that the added

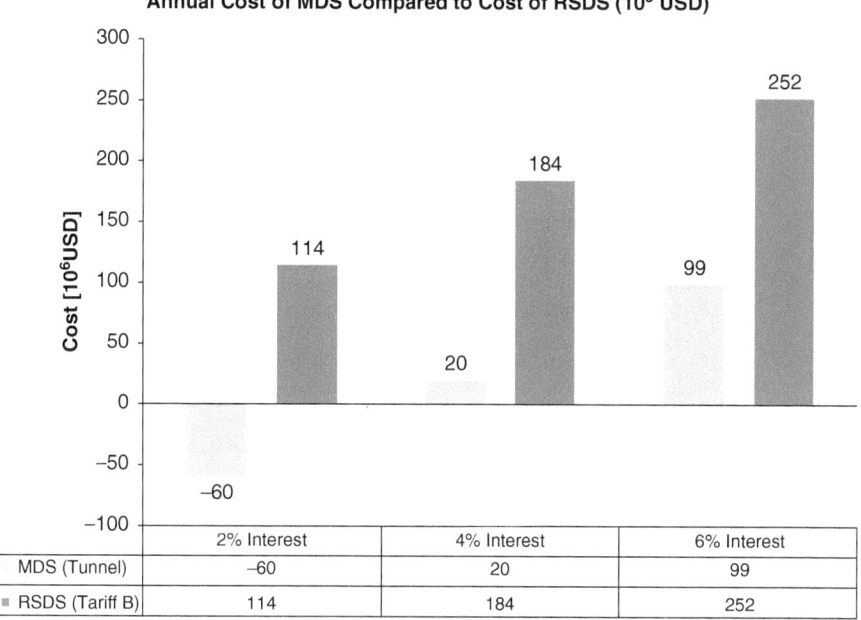

Fig. 14.6 Annual cost (*million USD*) of the Southern Mediterranean Sea–Dead Sea (MDS) Conveyance Project (from Ashkelon to Qumran) and the comparable costs of the Red Sea–Dead Sea (RSDS) Pipeline Project (tariff regime B)

Figure 14.6 presents the net annual cost (million USD) of the Ashkelon–Qumran (tunnel) Conveyance Project with a comparison to the comparable costs of the Red Sea–Dead Sea (Pipeline) Project under electricity tariff regime B (of Fig. 14.6). Figure 14.7 presents the same costs in USD/m^3 units (obtained by dividing the annual costs by 1,150 m^3/year – the annual quantity of brine discharge). The tables reveal the cost advantage of the Mediterranean over the Red Sea as a source of seawater conveyance. It also shows that conveying seawater from Ashkelon to the Dead Sea (around Qumran), using the elevation difference to generate hydroelectricity, is a profitable operation at 2 % interest rate.

To sum up, because of the shorter distance, the Ashkelon–Qumran Water Conveyance Project (via tunnel) is more cost effective than the Red Sea–Dead Sea Pipeline Project. At a 2 % interest rate, the Ashkelon–Qumran (tunnel) Project is profitable – the hydropower profits more than compensate for the tunnel costs (construction and operation).

---

cost associated with this disadvantage is much smaller than that of the Red Sea–Dead Sea Project. The course of a tunnel from Ashkelon to the northern Dead Sea intersects the mountain aquifer, and the exact route would need to be determined in order not to potentially harm this sensitive and important water source.

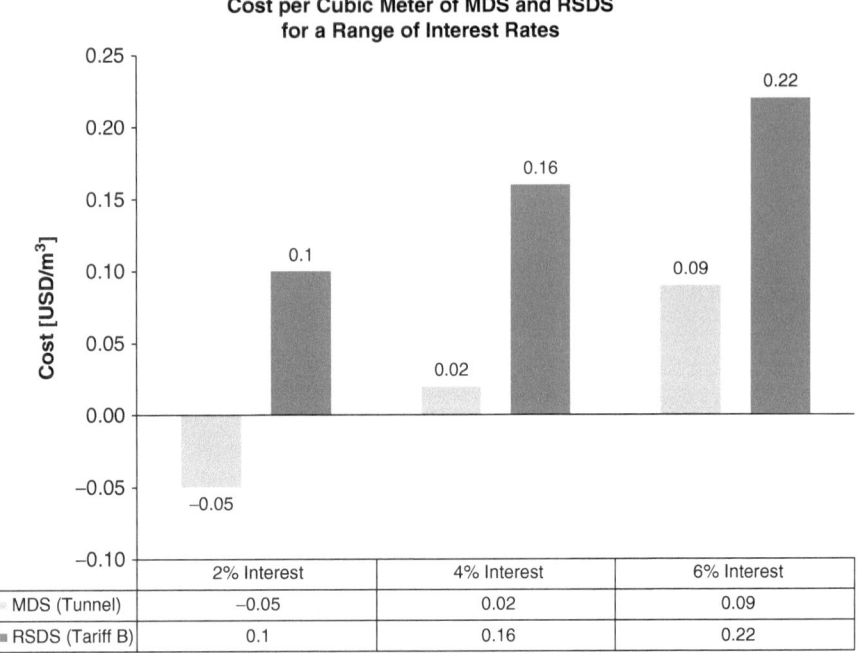

Fig. 14.7 The costs of Fig. 14.6 expressed in USD/m³ units (Obtained by dividing the annual costs of Fig. 14.6 by $1{,}150 \times 10^6$ m³/year – the annual quantity of brine discharge)

14.3.4   Northern Route (Atlit → Naharayim-Bakura → Dead Sea)

The distance from Atlit to Naharayim–Bakura is 65–70 km (Fig. 14.5). The possibility of conveying seawater along this route is ruled out because the course extends over fertile valleys and sensitive aquifers, and the risk for damage from leakage is high. Therefore, this alternative involves desalination along the Mediterranean coast near Atlit and conveyance of the desalinated water (via pipeline) to Naharayim–Bakura, where the Yarmuk flows into the Jordan River. The location of the eastern outlet (Naharayim–Bakura) opens up opportunities to combine this option with partial restoration of the lower Jordan River; from Naharayim–Bakura, the water flows along the lower Jordan River to the Dead Sea, performing a dual goal: partially restoring the lower Jordan River and stabilising (or halting the decline of) the Dead Sea level.

Allowing for a pumped storage reservoir near Kaukab-el-Houa (Belvoir), at 300 m elevation (more than 500 m above Naharayim–Bakura), and exploiting the Israeli peak-load electricity tariffs to pump water during low demand (nights) and generate hydropower during high demand, it was found that the profit of the hydropower plant is sufficient to cover the entire cost of conveying the desalinated water from Atlit to Naharayim–Bakura.[8] The total cost of the desalinated water in Naharayim–Bakura is therefore the desalination cost, which will soon approach USD 0.5/m³ (see Tsur 2014).

From Naharayim–Bakura, the water flows to the Dead Sea along the lower Jordan River route. The total cost of water discharge in the Dead Sea of the Atlit–Naharayim Project is therefore solely the cost of desalination of about USD 0.5/m³, which is higher than the comparable costs under the Red Sea–Dead Sea (pipeline) or Mediterranean–Dead Sea (tunnel) Projects (see Figs. 14.3 and 14.5). However, the water flow from Naharayim–Bakura to the Dead Sea contributes to the (partial) restoration of the lower Jordan River, which has an economic value of its own (see Gafny et al. 2010 for a description of the current situation of the lower Jordan River and restoration options). Estimates of regional (Israel, Jordan and the Palestinian Authority) willingness to pay (WTP) for partial restoration of the lower Jordan River were recently calculated by Becker et al. (2014). Considering different flows (220×10^6 m³/year and 400×10^6 m³/year) and two levels of water quality (high-quality natural and recycled), these authors estimate the WTP between USD 0.23/m³ and USD 0.87/m³, with the good quality (natural) water receiving the higher values (see Becker et al. 2014, Tables 2–3).[9] Based on these findings, it is not implausible that the benefit due to (partial) restoration of the lower Jordan River is high enough to cover most (if not all) of the cost of the Atilit–Naharayim–Bakura Project, leaving very little (if any) cost to the Dead Sea reclamation.

To sum up, high damage due to risk of leakage rules out the possibility of conveying seawater from the Mediterranean (near Atlit) to Naharayim–Bakura and requires desalination along the Mediterranean coast. The cost of desalination is approaching USD 0.5/m³. The cost of conveying the desalinated water from Atlit to Naharayim–Bakura can be fully covered by the profits of a pumped energy plant near Kaukab-el-Houa (Belvoir). From Naharayim–Bakura, the (desalinated) water flows to the Dead Sea along the lower Jordan River, contributing to the partial restoration of the lower Jordan River along the way. Preliminary estimates of the benefit (in terms of WTP of the local population) associated with partial restoration of the lower Jordan River range between USD 0.23/m³ and USD 0.87/m³, with the good quality water receiving the higher values. The cost left for the Dead Sea reclamation is the cost of desalination (USD 0.5/m³) minus the benefit of lower Jordan

---

[8] The cost of conveyance from Atilt to Naharayim–Bakura was calculated based on Coyne et Bellier's (2014) data with the appropriate modifications needed to fit to the different situation. These data, together with the peak-load pricing schedule of the Israeli grid, were used to calculate the profit of the pumped energy plant near Koukab-el-Houa (see Appendix of Allan et al 2014 for details).

[9] These estimates are obtained by dividing the total WTP corresponding to scenarios S3 and S4 by the restoration flow of 400×10^6 m³/year (see Becker et al 2014, Tables 2–3). It should be noted that these estimates are based on WTP of the local population and do not include WTP for Dead Sea reclamation (only lower Jordan River restoration was considered). Adding WTP of the international community and WTP for Dead Sea restoration (on which no data are yet available) will increase the WTP for allocating recycled water for environmental restoration of the lower Jordan River and the Dead Sea.

River restoration. Given estimates of the latter, the cost of Dead Sea reclamation associated with the northern Mediterranean route is negligible. However, performing desalination at a scale needed to stabilise the Dead Sea (above 700×10^6 m³/year) may not be feasible, implying that this project may not suffice on its own and should come in combination with other options.

14.4   Dead Sea Reclamation Based on Recycled Water

This alternative is based on the evolution of the following processes:

(i) The population of the region will continue to grow more or less along the projections of Fig. 14.2.
(ii) The supply of potable water (from natural sources, desalination or importation) will accordingly grow to provide at least the quantity deemed necessary for basic human needs – about 100 m³/year per person (Gleick 1996).
(iii) Environmental standards require appropriate treatment of domestic sewage, disregarding its outlet and whether or not it will be reused. Accordingly, each cubic metre (m³) allocated to households and industrial use will be collected, treated and become available for reuse (mainly in irrigation and environmental restoration). Under current recycling technology, each m³ allocated for domestic use provides between 0.6 m³ and 0.65 m³ of treated (recycled) water (Cohen et al. 2008).
(iv) Water will be priced according to its cost of supply (see Tsur 2009). Pricing water in this way will affect farmers' choices of crops (e.g. away from water-thirsty crops) and will induce them to switch irrigation water away from expensive natural (potable) sources into marginal (recycled, saline) water.

The population projections of *(i)* are reasonable because they are based on 60 years of actual data and account for the anticipated decline in the population growth rate due to economic development. Processes *(ii)*–*(iv)* have been progressively implemented in Israel: Firstly, five large-scale desalination plants, with accumulated capacity of 600×10^6 m³/year (more than 85 % of domestic water consumption in Israel), have been built in the last decade, and plans to increase desalination capacity in the future match the population growth projections (see Israel Water Authority 2012; Tsur 2014). Secondly, virtually all domestic water is collected, treated and made available for reuse in irrigation and environmental restoration (Israel Water Authority 2012). Thirdly, a series of water pricing reforms in the domestic and irrigation sectors have increased the efficiency of water use by reducing leakage, changing the mix of irrigated crops and inducing farmers to switch from natural (potable) water to recycled water (Tsur 2014; Kislev 2011). These measures have led, inter alia, to a 10 % reduction in domestic water consumption (about 100×10^6 m³/year – equivalent to a large-scale desalination plant) and induced farmers to switch natural water quotas (which become more expensive) for recycled water (see Tsur 2014). Israel currently produces more than 450×10^6 m³/year of recycled water and plans to increase this quantity to 930×10^6 m³/year by 2050 (Israel Water Authority 2012).

Jordan and the Palestinian Authority have already begun to apply similar measures and will (sooner or later) catch up with Israel's policy. The only exception is Jordan's ability to implement *(ii)* via desalination, as Jordan's main population centre (Amman) is some 1,000 m above sea level and 300 km away from its only sea access (the Gulf of Aqaba); thus, desalinating in Aqaba and conveying to Amman can be prohibitively expensive. However, there exist other, more economical ways to increase the supply of drinking water in Jordan in tandem with its population growth (see Tsur 2014).

Observing the population projections (Fig. 14.2), we see that by 2050 the region's population will exceed 30 million. Under Assumptions *(ii)–(iv)*, this population size will be capable of producing more than $2,000 \times 10^6$ m^3/year of recycled water. The bulk of the recycled water will be allocated for irrigation (as farmers switch from the more expensive natural water to the cheaper recycled water), but some of the recycled will be allocated for environmental restoration, including lower Jordan River restoration and Dead Sea reclamation. This requires conveyance of recycled water from the treatment plants to the upper end of the lower Jordan River (near Naharayim–Bakura) as well as compensating farmers for reallocating the water away from irrigation.

Mekonen (2013) calculated the cost of conveying recycled water from the Jerusalem–Ramallah area to Naharayim–Bakura, while using the elevation difference (of about 1000 m) to generate electricity. Mekonen (2013, Table 16) calculated the conveyance cost at USD 0.19/m^3 and the hydroelectricity profit at USD 0.12/m^3. The required compensation to farmers (under which farmers are indifferent between receiving the recycled water or the compensation) was estimated at USD 0.26/m^3. The net cost of using the recycled water for (partial) restoration of the lower Jordan River and the Dead Sea (conveyance minus hydroelectricity profit plus compensation to irrigators) is therefore USD 0.33/m^3. As was noted above, the associated benefit (based on WTP to restore the lower Jordan River) was estimated by Becker et al. (2014) between USD 0.23/m^3 and USD 0.87/m^3. The cost of using recycled water for lower Jordan River and Dead Sea restoration (USD 0.33/m^3 as estimated by Mekonen (2013) falls at the lower half of the benefit range. Based on these preliminary calculations, we conclude that in 3 to 4 decades, allocating 400×10^6 m^3/year of recycled water for partial restoration of the lower Jordan River and the Dead Sea is likely to pass a cost-benefit test.

Stopping the Dead Sea decline requires increasing its inflow by $700–800 \times 10^6$ m^3/year (TAHAL and GSI 2011), implying that an additional inflow of $300–400 \times 10^6$ m^3/year (in addition to the 400×10^6 m^3/year of recycled water) is needed. This additional inflow can come from a mini Red Sea–Dead Sea Project that will desalinate 300×10^6 m^3/year at Aqaba (100×10^6 m^3/year) and near the Dead Sea (200×10^6 m^3/year) and will convey the 367×10^6 m^3/year brine discharge (1 m^3 of seawater generates 0.45 m^3 of desalinated water and 0.55 m^3 of brine) to the Dead Sea.[10] The combination

---

[10]This mini Red Sea–Dead Sea Project was suggested by the study of alternatives team under CA1 (see Allan et al 2014) as a contribution to a comprehensive solution for Jordan's severe water

of the recycled water and the brine from the mini Red Sea–Dead Sea Project will be sufficient to stabilise the Dead Seat at its current level.

An advantage of this alternative is that it does not require investment in a large-scale conveyance project, such as those considered in the Red Sea–Dead Sea and Southern Mediterranean–Dead Sea Projects.[11] Moreover, the 367×10^6 m^3/year of brine discharge in the Dead Sea is below the 400×10^6 m^3/year flow considered safe (i.e. unlikely to give rise to gypsum crystallisation, stratification or algae bloom) by the Dead Sea study (TAHAL and GSI 2011). Disadvantages are the potential risks associated with letting 400×10^6 m^3/year of recycled water flow into the Dead Sea (e.g. effect on algae bloom), and these risks will need to be investigated and are likely to have ramifications regarding associated recycling technologies and costs.

To sum up, population growth will require increasing the supply of potable water to satisfy the needs of the growing population. As the natural water sources are already fully exploited, this will require better management of existing water sources (e.g. by pricing water to reflect true costs of supply) and increasing the supply of potable water from an alternative source such as desalination. Environmental standards require appropriate treatment of all sewage, implying, given the current recycling technology, that 60 to 65 % of the total domestic water consumption will be available for reuse, mostly in irrigation and environmental restoration. Given the population projections of Fig. 14.2, by 2050 the supply of recycled water will exceed $2,000 \times 10^6$ m^3/year. About 400×10^6 m^3/year of this supply can be allocated for the joint purpose of partial restoration of the lower Jordan River and the Dead Sea. The additional $300–400 \times 10^6$ m^3/year required to stabilise the Dead Sea at its current level may come from a mini Red Sea–Dead Sea Project that will desalinate 300×10^6 m^3/year at Aqaba (100×10^6 m^3/year) and near the Dead Sea (200×10^6 m^3/year) and will discharge 367×10^6 m^3/year of brine in the Dead Sea. The mini Red Sea–Dead Sea Project will serve the dual purpose of alleviating the shortage of potable water in the region (mainly in Jordan) and contributing to the stabilisation of the Dead Sea.

The cost of recycled water is estimated at USD 0.45/m^3 (the cost of conveyance to Naharayim–Bakura of USD 0.19/m^3 plus USD 0.26/m^3 compensation to irrigators). The benefit due to partial restoration of the lower Jordan River was estimated between USD 0.23/m^3 and USD 0.87/m^3, where restoration by recycled water is receiving the lower values. The net cost left for the Dead Sea reclamation is therefore below USD 0.22/m^3 (= USD 0.45/m^3 minus USD 0.23/m^3). The costs of the brine discharge from the mini Red Sea–Dead Sea Project are comparable to the costs of the full-scale project reported in Figs. 14.3 and 14.4.

---

scarcity problem (see discussion in Tsur (2014). The brine discharge in the Dead Sea is a by-product that contributes to stabilising the Dead Sea.

[11] The infrastructure investment required by the Red Sea–Dead Sea Project was estimated above USD 10 billion (Coyne et Bellier 2014).

14.5   How to Cover the Cost of Dead Sea Reclamation?

Using any of the alternatives discussed earlier to halt the decline of the Dead Sea level or to restore its historical state inflicts costs (except for the Southern Mediterranean Project under 2 % capital cost). This raises the questions stated in the title of this section. The problem arises because environmental amenities, such as the Dead Sea reclamation, have public good characteristics, and as such market mechanisms fail to allocate them properly. This feature often (though not always) implies that the value of an amenity cannot be inferred from market prices; hence, the need to use alternative approaches to obtain WTP measures, such as the contingent valuation method used by Becker et al. (2014), complicates the regulation needed to correct the failure. In the present context, the market failure arises because the benefits of reclaiming the Dead Sea (stabilising or restoring) stem from many reasons and affect different groups, some more directly (e.g. the hotels that will suffer less from deteriorating roads) and some indirectly (e.g. present and future pilgrims who aspire to see the Dead Sea ecosystem as it was during times of prophecy).

The issue of who should pay for environmental restoration (those who perpetrated the damage, or who stand to benefit from the restoration, or who suffer from the damage) and how to extract the correct sum from the different groups (polluters, beneficiaries, victims) is central to any environmental policy (see discussion in Goulder and Parry 2008, and the references therein). We propose here a simple mechanism to cover the costs of a Dead Sea reclamation project, based on a widely used environmental policy principle known as the polluter pays principle. The basic idea is to levy a surcharge on any cubic metres of water that would have reached the Dead Sea had it not been extracted or diverted upstream. This applies to diversions from the Jordan River (including Lake Tiberias), from the Yarmuk River (by Jordan) and from side wadies (tributaries) that flow into the Jordan River or directly into the Dead Sea (e.g. Zarqa, Mujib). It also applies to the 262×10^6 m³/year diversions (evaporation) of the Israeli and Jordanian potash industries (Zbranek 2013).

The exact surcharge rate will vary across the alternatives based on the cost of each alternative. We calculate the range of surcharge rates corresponding to the costs of the alternatives discussed above. As noted in the introduction, the total upstream diversion based on historical flows is about $1,500 \times 10^6$ m³/year (TAHAL and GSI 2011). Allowing for a decline in average precipitation due to climate change (Weinberger et al. 2012), we assume that total diversion today is about $1,300 \times 10^6$ m³/year, of which 460×10^6 m³/year is diverted from the Yarmuk River, mostly by Syria (which is excluded from the mechanism for a number of reasons, including the fact that it is not a Dead Sea riparian). The remaining diversions to be taxed are therefore about $1,100 \times 10^6$ m³/year, accounting for about 850×10^6 m³/year of upstream diversions plus the 262×10^6 m³/year consumed by the potash industries. This quantity is similar to the $1,150 \times 10^6$ m³/year to be discharged into the Dead Sea by the (full-scale) Red Sea and Southern Mediterranean Sea Projects. The surcharge per cubic metre needed to cover the costs of these projects is therefore equal to the costs per cubic metre presented in Fig. 14.7. Assuming a 2 % interest rate, which is justified by the low ecological discount rate to be used in environmental projects of this sort (see Gollier 2010), the required surcharge (which will raise the annual proceeds needed to cover the annual costs) is zero for the Southern Mediterranean Project and USD $0.1/m^3$ for the Red Sea (pipeline, tariff regime B) Project.

Regarding the Northern Mediterranean Project, the cost of desalination and conveyance to Naharayim–Bakura is about USD $0.5/m^3$ (recall that the conveyance cost is covered by the profit of the pumped energy operation). From there, the water flows to the Dead Sea, partially restoring the lower Jordan River along the way. The Dead Sea reclamation cost is therefore the USD $0.5/m^3$ minus the benefit associated with the lower Jordan River restoration. The latter benefit, as discussed above, lies in the upper half of the USD $0.23/m^3$–USD $0.87/m^3$ span estimated by Becker et al. (2014) (the higher values refer to restoration with good quality water and the lower values to restoration with recycled water). Assuming also that the flow of water under this alternative will be the flow needed to stabilise the Dead Sea at its current level (700–800×10^6 m^3/year), we conclude that the surcharge needed to cover the Dead Sea reclamation costs should range between zero and USD $0.1/m^3$.

Regarding the alternative of combining recycled water and a mini Red Sea–Dead Sea Project, it was found that the benefit associated with using the recycled water for the lower Jordan River restoration will outweigh the costs of the recycled water, leaving only the costs of the mini Red Sea–Dead Sea Project. The USD/m^3 cost of the mini project will be similar to that of its large (full)-scale counterpart, which was found to be about USD $0.1/m^3$ (see Fig. 14.4). However, the mini Red Sea–Dead Sea Project will discharge only 360×10^6 m^3/year of brine into the Dead Sea, which is less than a third of its large (full)-scale counterpart considered above. Accordingly, the annual costs will be about a third of the annual costs of the large (full)-scale project, and the surcharge will accordingly be about a third of that under the large-scale project, i.e. about USD $0.03/m^3$ (recall that the surcharge is levied on the same quantity of diverted water as under the large-scale project).

To sum up, following the logic of the polluter pays principle, we offer a mechanism to finance a Dead Sea reclamation project by levying a surcharge on all diversions that otherwise would have reached the Dead Sea. These diversions are estimate at about $1,100 \times 10^6$ m^3/year (850×10^6 m^3/year of upstream diversions plus 262×10^6 m^3/year consumed by the potash industries). The surcharge rate is calculated such that the annual proceeds cover the cost of the reclamation project. Because the costs vary across alternatives, so will the surcharge rates. Under a 2 % capital cost (consistent with ecological discount rate), the surcharge needed to cover the cost of the Red Sea and Southern Mediterranean Projects is USD $0.1/m^3$ and zero, respectively. The surcharge needed to cover the costs of the Northern Mediterranean Project is between zero (if the subtracted lower Jordan River restoration benefit is low) and USD $0.1/m^3$ (if the subtracted lower Jordan River restoration is high). The surcharge needed to cover the costs of the alternative combining recycled water and a mini Red Sea–Dead Sea Project is about USD $0.03/m^3$. The feasibility of imposing these surcharge rates determines the feasibility of reclaiming the Dead Sea.

14.6   Concluding Comments

The average supplies of natural water available on a sustainable fashion in the water basin feeding the Dead Sea (comprising Israel, Jordan and the Palestinian Authority) will soon drop below 100 m³/year per person – the quantity deemed necessary for basic human consumption. Upstream diversions have deprived the Dead Sea of more than 90 % of its historical inflow, leading to progressive decline of its water level, which currently exceeds one metre per year on average. Stabilising the Dead Sea at its current level requires increasing the inflow by 700 to 800×10^6 m³/year, while restoring historical levels requires above $1,100 \times 10^6$ m³/year. We addressed four alternatives to stabilise or restore the Dead Sea: a large-scale Red Sea–Dead Sea Project, examined by Coyne et Bellier's (2014) feasibility study; two Mediterranean Sea–Dead Sea Projects (examined in the study of alternatives to the Red Sea–Dead Sea project); and an alternative based on recycled water and a mini Red Sea–Dead Sea Project (also examined in the abovementioned study of alternatives). We evaluate the costs associated with each alternative and offered a mechanism to pay for their implementation, based on a surcharge levied on all upstream diversions (including water consumed by the potash industries).

The Southern Mediterranean Project was found to be the most economical, in that it is a profitable project (the hydroelectricity profits more than compensate for the infrastructure and operating costs), thus requires no surcharge on upstream diversions. The full-scale Red Sea–Dead Sea Project was found to be the most expensive one, and financing it would require a surcharge of about USD 0.1/m³ on all upstream diversions (including the water consumption by the potash industries). Both projects are capable of restoring the Dead Sea level to its historical state. However, they should be implemented gradually, and discharge flows above 400×10^6 m³/year are currently considered risky in terms of possible damages due to stratification, gypsum crystallisation or algae blooms (TAHAL and GSI 2011).

The Northern Mediterranean Project involves desalination at the coastline, conveyance to Naharayim–Bakura, while exploiting the elevation difference to generate hydroelectricity and letting the water flow to the Dead Sea along the lower Jordan River route. The profit from the pumped energy plant covers the conveyance cost from Atlit to Naharayim–Bakura. A by-product of this alternative is a partial restoration of the lower Jordan River, and the ensuing benefit is sufficient to cover all or most of the desalination cost. The costs of the stabilisation of the Dead Sea level are therefore negligible. However, desalinating $700–800 \times 10^6$ m³/year (the minimal flow needed to stabilise the Dead Sea at its current level) along the northern Mediterranean coast may not be feasible, implying that this alternative should be combined with other alternatives.

The fourth alternative considered was built on the evolution of the following three ongoing processes: population growth, increased supply of potable water by desalination and reuse of domestic water after appropriate treatment. Over time (3–4 decades), these processes will give rise to a regional supply of recycled water above $2,000 \times 10^6$ m³/year, which will be available for reuse in irrigation and environmental restoration. We predict that $300-400 \times 10^6$ m³/year of the recycled water could be allocated for the purpose of partially restoring the lower Jordan River and the Dead Sea. Estimates of the benefit associated with the partial restoration of the lower Jordan River suggest that the associated benefit is sufficient to cover the costs of the recycled water (conveyance cost plus compensation to irrigators). The residual costs of the recycled water to the Dead Sea reclamation are therefore negligible. Stabilising the Dead Sea, however, requires additional flow of 300×10^6 m³/year to 400×10^6 m³/year. This additional flow can come from a mini Red Sea–Dead Sea Project that will desalinate 300×10^6 m³/year at Aqaba (100×10^6 m³/year) and near the Dead Sea (200×10^6 m³/year), while conveying and discharging the brine (367×10^6 m³/year) into the Dead Sea (the purpose of the desalination is to alleviate Jordan's severe water shortage problem). Since the cost of the mini Red Sea–Dead Sea Project is about a third of its large (full)-scale counterpart, the surcharge on all diversions (which remain unchanged) needed to finance this project will be USD 0.03/m³ – about a third of the surcharge of the large-scale project.

References

Allan JA, Malkawi AIH, Tsur Y (2014) Red Sea – Dead Sea water conveyance study program. Study of alternatives. Final report. Executive summary and main report. World Bank Publications, Washington, DC

Becker N, Katz DL (2009) An economic assessment of Dead Sea preservation and restoration. In: Lipchin C, Sandler D, Cushman E (eds) The Jordan River and Dead Sea: cooperation and amid conflict. Springer, Dordrecht, pp 275–296

Becker N, Helgeson J, Katz DL (2014) Once there was a river: a benefit-cost analysis of rehabilitation of the Jordan River. Reg Environ Chang 14:1303–1314. doi:10.1007/s10113-013-0578-4

Beyth M (2006) Water crisis in Israel. In: Leybourne M, Gaynor A (eds) Water: histories, cultures, ecologies. University of Western Australia Publishing, Crawley, pp 171–181

Beyth M (2007) The Red Sea and the Mediterranean – Dead Sea canal project. Desalination 214:365–371. doi:10.1016/j.desal.2007.06.002

Cohen A, Feiman D, Ladal M (2008) Sewage collection and treatment for irrigation: national survey 2006/2007. http://www.water.gov.il/Hebrew/ProfessionalInfoAndData/Water-Quality/DocLib1/NationalSurvey2006-2007.pdf. Accessed 14 Jan 2015

Coyne et Bellier (2014) Red Sea – Dead Sea water conveyance study program, feasibility study. Final feasibility study report. http://siteresources.worldbank.org/EXTREDSEADEADSEA/Resources/5174616-1416839444345/RSDS-Summary_of_Final_FS_Report.pdf. Accessed 14 Jan 2015

Falkenmark M, Lundquist J, Widstrand C (1989) Macro-scale water scarcity requires micro-scale approaches: aspects of vulnerability in semi-arid development. Nat Res Forum 13:258–267. doi:10.1111/j.1477-8947.1989.tb00348.x

Gafny S, Talozi S, Al Sheikh B et al (2010) Towards a living Jordan River: an environmental flows report on the rehabilitation of the Lower Jordan River. https://www.globalnature.org/bausteine.net/file/showfile.aspx?downdaid=7273&domid=1011&fd=0. Accessed 14 Jan 2015

Gleick PH (1996) Basic water requirements for human activities: meeting basic needs. Water Int 21:83–92. doi:10.1080/02508069608686494

Gollier C (2010) Ecological discounting. J Econ Theory 145:812–829. doi:10.1016/j. jet.2009.10.001

Gollier C (2013) Pricing the planet's future: the economics of discounting in an uncertain world. Princeton University Press, Princeton

Goulder LH, Parry WH (2008) Instrument choice in environmental policy. Rev Environ Econ Policy 2:152–174. doi:10.1093/reep/ren005

Israel Water Authority (2012) Long-term master plan for the national water sector part A – policy document version 4. http://www.water.gov.il/Hebrew/Planning-and-Development/Planning/ MasterPlan/DocLib4/MasterPlan-en-v.4.pdf. Accessed 17 Feb 2015

Kislev Y (2011) The water economy of Israel. Taub Center for Social Policy Studies in Israel, Jerusalem

Malkawi AIH, Abu Arabi M, Khasawneh M (2010) The Red Sea – Dead Sea conveyance system: bridging the water shortage and prospects for desalination. Paper presented at the 1st international nuclear and renewable energy conference, Amman, 21–24 March 2010

Mekonen S (2013) Economic alternatives for rehabilitation of the Lower Jordan River. M.Sc. thesis (in Hebrew)

MWI (2009) Water for life: Jordan's water strategy 2008–2022. http://www.irinnews.org/pdf/jordan_national_water_strategy.pdf. Accessed 14 Jan 2015

Rawashdeh S, Ruzouq R, Al-Fugara A et al (2013) Monitoring of Dead Sea water surface variation using multi-temporal satellite data and GIS. Arab J Geosci 6:3241–3248. doi:10.1007/s12517-012-0630-6

Salameh E, El-Naser H (2000) Changes in the Dead Sea level and their impacts on the surrounding groundwater bodies. Acta Hydrochim Hydrobiol 28:24–33. doi:10.1002/(SICI)1521-401X(200001)28:1<24::AID-AHEH24>3.0.CO;2-6

TAHAL, GSI (2011) Dead Sea study (final report). http://siteresources.worldbank.org/INTREDSEADEADSEA/Resources/Dead_Sea_Study_Final_August_2011.pdf. Accessed 14 Jan 2015

Tsur Y (2009) On the economics of water allocation and pricing. Ann Rev Resour Econ 1:513–536. doi:10.1146/annurev.resource.050708.144256

Tsur Y (2014) Closing the (widening) gap between natural water resources and water needs in the Jordan River Basin: a long term perspective. Water Policy. doi:10.2166/wp.2014.129

United Nations (2011) World population prospects: the 2010 revision, volume 1: comprehensive tables (table A.17). http://www.un.org/en/development/desa/population/publications/pdf/trends/WPP2010/WPP2010_Volume-I_Comprehensive-Tables.pdf. Accessed 14 Jan 2015

Vardi J (1990) Mediterranean – Dead Sea Project – a historical review. Geol Surv Israel 90:31–50

Weinberger G, Livshitz Y, Givati A et al (2012) The natural water resources between the Mediterranean Sea and the Jordan River. Israel Hydrological Service, Jerusalem

Zbranek V (2013) Red Sea – Dead Sea water conveyance study program reports: chemical industry analysis study. http://siteresources.worldbank.org/INTREDSEADEADSEA/Resources/Chemical_Industry_Analysis_Study_Final_Report.pdf. Accessed 14 Jan 2015

Chapter 15
Jordan's Shadow State and Water Management: Prospects for Water Security Will Depend on Politics and Regional Cooperation

Valerie Yorke

Abstract Over two decades, many have regarded the idea of a Red Sea–Dead Sea (RSDS) Conveyance Project to save the Dead Sea as a golden opportunity for Jordan. It held promise of providing the Kingdom with desalinated water to meet its long-term needs. Now, with the project's potential abandonment, Jordan needs to consolidate and accelerate reforms to close its deficits and improve water sector sustainability until a long-term bulk solution is found. Addressing how these challenges might be dealt with, this chapter focuses on politics. Analysis of Jordan's water reforms over 20 years shows that the sector's limited ability to achieve its goals is rooted in a wider problem – the Kingdom's organisation of political power. The chapter explores how an evolving "political compact" between Throne and people, underpinned by patronage, permitted an increasingly powerful neo-patrimonial, anti-reformist elite – a resilient "shadow state" – to influence policies and control the economy including in due course water resources. The water problem cannot therefore be remedied only through improved water management. Taking account of the link between political dynamics and governance, the chapter sets criteria for implementing reforms that, if met, would gradually free policymaking and institutions from shadow-state influence, providing context for effective water solutions. A final section, addressing the need for a nationwide coordinated approach to comprehensive water reforms that could provide a path to water security, discusses how Jordan might accelerate policies underway, implement deeper reforms and pursue further options – locally and regionally – to address medium- and long-term challenges.

This paper draws on research for, and the analysis and data in the author's study, *Politics matter: Jordan's path to water security lies through political reforms and regional cooperation*, April 2013, commissioned by NCCR-Trade Regulation, World Trade Institute, Bern.

V. Yorke (✉)
NCCR Trade Regulation/World Trade Institute, University of Bern,
Hallerstr. 6, 3012 Bern, Switzerland
e-mail: v.yorke@btinternet.com

© The Author(s) 2016
R.F. Hüttl et al. (eds.), *Society - Water - Technology*, Water Resources
Development and Management, DOI 10.1007/978-3-319-18971-0_15

Keywords Jordan water scarcity • Water governance shortcomings • Politics of water • Shadow state networks • National and regional solutions • Cooperation • Water security • Water demand • Water supply • RSDS Conveyance Project

15.1 Defining the Problem

Jordan faces a deepening water crisis, aggravated over decades by climate-change impacts, regional conflict, inflows of migrants and poor governance. Its people are among the most water-deprived worldwide, with 145 m³/year per person (2008) (MWI 2009, p. 3–1), a level projected to fall as the gap between demand for water and renewable and financed non-renewable supplies widens. Few in-country options remain to develop new water. Meanwhile, with more than 80 % of available annual supply of up to 900×10^6 m³ depending on unsustainable abstraction from groundwater aquifers and on diminishing transboundary surface flows, Jordanians could face absolute water poverty by 2025, with only 90 m³ per person (USAID 2012, p. vii). MWI statistics for 1994–2010 (Fig. 15.1) also reveal continuing structural distortions in water use: agriculture remained the dominant consumer (66 %), whilst contributing only 3.5 % of GDP (MWI and WRG 2011, p. 23); the municipal sector modestly increased its share (30 %) and industry/tourism use remained low (4.5–7 %). Behind these data lie dangerous trends, namely continuing over-pumping of groundwater aquifers; continuing inefficient water use in terms of productivity and diminishing scope to cope with climate-change impacts.

15.1.1 Demand Exceeds Supply

In terms of numbers, the scale of the challenge facing Jordan to 2025 and beyond of securing water to provide for its expanding population and meet its economic growth aspirations is crystal clear. Recent demand and supply projections in studies using different methodologies all show substantial deficits (MWI and WRG 2011; Coyne et Bellier 2012; USAID 2012). These differ in size, however. In order, therefore, to reflect the extent of the problem as realistically as possible, alternative projections – based on the MWI's annual water budget of 2012, but stripped of outdated or uncertain assumptions (Yorke 2013, pp. 25–27) – are provided at Fig. 15.2.

They show widening deficits, reaching 630×10^6 m³/year by 2025, the implications of which are well recognised: Jordan needs a costly new bulk supply source for the long term and meanwhile demand will likely continue to be met by over-abstracting valuable low-cost renewable supplies.

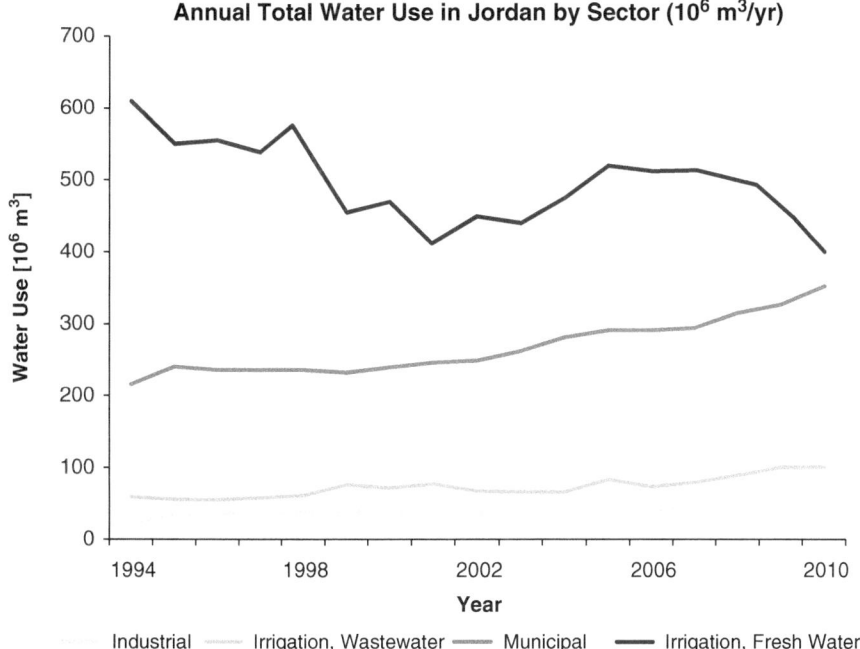

Fig. 15.1 Total water use in Jordan by sector since 1994 (10^6 m³) (Source: MWI Water Budget 2009/2010, MWI Water Resources Directorate open files, 2011. Cited in USAID (2012, p. viii))

15.1.2 Unpredictable Transboundary Flows

Seen through the lens of shared resources, the problem becomes more serious. Arid Jordan is already vulnerable to multi-year drought (JME 2009, p. 31). Its geo-strategic location and the legacy of regional conflict combined with heightened regional instability add to uncertainties. This is because a significant proportion of the surface and renewable groundwater providing most of Jordan's water derive from transboundary flows. Jordan is located *downstream* from Syria and Israel, with whom waters of the River Yarmouk and River Jordan respectively are shared (Yorke 2013, pp. 120–123). Aggravating matters, bilateral agreement on *joint management* of cross-border surface supplies[1] is either non-existent (1987 Jordan-Syria accord) or inadequate (Jordan-Israel Peace 1994, Annexes II, IV) and Jordan receives less than its rightful shares (SFG 2011, p. 26; MWI 2012). Meanwhile, fossil water

---

[1] The problem also surrounds groundwater. Of eleven renewable groundwater reservoir basins, four are shared with Syria (Yorke 2013, pp. 16–17).

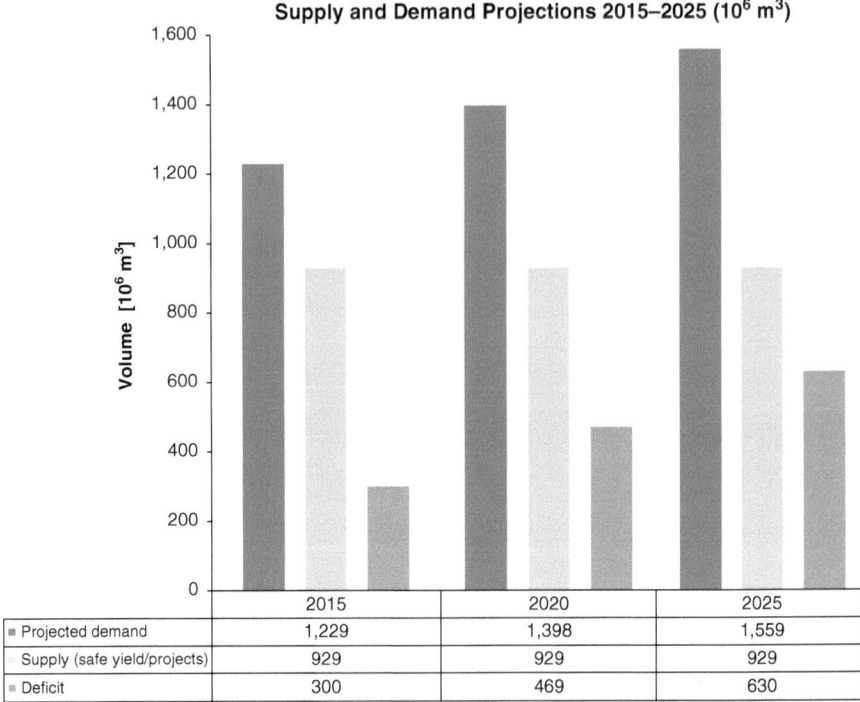

Fig. 15.2 Author's summary of supply (safe yield/projects underway) and demand projections, 2015–2025 (10^6 m³) (Source: Cited in Yorke (2013, pp. 28–29). Based on MWI (Jordan Ministry of Water and Irrigation) (2012)). Notes: Overall supply of 929×10^6 m³ to 2025 comprises (1) sustainable supply calculated at 776×10^6 m³ (total supplies of 892×10^6 m³ in 2010 minus over abstraction of 76×10^6 m³ minus 40×10^6 m³ irrigation use of Disi resources); (2) new water supply: Disi-Amman conveyor (100×10^6 m³/year), As Samra Wastewater Plant Extension (45×10^6 m³/year), Kufranjah dam (5×10^6 m³/year), small wastewater plants (3×10^6 m³/year). Projections exclude a possible 100×10^6 m³ desalination plant near Aqaba and supply side efficiency projects. Demand projections are based on MWI statistics

beneath Jordan's border with Saudi Arabia accounts for 8 % of overall supply,[2] but the two states have no accord. Inevitably, absent treaty-based or more formal agreement, inadequate water governance fuels bilateral suspicions.

The question arises whether this *modus vivendi* on water use is sustainable. According to the World Bank, the Middle East is especially vulnerable to climate change (World Bank 2007). With average temperatures expected to rise and precipitation levels to decline, Jordan's freshwater availability is projected to fall by 15 % by 2020 (SFG 2011, p. 92). Climate change-induced water stress, impacting agriculture, economic growth, health and food security, could potentially excite unrest and fuel tensions over water with neighbours.

---

[2] Disi resources have been used since the early 1980s for municipal and industry purposes in Aqaba and later for agriculture; since 2013 100×10^6 m³/year have been pumped to supply Amman.

15.1.3   Governance Shortcomings

Jordan's water governance record provides a third perplexing perspective. With vital national and human security interests at stake, Jordan has published strategies and introduced donor-supported reforms, targeting demand management, over more than two decades. But, these efforts have been insufficient to narrow worsening water deficits, protect groundwater and pre-empt the disaster to which the country is heading. Consider the balance-sheet during the two decades to 2010.

15.1.3.1   Reducing Unsustainable Highlands Groundwater Abstraction

After 1997 demand management focused on reducing groundwater use for irrigated agriculture especially in the Highlands, where it consumes up to half of abstracted water. The landmark Groundwater By-Law No 85 (2002) sought to control abstraction through block tariffs, penalising the abuse of licensed withdrawal levels and closing illegal wells (FOEME 2010).[3] These moves proved insufficient – in 2010, withdrawals from renewable aquifers still exceeded safe yield by an average 55 % (and by 176 and 215 % in the Amman-Zarqa and Azraq basins, according to (GIZ 2011; Subah and Habjoka 2011; MWI 2012). Water institutions acting alone lacked the capacity – both authority and back-up from the ministries of interior and agriculture – to enforce regulations on drilling and private sales and to effectively monitor and close wells. Compounding matters and in the knowledge of donors, two major donor-funded initiatives, the Water Demand Management Unit and the USAID's Instituting Water Demand Management failed to target agriculture (Yorke 2013, p. 43).[4]

15.1.3.2   Encouraging Efficient Water Use

From 1997, the importance of pricing and incentives to encourage more efficient water use and to raise revenue was also recognised. Reflecting Jordanian-donor collaboration, Jordan's Water Strategy (JWS) of 2009 called for cost-reflective tariffs for the long term, and the USAID sought to raise awareness on the need to change the way water is valued and allocated.

However, political, historical and social factors limited the sector's capacity to implement its plans, according to officials and water specialists interviewed by the author (Yorke 2013, pp. 43–44). Authority to alter water tariffs or remove tariffs on

---

[3] The By-Law created a 150,000 m³/year quota of free abstraction for each well, with block tariffs for further amounts abstracted. A 2004 amendment, providing for changes in these tariffs in 2008, was not implemented.

[4] According to the USAID Office of Inspector General's 2011 audit, USAID/Jordan had no project focusing on agriculture's water use and *conditions precedent* no longer required the government to address water use in agriculture. Jordan officials had also requested the mission to drop the condition on closing 50 illegal wells.

inter alia agricultural imports lies with the cabinet – not with the sector; and historically, low tariffs have helped to protect the poor and farmers working at the margins. Meanwhile, government focus on *new* supplies to deliver water security silenced debate on the need for efficient and fairer allocation; while access to "virtual water"[5] helped to disguise the scale of the crisis and the need for increased demand management (Allan 2003).

With these constraints, the 1997 tariff system, with its 2002 By-Law revisions, remained unchanged, despite USAID making cash transfers for GOJ projects conditional on agreed steps towards full cost recovery, including raising tariffs (USAID 2011, p. 8). Agriculture continued to pay less for water than the municipal and industry sectors – and below the cost of delivery. Low prices and import tariffs worked against commitments to improve end-use efficiency reduce waste and increase conservation. They distorted allocations so farmers lacked incentive to switch to higher-value water-efficient crops. Agriculture's share of renewable water declined, but not significantly; and overall, the Highlands area given to irrigation increased (Hagan 2008, p. 32).

15.1.3.3   Reducing Non-Revenue Water (NRW)

The high level of NRW – water supplied that raises no revenue – represents huge waste of low-cost water supply (Salameh 2007; MWI 2009). With an average 45–47 % of water pumped through supply works lost to NRW, an estimated 137×10^6 m^3 of the total 320×10^6 m^3 municipal allocation would have been lost in 2009 (Denny et al. 2008, p. 10; MWI 2009, pp. 3–2). The government has reiterated the need to reduce NRW – with the JWS aiming to reduce the level to 25 % – through addressing leakages, illegal pumping and meter deficiencies. Despite huge expenditure on infrastructure, corporatisation of utilities and donor projects to improve efficiency, reductions have not been significant.

One reason has been the slow pace of privatisation owing to political opposition. The government favours private-sector involvement in water and wastewater services in order to attract investment and implement reforms to improve financial sustainability. But by 2011, only three utility companies, owned by the Water Authority of Jordan (WAJ), had been established and their operations corporatised.[6] Moreover, results were mixed. The Aqaba Water Company (AWC) reduced NRW levels significantly, but Miyahuna (Amman) and the Yarmouk Water Company (YWC) recorded continuing high losses – 35.3 % and 40.8 %, together accounting for an estimated average annual loss to NRW of 77×10^6 m^3 of a total 80×10^6 m^3 for the three companies (USAID 2012). Poor performance was reportedly mainly due to distribution practice of undersupply and rationing, since intermittent flows

---

[5] The term coined by Professor Tony Allan refers to water used to produce grain and food imports.

[6] In 2007, the original Amman private operator was substituted by Miyahuna, a company owned by WAJ with private involvement. The AWC and YWC were similar companies established in 2004 and 2011 respectively.

damage pipes. The second reason cited to the author was the loss of resources from the grid and illegal wells, through *theft* – resulting from lax enforcement and lenient penalties.

15.1.3.4   Reforming Water Institutions

Decision-makers have long recognised that developing sector institutional capacity to manage and deliver efficient operations would be a pre-requisite for coping with the water crisis. Progress was blocked, however, due to the sector's inability to sort the overlapping responsibilities of the MWI and the government-owned WAJ and Jordan Valley Authority (JVA) – all three with ties to traditional constituencies, and the conflicts of interest these generated: MWI planning and projects were hampered by a lack of overall responsibility for supplies; the WAJ's dual supply and retail roles complicated water management but their separation remained politically constrained; lack of agreement over use of multiple data sources hindered sector planning; and over-staffed institutions suffered from low morale and a "brain drain".

The above record demonstrates that despite some achievements, efforts were *ad hoc* and *partial* and failed to set in train transformative reforms that could significantly change water outcomes. The sector lacked capacity *on its own* to fulfil its goals in key areas, was inhibited by factors *outside* the sector and disadvantaged by weak organisation. Overall, this suggests Jordan's water crisis is not only one of supply and demand, but also one of national governance. Understanding why this is so requires consideration of a wider structural problem.

15.2   The Shadow State

15.2.1   The Politics of Co-option

The longevity of Hashemite rule and Jordan's present governance problems are linked. Both are rooted in the distinctive organisation of power to which state formation and Hashemite rule gave rise, and in its restructuring over time (Yorke 1988, 1990). In Jordan, as in other Middle East states formed on the back of colonial legacy and weak institutions, personalised leadership based on patron-client relations, underpinned by patronage and lubricated by outside rents, remains key.[7] Although the Kingdom has a constitution and formal executive, legislative and judicial structures over which the king enjoys broad powers, Hashemite rule has been underpinned by parallel informal neo-patrimonial structures, which bind military and civilian elites into a web of support, and by a costly "social contract" between Throne and people. Patronage underpins the system: privileges, benefits and cheap

---

[7] A. Peters and P.W. Moore provide a valuable account of rentierism (Peters and Moore 2009, pp. 256–285).

services – today, including access to water – are exchanged for allegiance. Over time, however, the political influence and sense of entitlement accompanying privileges and benefits transferred has eroded the monarchy's autonomy.

In the face of changing circumstances – the challenges of state formation, pan-Arabism, nationalism, war, migration and development, and the economy's downturn in the 1980s – the Hashemites pursued imaginative co-option policies to consolidate power and maintain stability. In the process, the organisation of power and its location would alter, with consequences for decision-making, and the strength of the king's role in it. What amounted to an evolving political bargain between Throne and people permitted influential traditional elites and businessmen, enjoying a variety of privileges, to become embedded in a web of power – or *shadow state* (Tripp 2007, pp. 4–5), whose influence today penetrates the core elite, military and intelligence, bureaucracy and parliament. Understanding the power dynamics of these "shadow" networks whose growing influence afforded special interest groups opportunity to resist change and influence policies to protect their interests at the expense of those of the state (Muasher 2011), sheds light on the water governance deficits of the last two decades.

Transjordan's creation in March 1921 under Amir Abdullah was the result of an historic opportunity seized by the British to contain French expansion in the region. Transjordan was not a separate political entity and its people's heterogeneity – Transjordanians of East Bank origin but also from Arabia and from Palestine before 1948 – complicated the forging of a national identity. Palestinians would come to comprise more than half the population. The presence of non-Arab Circassians, Arab Christians and substantial Iraqi, and today Syrian, communities, as well as rivalries between tribes complicate the picture. Faced with inevitable disintegrative pulls, King Abdullah I and his successor, King Hussein, sought to build a common identity of interest between the Hashemites and their subjects.

The pursuit of a complex process of co-option to keep in with supporters and win the acquiescence of potential opposition would be key to consolidating Hashemite rule, which became embedded in patron-client relations, laying the basis of the "shadow state". Lacking central power, Abdullah courted the tribes with arms, funds and employment in the military and extended patronage to merchants in order to extend state authority – policies facilitated by British subsidies. In the turbulent 1950s and 1960s, the political bargain took on new forms: with flows of unconditional US financial assistance replacing British funds (Peters and Moore 2009),[8] Hussein responded to the growing demands of East Bank tribes by expanding the army and distributing members of tribes and minorities, his key supporters, through the administration and political structures, while drawing others into state. He rewarded merchants with public-sector jobs and contracts, protected the interests of and provided jobs for Transjordanian and Palestinian critics, and met the needs of Palestinian refugees through US-funded building of infrastructure. Arab aid flows during the 1970s' oil boom contributed to funding growing state largesse. Thus the king was able to subordinate Transjordanian traditional elites – and potential challengers amongst them – to the influence of the state and give the Palestinians a stake in the

---

[8] Military spending increased by 74 % in 1955–1960 and tripled in 1961–1975.

country. Stability of Hashemite rule was enhanced through crafting a social balance where social groups offset each other's influence. Cohesion appeared to prevail.

15.2.2   Politics of Water

It is in the context of these Hashemite regime maintenance policies and evolving organisation of power to which they led that the role of water provision today – for agricultural and domestic purposes – is best understood. With growing water scarcity, water has become a *political* matter – serving as a political instrument to maintain support, a source of personal wealth and influence for elites to be won and protected, and a cheap service as part of the "social contract" with the people.

The critical importance to King Abdullah I and his successors of co-opting powerful traditional and merchant landowners would from the start put agriculture at the centre of patronage politics, with forms of exchange around water in later years becoming an integral part and benefiting a widening constituency. An early indication came with British-designed land reforms that benefited Abdullah's landowning supporters – tribal leaders, peasants, and merchants (Fischbach 2000, pp. 209–212). In the process, they assisted many sheikhs to increase their holdings and influence (Alon 2009, pp. 127–128). By the turbulent 1950s, King Hussein was able to continue to fund privileges for tribal farmers whilst simultaneously financing development, urbanisation and services for the population – hugely expanded as a result of the 1948 war with Israel: US-funded water projects were key in settling Palestinian refugees in the Jordan Valley, but land irrigation projects also benefited tribal favourites and their followers (Peters and Moore 2009, p. 271). Landowners benefited from *political* rewards, too, as Hussein sought to consolidate loyalties in the wake of pan-Arabist pressures in the 1950s and the Palestinian rebellion in 1970/1971. Members of landowning families were appointed to key army and government posts – those of prime minister and defence minister, as well as in the agriculture, water and planning ministries (Yorke 2013, p. 67).[9] With this privileged access, traditional elite groups became entrenched. Embedded in politics and the administration, they were well-placed to advance their interests and preferred water policies in the Jordan Valley and Highlands.

Later in the 1980s, when the Hashemites sought to compensate key constituencies for economic setbacks as a result of the faltering economic boom and the subsequent IMF-imposed structural adjustment program, political liberalisation brought new kinds of pay-offs providing opportunity for these elites to consolidate their stakes in farming and in access to renewable and fossil groundwater. Henceforth influential landowners and farmers in the Northern Highlands and in Disi in the south, for example, were able to lobby inter alia for access to water for their farms,

---

[9] Hazza Al-Majali was twice prime minister (1955, 1959) and minister of agriculture (1950–1951) and Field Marshall Habis Al-Majali, his cousin, was chief of staff (1958–1975), defence minister (1967–1968), member of the Upper House 1967, 1984, 1989, 1993, and 1997. Adnan Badran was prime minister in 1989.

using their public positions in parliament and politics and links to the centre for private gain – all with grave implications for the water sector.

15.2.2.1 Northern Highlands

In the Northern Highlands where Jordan's population has historically been concentrated, farming pre-dates the state's formation. Thus, from the 1920s, the political exchanges around land were key to binding Highland farmers to Amir Abdullah and to underpinning state consolidation. They laid the foundation for today's patronage state which would take on its own logic.

As noted earlier, Transjordanian landowners would increase their political influence and come to dominate land ownership and agricultural markets. Early on, larger Highland landowners had benefited from British land reforms, and many still today maintain the legal rights acquired then (Alterman and Dziuban 2010). With land their main source of wealth, they cooperated with merchants in parliament to influence government resource policy. Meanwhile, the Hashemites – in confronting pan-Arabist and Palestinian nationalist challenges – extended generous privileges to these key Highland constituents. They benefited from light regulation of their farming affairs, from government investment in agriculture to provide jobs and food for the expanding population and from the establishment of well fields to exploit groundwater aquifers (Hagan 2008, p. 16). Agricultural expansion in the 1960s and 1970s paved the way for a huge increase in irrigated agriculture – in later decades using groundwater pumped from depleting aquifers (Hagan 2008, p. 32). As a result, in 1997, when the government launched its groundwater strategy to protect the sustainability of vital renewable resources, powerful landowners resisted. By-Law No 85 was eventually passed in 2002, but its provision to close illegal wells was not implemented – under a 2004 amendment, illegal wells could register for inclusion in the MWI monitoring program. Quotas and bulk tariff rates to encourage efficiency were introduced, but these remained virtually unchanged thereafter, and, until recently, fees, even for legal wells, were often not collected. Astonishingly, farmers reportedly used groundwater for new plantings of low value olive trees (per m^3 of irrigation water used) in order to enhance land values (USAID 2012, p. 34).

15.2.2.2 Southern Highlands

In the 1980s, the government started large-scale farming around the Disi aquifer in the south in an effort to make Jordan self-sustainable in wheat and barley. Private landowners and commercial elites would benefit: the government soon sold its farm to the Al Masri family and offered cheap land near the aquifer to private farm companies, whose shareholders included former government officials (Yorke 2013, pp. 69–70).[10]

---

[10] 25-year contracts provided cheap land and allowance of 1,000 m^3/dunum/year to the companies, which in return would use 50 % of land to grow wheat and barley for government purchase (1 dunum = 1,000 m^2).

Twenty years later public controversy surrounded Disi farming over its inefficient use of up to 65×10^6 m^3/year (equivalent to almost a third of annual municipal supplies in 2007) of precious non-renewable drinking water.[11] But *private* criticism, relayed by interviewees to the author in 2010, focused on the "shadow state". Disi farming was viewed as symptomatic of the political power wielded by vested interests and the government's lack of political will to act in the state interest to curb corrupt practice. Of special concern was rent-seeking by political and business elites through securing policies to promote their farming interests and to provide immunity from those that might undermine them.[12] Critics argued that government failure to allow the companies' licenses to expire in 2002 and to enforce its decision *not* to renew the companies' contracts in 2012 reflected the influence of powerful groups with ties to ruling circles and the ministries (Namrouqa 2012).

15.2.2.3   Resilience of the Shadow State

This analysis indicates that the water governance problem is part of a wider problem – the organisation of political power which the regime has shaped and altered over time to maintain security, which has politicised water, today controls resource allocation and inhibits regional cooperation, and in which anti-reformists are entrenched. The water problem cannot therefore be addressed through technical solutions alone. A pre-requisite will be political reforms that reduce the influence of "shadow state" networks over institutions, the economy and policies.

The task will not be easy. The web of power emerging from carefully choreographed co-option to protect Hashemite longevity has proven remarkably resilient (Yorke 2013, pp. 71–75). Despite his significant executive powers, King Abdullah II's reform initiatives after 1999 met resistance from conservative elites (Muasher 2011, pp. 7–23). In the event, he chose to back down rather than risk forfeiting the loyalty of a key constituency. Royal room for manoeuvre was constrained by an absence of sufficient "bottom-up" pressure for reform.[13] Post-1989 the Throne's new partnership with the people based on the National Charter permitted political parties but constrained political activity,[14] whilst electoral gerrymandering and the

---

[11] Water experts, Royal Water Committee members and donors shared criticisms: grains that the farms were contracted to produce could be grown more efficiently in the north using recycled water; farms were violating contracts, employing expatriate rather than local workers, and producing water-intensive produce for export rather than cereals for home consumption.

[12] Criticism focused on former government officials benefiting from the companies, on one serving prime minister using his public office to freeze a government tax claim on a company of which he was shareholder, and on the government for ignoring contract violations re growing water-thirsty crops.

[13] Since 1989 unrest has been frequent as a result of neo-liberal policies eating into the patronage state, with quasi-privatisation benefiting those linked to the palace but reducing state sector jobs particularly unpopular.

[14] The National Charter led to the legalisation of parties, but did so in return for acceptance of the constitution and the powers of the monarchy, thus placing boundaries on political activity and marginalising opposition.

"one-man one-vote" system ensured over-representation of traditional elements at the opposition's expense. Moreover, reformists themselves representing the plurality of a fragmented society – sharing goals perhaps, but with distinct orientations – have proven no match for anti-reformists of a centralised "shadow state".

Meanwhile, U.S. and Arab Gulf allies have played a key role in shaping this reality through funding the security and patronage networks underpinning Hashemite rule. In pursuit of their strategic goals, they provided flows of financial and technical assistance to underpin state largesse – permitting Jordan to develop without difficult reforms and cushioning it from the adverse socio-economic impacts of overspending. Donors thus bear some responsibility for the adverse consequences, felt inter alia in the water sector to which much aid has been directed but whose slow pace of reform reflects the workings of the patronage state.

However, the combined effects of the Arab Spring and global financial crisis provided a wake-up call – exposing weaknesses in the monarchy's compact with the people and the contradiction between centralised power and popular demands for democratic reforms.

The Hashemites have historically taken dramatic steps to cope with security threats to the monarchy and state: In 1988, King Hussein severed ties with the West Bank and the next year in response to economic austerity reinstated parliamentary life, both steps contributing to a restructuring of relations with his people. King Abdullah has shown that he, too, is aware that politics cannot stand still if ongoing socio-economic crises are to be dealt with, and that the current patronage state is unsustainable.

15.3 Proposed Actions

15.3.1 Building Water Security Through Political Reforms

Jordan's government, donors and water specialists publicly recognise that developing water security will be a pre-condition for generating the needed levels of economic growth to meet future energy, food and financial challenges and that failure risks domestic instability, a message strongly delivered in the government's 2005 National Agenda (JMGP 2005). Commitment to water reforms is strong. Yet, despite the fast-deteriorating water situation, Jordan's leaders and their counterparts abroad avoid *publicly* recognising "the elephant in the room", namely that Jordan's water strategy is *politically* challenged:

> What is missing is frank talk about how the Kingdom's distinctive power dynamics (i) contribute to distorted allocations and unsustainable use; (ii) foster a culture of competition for, rather than conservation of, scarce water resources; and (iii) inhibit efficient and democratic water governance necessary for economic growth, for putting in place adaptive capacity to deal with climate change and for laying the foundation for regional cooperation over shared water resources and new supplies (Yorke 2013, p. 77).

It follows that the path to sustainable water management will involve both addressing political factors adversely affecting water governance and exploiting opportunities to build well-functioning political and economic systems to improve the context for implementing effective water solutions. Meeting the following criteria will likely determine success:

Recognising Poor Water Governance Reflects a Wider Political Problem
While "shadow state" elites dominate centralised unaccountable structures, reforms will likely continue to show imbalances with the privileged benefiting disproportionately. More representative parliaments would give Jordanians greater political say in, and parliamentary scrutiny over, policies and thus render more likely their acceptance of difficult reforms.

Recognising How Non-water Policies Impact on Water Outcomes and How Poor Water Governance Affects Other Sectors This will involve national leaders making consideration of how to use scarce water a priority in national planning and investment decisions and aligning its governance with cross-sector policies chosen efficiently and effectively to promote economic growth. Coherent decisions require policymakers to understand how water management affects their sectoral responsibilities and to tap into water-sector expertise, calling for a central single water data set.

Improving Accountability and Transparency, and Curbing Corruption Good governance across the board will be a pre-condition for improving water management at home, and both will be necessary for building the level of regional cooperation required to manage and protect shared water resources and for attracting foreign investment for water and other projects. Governments that are accountable are incentivised to implement the reforms needed to deliver services in line with public and investor expectations.

Recognising Reforms Must Be Inclusive, Results-Oriented and Well-Planned Jordanians will be more likely to "buy into" difficult reforms if they can perceive tangible benefits, and to stay the course if they have been consulted.

Restructuring Donor Collaboration If donors are to support an extended reform process, they need reassuring that national leaders are shouldering responsibility in meeting reform-related conditions negotiated. Revised arrangements with donors for cash transfers and their *conditions precedent* (especially on water use in agriculture) will be key.

All this suggests a bold political approach linking political and governance reforms to effective water solutions is needed. King Abdullah II's response to the Arab Spring (King Abdullah II Ibn Al Hussein 2012)[15] provides unprecedented opportunity, when regional politics permit, for Jordan to implement with international support the needed domestic, regional and international policies to make this happen. The key challenge will be whether Jordan's roadmap of political reforms[16] can translate into politico-economic transitions capable of delivering the democratic freedoms and water-secure future to which Jordanians aspire. This paper's analysis and above criteria suggest success will be contingent on: the gradual but systematic implementation in due course of historic structural change away from the rentier system supporting Jordan's unaffordable patronage state to a participatory democracy based on the rule of law, transparency and accountability; and on all Jordanians – leadership, elites and public – "buying into" and putting their weight behind a difficult transition through agreeing cross-sector reforms serving the collective interest and a strategic agenda for implementing political, economic, legal and institutional aspects in which water is prioritised (Yorke 2013, pp. 80–93).

Who would drive these processes? The King has recognised the need for "top-down" leadership to facilitate "bottom-up" reform. By calling on Jordanians to embark on a process of "self-transformation", he has pointed the way through uncharted waters. While offering guidance he has called on Jordanians – government, parliament, administration, NGOs and citizens – to assume responsibility for agreeing a vision for Jordan's future and to drive through inclusive reforms to implement it. In this context, there is need for "political champions" amongst the elite to recognise win-win outcomes for all and to bring their followers with them. Donors must also play their part, with robust support for such political change.

15.3.2 National and Regional Solutions

On this paper's analysis and evidence only a political transition based on a restructured partnership between Throne and people will pave the way to the institutional, legal and economic reforms needed for Jordan to sequence implementation of effective water solutions. Meanwhile, Jordan needs as a matter of priority to support on-going ministry efforts to transform its water sector. "Business-as-usual" is not an option. Water management shortcomings render unachievable both the narrowing of water deficits and laying foundation for a sustainable water-secure future. Reforms need to move further and faster than planned. Ideally, they would be facilitated by

---

[15] In 2012–2013, the king outlined goals for "self-transformation and progressive reform": fair parliamentary elections, a law guaranteeing broad representation, a Parliament based on political parties and governments drawn from that Parliament. In 2013, the king distributed three papers on political transition.

[16] Constitutional changes implemented in 2012 and elections held in 2013, to be followed by changes in the electoral law, and formation of governments comprising elected parliamentarians.

timely pursuit of policies to meet the above criteria, which would start to free institutions from patronage politics and empower them. But they cannot wait.

How might this transformation be brought about? In view of the governance shortcomings identified earlier, focus on an integrated cross-sectoral approach incorporating domestic and regional options, which in different combinations could assist Jordan to meet its water challenge in the absence of a large-scale RSDS Conveyance Project offers the best way forward (Yorke 2009, 2013, pp. 94–132). Priorities for immediate implementation – some already embarked upon by the government – are identified below that would meet the criteria earlier set out and assist Jordan through small steps to narrow its short- and medium-term deficits; to protect its aquifers; to extend the period before a bulk solution is required and to identify long-term options for that supply. It is suggested these solutions be pursued in parallel, planned by national decision-makers in consultation with the public, coordinated inter-ministerially, and implemented with an agreed time-line.

15.3.2.1   Transforming the Water Sector

A necessary start point for the new approach would be the formal adoption of *integrated planning* across government and sectors. This would provide means to ensure that strategically important but scarce water is prioritised, its true cost accounted for and its management aligned with cross-sector implementation of the policies chosen to fulfil growth objectives. The resulting plan, with time-line and benchmarks, would provide frame for the required "step change" in water management.

A switch to integrated planning, reportedly underway, will be technically challenging. Decision-makers will face difficult choices given the food and energy needs of an expanding population, the need to attain higher growth levels and climate change impacts on water availability. It is suggested therefore that as part of the political transition, the king, prime minister and cabinet formally mandate this switch, thereby providing for prioritisation of water in investment decisions and optimisation of its use for future water security. This "game-changing" move would signal to "shadow state" members that new rules of the game will henceforth apply to water allocation and counter public scepticism on government intentions. At the same time, the national dialogue underway would provide opportunity for stakeholders to identify and agree "win-win" solutions that take availability, costs and the needs of the poor into account. These could be fed into planning. Donors could be expected to collaborate with the government to develop institutional capacity to roll out the plan. They could fund, if Jordan wishes: analysis of the true value of water, support for compilation of comprehensive data in a single centralised source, consultancy on options for restructured tariffs targeting subsidies more fairly and for trade policy adjustments on lifting import tariffs on water-thirsty crops (Yorke 2013, pp. 96–97). Power realities in the kingdom will change only slowly, however. The following discussion with proposals for domestic and regional solutions therefore identifies how these might be facilitated by the wider political transition, sequenced over time to neutralise resistance of opponents and maximise stakeholder cooperation, and be supported by regional and international players.

15.3.2.2 Exploiting Scope for Indispensable Demand and Supply-Side Efficiency Measures

Officials and donors recognise that opportunities exist to improve efficiency on both demand and supply sides. Some would be politically difficult to exploit, but all could potentially find sufficient support to drive them through and are technically and economically feasible. As a package they could bring short to long-term benefits – conserving low-cost supplies, protecting aquifers, enhancing adaptive capacity to climate change, raising additional revenue and making additional water available at lower cost than development of new sources:

Non-Revenue Water Reduction

A well-recognised opportunity lies in addressing inefficient supply in the municipalities and improving performance to standards that would reduce NRW at the Miyahuna and YWC networks. Politically, the government is committed to NRW reduction and Jordanians understand that failure to address the public health problem of leakages could result in political unrest. In addition leakage reductions and pressure controls at supply networks would reportedly be one of the most cost-effective measures to increase water supply and raise revenue (MWI 2009; MWI and WRG 2011). Failure to improve the Amman network (Miyahuna) would raise the cost of Disi water from JD $0.85/m^3$ to around JD $1.75/m^3$. Technically, leakage reduction would be a low cost-high volume move (Salameh 2007, p. 9). *Technical leakages* account for around 50 % of water supplies lost to NRW, but experts say network rehabilitation and improved management could reduce these losses by 50 %. On this basis, starting at Miyahuna and YWC, an annual saving of $19.5 \times 10^6 \, m^3$ (based on current production) could be achieved (USAID 2012, p. 32). A country-wide program would require further analysis, but water volumes of around $40 \times 10^6 \, m^3$ annually could potentially be saved by 2025, a significant contribution to projected requirements. *Theft* should be tackled in parallel, with the WAJ coordinating closely with the police on law enforcement, and a public education program launched.

The USAID is likely to support NRW projects, but the high cost of infrastructure makes essential a new US-Jordan understanding on USAID cash transfers to infrastructure projects and their ties to the fulfilment of conditions precedent.[17] Inter alia, conditions would need to cover government moves to rein in theft and shut illegal wells, the extension of PSP to more utilities with tighter performance-based management contracts and tariff restructuring.

---

[17] Echoing views expressed to the author in 2010, the (USAID 2011) found USAID/Jordan had not developed *conditions precedent* to ensure sustainability of its program activities since 2010. Criticism focused on the failure to address agricultural water use.

Cross-sector Highlands Water Strategy

A second opportunity lies in rethinking agriculture in the *Highlands*, where use of strategically important, low-cost groundwater is controversial. High-volume water use and low productivity point to scope for increasing agriculture's efficiency and contribution to national wealth, for conserving renewable low-cost supplies through sustainable use and for arresting aquifer depletion[18] whilst enhancing adaptive capacity.[19] Jordan's political transition and the increased public and parliamentary scrutiny it will bring has put leaders on notice to pursue policies serving the national interest rather than their own vested interests, providing political space for the Prime Ministry and governing institutions to intervene with the required approach to Highlands groundwater management.

Driven by national leaders, this would need to maximise the use of available legal, financial, technical and administrative levers to ensure groundwater is used efficiently and sustainably. It would build on the 2002 By-Law No 85 and would need to involve stakeholders – as did the Highland Water Forum initiative – in its design and implementation and to secure donor funding. A "carrot and stick" approach to win the cooperation of landowners and farmers, or cajole them, to change behaviour[20] and to assist the poor to adjust might include:

- *Enlisting tribal interlocutors to explain the situation and solutions to water users in a culturally sensitive manner.* Shadow state members, identifying with the national interest to reduce over-abstraction, would make excellent "political champions" in a collective effort to optimise water use and productivity, according to one interviewee. Measures to improve on-farm efficiency, raise government revenue and avoid social hardship might include:
- *Incentives to encourage farmers voluntarily to change crops and production methods or to cease farming in order to reduce groundwater use.* Consensus has grown among political elites and landowners on the need for suitable crop patterns. But crop transition will be long term, requiring steps to provide "win-win" outcomes for wealthy and poor farmers and workers. Interventions to ease the transition require study but might include: (a) subsidies, loans and debt forgiveness (Denny et al 2008, p. 11) where farms are profitable and farmers plan behaviour shifts to reduce groundwater use and to generate increased productivity; (b) retirement packages and alternative income opportunities for those unable

---

[18] According to a U.S. Geological Survey report cited in (USAID 2012, p. 7), groundwater abstraction levels have declined over recent years. But they still exceed safe-yield.

[19] Highlands irrigated agriculture is three times less productive than in the Jordan Valley, less profitable and accounted for over half (207×10^6 m³/year) of abstracted renewable groundwater, or nearly 25 % of total resources allocated in 2010 (Subah and Habjoka 2011).

[20] The government has been reluctant to press for demand management in the past because this involved confronting influential landowners resisting reforms. Ministry of Agriculture's conservative policy is best understood in this context, though it also strongly champions the poor. The Ministry's cap for overall irrigation use set at 700×10^6 m³/year may be politically-doable, but is arguably too high (510×10^6 m³ in 2010).

to farm at profit. Compensation could be offered in return for well closures. Agribusiness development, eco-tourism, wind and solar projects and expanded rain-fed agriculture could provide reassurance for those who risk losing jobs and opportunity for landowners to diversify investments.

- *Parallel inducements to encourage farmers to reduce groundwater abstraction* will be essential: (a) phased increases in extraction charges to encourage farmers to shift to higher value, water-efficient-crops or to move to alternative employment and to raise revenue; (b) removal of quantitative restrictions and customs duties on imported water-intensive crops to encourage farmers to grow higher-value, less-thirsty crops to enjoy comparative advantage; (c) accelerated enforcement of the 2002 By-Law No 85 (calling for closure of an estimated 1,000 illegal wells and restrictions on extraction from 2,799 licensed wells).
- *USAID financial support.* Although USAID projects have not focused on agriculture, this is changing following the highly critical 2011 USAID Audit. USAID could financially facilitate cross-sector studies on crop diversification and the above incentives, but make inducements subject to tightened *conditions precedent* for cash transfers.

Wastewater Reuse

Third is the opportunity to increase significantly the levels of treated wastewater for industry and agriculture in order to reduce demand for groundwater in the medium to long term. The option appears feasible: USAID consultants recommend that the agency fund over 5 years the building of medium-sized wastewater treatment plants (WWTPs) (USAID 2012, p. xvi). The option is politically attractive on account of its environmental and public health benefits and is technically feasible.

Expanding wastewater treatment is likely to be key to reducing highlands groundwater use and to long-term plans to release surface water (used in irrigation) for municipal use. Research is required on what volumes of treated wastewater might realistically be produced. This article's projections foresee around 60×10^6 m^3/year of additional treated wastewater production by 2020 and 85×10^6 m^3/year by 2025[21] – a significant contribution to closing projected deficits.

15.3.2.3 Pursuing Affordable Regional Desalination

Jordan has long recognised that water solutions cannot be exclusively national. This is because even when improvements in demand and supply efficiency and the remaining feasible *new* water options (wastewater re-use and brackish water

---

[21] Figures exclude supplies from As Samra extension for which finance is committed.

desalination) are taken into account, significant additional supplies will be required to narrow remaining deficits – whether from desalination, regional conveyance or transboundary solutions; and if these projects are to succeed, they will require international involvement – funding, provision of data, legal frameworks and guarantees. Meanwhile the global financial crisis, climate change, Arab uprisings and conflict in Syria and Iraq have transformed the political landscape, pushing Jordan's allies to reshape Middle East policies in which quid pro quos in exchange for the political and material support they extend will play a part. In short, past circumstances underpinning Jordan's rentier state are falling away and unilateral "business-as-usual" water policies based on it no longer hold. National leaders recognise the need radically to rethink how they address bulk-water supply and regional cooperation, if Jordan is to find a path to a water-secure future.

The government's preferred option has been desalination on its own coast. In December 2012, in a changed approach reflecting national interest, it refocused on the long-studied RSDS Conveyance Project, having decided to "scale down" its ambitious but unaffordable Jordan Red Sea Project (JRSP) in favour of a 70×10^6 m³/year desalination plant near Aqaba (Namrouqa 2012). The plant, which experts say is technically feasible if built north of the airport, will be designed to meet requirements for growth in the south over the medium term and – as agreed in December 2013 with Israel and the PA – transboundary demand in Israel and Palestine (see below). With its regional cooperative aspects, funding is likely to be forthcoming. However, discharge of the brine is problematic: Environmentalists argue against piping brine to the Dead Sea. On the other hand, any plan to discharge brine in the Gulf of Aqaba would need a positive environmental assessment and a mandate from riparian states.[22] The alternative of distributing brine in the desert requires study.

Jordan hoped the new plant would serve as the first phase of the mega-RSDS Conveyance Project. This latter project by virtue of Israeli, Jordanian and Palestinian participation, would likely have attracted international funding, permitting Jordan to sell a proportion of the more than 370×10^6 m³/year of water earmarked for it at prices affordable to Jordanians. Moreover, though not publicly discussed by decision-makers, it could have secured these outcomes without the government having to implement politically difficult water and economic reforms that alternative solutions would require. The recent World Bank-funded RSDS feasibility study raised questions over the project's technical feasibility, however, and environmentalist opposition now points to its potential abandonment. Jordan is thus in need of a Plan B for its long-term bulk supply, which has led some to resuscitate the politically difficult option of conveyance of desalinated water from the Mediterranean coast. But there are regional alternatives to this, as discussed below.

---

[22] Jordan, Saudi Arabia, Egypt and four other riparian states are members of the Regional Organisation for the Conservation of the Environment of the Red Sea and Gulf of Aden which is committed to conservation of coastal and marine environments.

15.3.2.4 Intensifying Regional Diplomacy to Manage Shared Resources

Jordan's water policy for decades has reflected asymmetry of power with its neighbours and its downstream riparian status. Lacking leverage to enforce its rights over shared surface waters, it has preferred unilateral options to secure new supply. However, in view of the scale of the water crisis, Jordan has strong interest in reaping the benefits of regional cooperation, for it will be a pre-requisite for the success of its mix of short-, medium- and long-term water solutions. A multi-pronged approach is required to facilitate regional and transboundary cooperation in order to mitigate dangers to shared waters and to develop new resources. Albeit less powerful than its neighbours, Jordan may not be maximising what leverage it has. Smart diplomacy could lead to beneficial trade-offs over water, food and energy. Besides, insufficient attention to bilateral differences might permit these to fuel broader tensions.

Since it is in the region that medium- and long-term water security will be found, now is the time for Jordan to raise the profile of water in diplomacy and harness support for indispensable regional efforts to manage resources sustainably – within and inter-basin. Identified below are four areas for cooperation which would serve the interests of Jordan and its neighbours. Requiring national and regional action, they cover demand and supply aspects and could be effective in the short to long term. All are in the public domain and given the right political dynamics would be feasible.

Laying Foundations for a Regional Water Authority

The promotion of the idea of a Council for Water Resources in the Middle East made up of heads of government represents a start (SFG 2011, p. 20). The Council's purpose would be to support five countries – Jordan, Iraq, Lebanon, Turkey and Syria – to prepare for decisions around water through reaching consensus on principles of cooperation, developing guidelines for standardising measurements, interpreting and exchanging data, on measures to combat climate change and setting targets for sustainable water management. In pursuit of these goals, the new High Level Group chaired by HRH Prince Hassan bin Talal[23] provides a platform for fresh thinking on interdependencies; and it can drive regional interactions until leaders agree to raise these to institutional level – in the form of a Regional Water Authority. Such collaboration would likely generate momentum behind the strategies discussed next, which if successful, would positively influence Jordan's water outcomes.

---

[23] The group comprises a former Turkish foreign minister and Lebanese finance minister and will in due course include Iraqi and Syrian representatives.

Pursuing Transboundary Cooperation to Close the Medium-Term Deficit

Jordan will need to maintain the quantities of transboundary water it receives, ensure it receives its rightful shares and find ways to increase them. This provides incentive for improving cooperation with Syria, Israel and Saudi Arabia. Present uneasy relations prevent cross-border supervision, the minimum required for successful water sharing. Moreover, current allocation accords[24] are fragile – complicated by mutual distrust and vulnerable to climate-change impacts. Regional politics are in flux, so it will be important for Jordan to maximise opportunities for diplomacy and cooperation over shared water – with regard to technology, water data or wider energy, trade and transit deals which would likely win international support.

It is too early to know Syria's political future, but its people will need to rebuild their country, a process from which Jordan could benefit when it renews relations. At that stage it will have strong incentive to prioritise water and press for a new accord that irons out differences over Yarmouk resources and provides for water sharing and joint cooperation over maintenance of water quality and pollution prevention. Experts suggest Jordan could push for a larger share than originally agreed – say 100×10^6 m³/year.

Twenty years after the 1994 Jordan-Israel Peace Treaty commitment to augment water supplies to Jordan by 50×10^6 m³/year, a formula has been found. Under the recent Jordan-Israel-PA accord, Jordan will sell Israel up to 50×10^6 m³/year from the proposed Aqaba desalination plant in exchange for Israel's sale to Jordan of an additional 50×10^6 m³/year to be released from Lake Tiberias in the north to supply Amman. The likelihood of access to international funding will reduce the costs to Jordan. The arrangement, which will contribute directly or through exchange up to 70×10^6 m³/year to Jordan's medium-term deficit, is an encouraging precedent for further joint projects.

One such would be a Jordan-Israel-PA application to UNESCO for World Heritage Site recognition for the Jordan River/Dead Sea – an idea already considered in Amman. An application, if it is to succeed, would need to specify what responsibilities the participants would assume for joint management of the shared water course flowing to the Dead Sea and agreeing plans for the sustainable use of waters allocated.[25] A successful application could provide unimaginable "win-win" outcomes for riparians, including mechanisms to cooperate over rather than compete

---

[24] In 1987 Syria agreed to supply Jordan with 208×10^6 m³/year but Jordan reportedly receives only $50–100 \times 10^6$ m³/year, and in drought years less. Experts say Syria has dug thousands of dams which deprive Jordan of its rightful share. Under the 1994 Israel-Jordan Peace Treaty, Israel is obligated to release 100×10^6 m³/year to Jordan, but actual releases are closer to $50–60 \times 10^6$ m³/year, reports the MWI *Water Budget*, 2012.

[25] Arab parties would need to consult with members of the Jordan River Initiative for Cooperation, which is committed to the Convention on the Law of the Non-Navigational Uses of International Water courses.

for precious river water, to protect the environment and to access political and financial support for projects to protect the Dead Sea and Jordan River. This degree of cooperation could encourage international parties to consider how they might support regional water supply projects for Jordan (and Palestine) through bulk conveyance, which are considered next.

Supporting Studies on Inward Transfer of Turkish Water

With Jordan's long-term need for bulk water, exploring the potential for the import of Turkish water (Gruen 2007; Wachtel and Liel 2009; IPCRI 2010; Yorke 2013, pp. 127–131) has become a priority as prospects for the mega RSDS Conveyance Project wane. Turkey would like to export water to the extent it has capacity (SFG 2011, pp. 39–41) and has historically favoured Middle East destinations for political reasons. The SFG's initiative to build cooperation around water between Turkey and the Arab states is therefore crucially important. Diplomacy and studies could keep open the option of water transfer to Jordan and its neighbours in the face of competing purchase offers from Mediterranean countries and lead to related deals on the transport of gas, oil and electricity to underpin growth. Given the potential contribution of Turkish water to Jordan's ability to meet long- and even medium-term needs, studies of the options deserve technical and financial support:

- *Transfers from Manavgat River:* Most promising would be Jordan's purchase of water conveyed from Turkey's Manavgat facility by tanker or floating bag. Imports of up to 400×10^6 m³/year would be significant and could be negotiated as part of a set of bilateral accords with Turkey or as a regional package involving Israel, or as a straightforward purchase of potable water for the medium term, pending a long-term solution. Conveyance by tanker is considered technically more reliable than by bag.[26]
- *Transfers from the Seyhan and Ceyhan Rivers:* Turkey's past interest in the idea of a pipeline to Syria and Jordan (Gruen 2007; Rende 2007; Wachtel and Liel 2009) has receded since development around the rivers has dented export capacity. Still, an estimated $1,000–1,500 \times 10^6$ m³ would be available to 2020, though flows would be intermittent. A modified pipeline in the context of a possible future Marshall-type plan for Syria's reconstruction with international guarantees and a legal framework might therefore appeal to all parties as well as attract the needed funding. The plan's self-sufficiency in terms of energy to convey the water would be an added plus. Scientific analysis to identify the scale of Turkey's future export capacity would be required as well as commitments by participating countries on joint monitoring and management.
- A third option would be an *undersea network of pipelines through the Mediterranean* to carry water, gas, oil and fibre optics (Personal interviews 2010;

---

[26] For discussion on Spragg bag technology, see www.internationalwaterlaw.org/blog/category/middle-east

(IPCRI 2010). Though politically easier to implement and less costly than other bulk water projects, the option is not technically feasible due to the Mediterranean's depth. Studies on how to overcome this difficulty and cost comparison with conveyance by transport or bag from the Manavgat are proposed.

15.4 Conclusion

Taken together, the above options for water solutions, if implemented, could assist Jordan to close the medium and long-term deficits projected in this chapter, conserve low-cost renewable resources for medium-long term use, protect aquifers, enhance adaptive capacity against climate change and find a bulk-supply solution as an alternative to a mega RSDS Conveyance Project. They involve: (a) demand and supply-side improvements at home – reduction and more efficient use of groundwater, conservation of these low cost renewable resources for medium-long term use, increased wastewater treatment for agricultural use and non-revenue water reduction; (b) building a desalination plant in Aqaba; (c) access to higher volumes of surface water from Syria and Israel and the conveyance of bulk supplies from Turkey. A pessimistic scenario would be one where Jordan fails to reach a favourable deal in the next couple of years with Syria on Yarmouk waters and no regional bulk-supply project, via imports from Turkey, is developed by 2025/2026.

Realistically, there is no set of local and regional policies that can guarantee the delivery of water to Jordan at affordable prices over the long term in sufficient quantity for it to fulfil its economic aspirations. Many factors beyond Jordan's control will inevitably influence prospects and access to water will remain highly political – domestically and regionally. What is clear is that finding a path to future water security will depend on Jordan accelerating a nationwide coordinated approach to parallel political, economic and water reforms and on a new imaginative regional diplomacy that boosts indispensable cooperation over water, economic and trade issues – the two are linked. International donors can be expected to support their strategic ally in its endeavour. The policy options discussed in this chapter could improve Jordan's water sector sustainability and through making additional supplies available relieve the pressures of water scarcity until long-term bulk supplies are found. The need to find a solution to the Arab-Israel conflict and to assist Syria move to a peaceful transition that would permit hundreds of thousands of refugees in Jordan to return will be determining factors.

References

Allan JA (2003) Virtual water eliminates water wars? A case study from the Middle East. In: Hoekstra AY (ed) Virtual water trade. Proceedings of the international expert meeting on virtual water trade. IHE Delft, The Netherlands, pp 137–145

Alon Y (2009) The making of Jordan: tribes, colonialism and the modern state. I.B. Tauris & Co. Ltd, London

Alterman JB, Dziuban M (2010) Clear gold: water as a strategic resource in the Middle East. A report of the CSIS Middle East program. http://csis.org/files/publication/101213_Alterman_ClearGold_web.pdf. Accessed 14 Jan 2015

Coyne et Bellier (2012) Red Sea – Dead Sea water conveyance study program. Draft final feasibility study report

Denny E, Donnelly K, McKay R et al (2008) Sustainable water strategies for Jordan. http://www.umich.edu/~ipolicy/PolicyPapers/water.pdf. Accessed 14 Jan 2015

Fischbach MR (2000) State, society and land in Jordan. Brill Academic Publishers, Leiden/Boston/Köln

FOEME (2010) Towards a living Jordan River: an economic analysis of policy options for water conservation in Jordan, Israel and Palestine. Friends of the Earth Middle East, Amman

Gruen GE (2007) Turkish water exports: a model for regional cooperation in the development of water resources. In: Shuval H, Dweik H (eds) Water resources in the Middle East: Israel-Palestinian water issues – from conflict to cooperation. Springer, Berlin, pp 157–164

Hagan RE (2008) Strategic reform and management of Jordan's water sector. USAID Jordan, Amman

GIZ (Deutsche Gesellschaft für Internationale Zusammenarbeit) (2011) The Highland Water Forum: a multi-stakeholder dialogue for sustainable groundwater management in Jordan. Amman.

IPCRI (2010) Water imports – an alternative solution to water scarcity in Israel, Palestine and Jordan? Palestine Center for Research and Information, Jerusalem

JME (2009) Jordan's second national communication to the United Nations Framework Convention on Climate Change (UNFCCC). Jordan Ministy of Environment, Amman

JMGP (2005) The national agenda 2006–2015: the Jordan we strive for. Jordan Ministry of Government Performance, Amman

Jordan-Israel Peace Treaty (1994) Treaty of peace between the state of Israel and the Hashemite Kingdom of Jordan (including texts of Annex II – water and Annex IV – environment), 26 October 1994. http://foeme.org/www/?module=regional_data&record_id=3. Accessed 3 Feb 2015

King Abdullah II Ibn Al Hussein (2012) Discussion Papers His Majesty King Abdullah II Ibn Al Hussein. http://kingabdullah.jo/index.php/en_US/pages/view/id/244.html. Accessed 27 Jan 2015

Muasher M (2011) A decade of struggling reform efforts in Jordan: the resilience of the rentier system. http://carnegieendowment.org/files/jordan_reform.pdf. Accessed 3 Feb 2015

MWI (2009) Water for life: Jordan's water strategy 2008–2022. http://www.irinnews.org/pdf/jordan_national_water_strategy.pdf. Accessed 14 Jan 2015

MWI, WRG (2011) Confidential Paper. Amman

MWI (2012) Water budget: projected demands and resources 2010–2025. Jordan Ministry of Water and Irrigation, Amman

Namrouqa H (2012) Southern agricultural companies still operating despite cancelled contracts – former minister. In: The Jordan Times. http://jordantimes.com/southern-agriculture-companies-still-operating-despite-cancelled-contracts. Accessed 9 Apr 2015

Peters A, Moore P (2009) Beyond boom and bust: external rents, durable authoritarianism, and institutional adaptation in the Hashemite Kingdom of Jordan. In: Studies in comparative international development, Springer science. Springer, Piscataway, pp 256–285

Rende M (2007) Water transfer from Turkey to water-stressed countries in the Middle East. Water resources in the Middle East: Israel-Palestinian water issues – from conflict to cooperation. Springer, Berlin, pp 165–173

Salameh E (2007) Towards a water strategy for Jordan. Royal Committee on Water, Amman

SFG (2011) The blue peace: rethinking Middle East water. Strategic Foresight Group, Mumbai

Subah A, Habjoka N (2011) The highland water forum: an experience in participatory water demand management. Paper presented at the economics of water demand management in Jordan workshop, Amman, 3 Dec 2011

Tripp C (2007) A history of Iraq. Cambridge University Press, Cambridge

USAID (2011) Audit of USAID/Jordan's design for sustainability in its water resources. U.S. Agency for International Development, Cairo

USAID (2012) Review of water policies in Jordan and recommendations for strategic priorities, final report. U.S Agency for International Development, Amman

Wachtel B, Liel A (2009) The "Peace Canal on the Golan" proposal: benefits and risks for regional water cooperation in the Middle East. In: Lipchin C, Sandler D, Cushman E (eds) The Jordan River and Dead Sea Basin: cooperation amid conflict. Springer, Dordrecht, pp 235–253

World Bank (2007) Making the most of scarcity: accountability for better water management results in the Middle East and North Africa. http://siteresources.worldbank.org/INTMNAREGTOPWATRES/Resources/Making_the_Most_of_Scarcity.pdf. Accessed 27 Jan 2015

Yorke V (1988) Domestic politics and regional security: Jordan, Syria and Israel – the end of an era? Gower Press/IISS, London

Yorke V (1990) A new era for Jordan? World Today 46:27–31

Yorke V (2009) Jordan water scarcity, strategy and alternative solutions: a politico-economic perspective. Paper presented at NCCR Trade Regulation Workshop, World Trade Institute, Bern, 27 November 2009

Yorke V (2013) Politics matters: Jordan's path to water security lies through political reforms and regional cooperation. Commissioned by NCCR-Trade Regulation, World Trade Institute, Bern

Chapter 16
Technologies, Incentives and Cost Recovery: Is There an Israeli Role Model?

Christine Bismuth, Bernd Hansjürgens, and Ira Yaari

Abstract This chapter focuses on water policy reforms and the introduction of a new water pricing policy in Israel. These reforms have to be seen in combination with measures to extend the available water sources that have been introduced in Israel in recent years. The effects of the Israeli demand and supply management policy on the use and availability of the water resources is investigated, and the potential transferability of the Israeli experiences to other countries in the region is examined. We also discuss the contribution of the water policy reforms as part of possible options and alternatives to the planned RSDS Conveyance Project.

Keywords Water policy • Water management institutions • Water pricing • Water use • Desalination • Wastewater treatment • Water consumption • Agriculture • Levies • Israel national water system

16.1 Introduction

Israel's water policies in recent years have presented its end users with two diametrically opposite modus operandi: from one point of view, Israel can be seen as a pioneer in agricultural water-saving technologies and water desalination projects (OECD 2011), and from the other view, Israel is wasting water, using the majority

C. Bismuth (✉)
Interdisciplinariy Research Group Society - Water - Technology,
Berlin-Brandenburg Academy of Sciences and Humanities,
Jägerstraße 22/23, 10117 Berlin, Germany

Helmholtz Centre Potsdam - GFZ German Research Centre for Geosciences,
Telegrafenberg, 14473 Potsdam, Germany
e-mail: bismuth@gfz-potsdam.de

B. Hansjürgens
Department of Economics, Helmholtz Centre for Environmental
Research – UFZ, Permoserstraße 13, 04318 Leipzig, Germany
e-mail: bernd.hansjuergens@ufz.de

I. Yaari
Robert H. Smith Faculty of Agriculture, Food and Environment,
The Hebrew University of Jerusalem, POB 12, Rehovot 76100, Israel

© The Author(s) 2016   253
R.F. Hüttl et al. (eds.), *Society - Water - Technology*, Water Resources
Development and Management, DOI 10.1007/978-3-319-18971-0_16

of the Jordan River's water resources and overexploiting the aquifers which are shared with the Palestinians (Amnesty International 2009).

In the 1990s and in the first decade of the third millennium, Israel was confronted with the impacts of a series of drought years. Several reports revealed that with respect to water, to continue with a "business as usual" policy was not possible. Israel reacted with the introduction of technical innovations in the water sector and a reform of water management institutions and pricing policies.

Some of these measures could be looked upon as being the adaptation of existing major water engineering projects (MWEPs), such as the Israel Water Carrier, or even being a second-generation MWEP.

Does the reform mean a shift away from the traditional technology-centred approach (MWEP logic)? And how have the reforms influenced the discussion around the Red Sea–Dead Sea Conveyance Project and its possible options?

16.2 The Evolution of Israel's Water and Agrarian Policy

From the first Zionist settlements, water in conjunction with land had been the driving force for the development of Israel's society and economy. Working the land was strongly linked to the Zionist ideology (Lipchin 2007). Modern agricultural techniques were introduced in Palestine not only by the Jewish settlers but also by Christian religious settlements from America and German Pietism movements (Kark 1983). A major influence stemmed not only from the introduction of modern agricultural techniques but also from the commercial marketing methods introduced by Baron de Rothschild who transferred all his properties and interests from his Palestinian settlements (*moshavim*) to the Jewish Colonization Association. Right from the beginning of Jewish expansion in the late nineteenth century, there was a difference between the subsistence system of the local Arab communities and the European model, based on cash crop plantations and shaping the Jewish settlements (Aaronsohn 1995). Jewish settlers did not come from traditional agrarian societies, but had a more urban background. From the commencement, the settlers invested not only in modern technologies and fertilisers, but they also improved the water supply devices within the settlements (*ibid.*). The strong ties and close communication between agricultural research and the farmer communities, most of whom had an excellent education, though not necessarily in agriculture, and the strong ties of the research communities with the Kibbutzim were determining factors for the high innovation rate within the Israeli agriculture. When compared to other agricultural societies, their pace of practical implementation in agricultural innovation has been considerably slower (Katz and Ben-David 1975).

Israel's Water Law of 1959 declared water as a public good, and its management was given into the responsibility of state institutions (Lipchin 2007). Any abstraction of water and any use of water require a permit. The requirement to monitor and measure the water was already included in the 1955 Water Measurement Law (Feitelson 2013). Since then, the water management system of Israel has been highly centralised. Its main purpose was subordinated to agricultural uses (Feitelson et al. 2007).

The water law also foresaw the institution of a "Water Commissioner" whose task was to monitor and allocate water rights. The Water Commission itself was supposed to be composed of two thirds of the general public, but the commission was dominated by representatives from the agricultural sector receiving 13 out of the 39 representative positions allocated (Plaut 2000). Until 2000, the Minister of Agriculture was the supreme statutory authority for the formulation of Israel's water policy. Also the Knesset Finance Committee, which was supposed to supervise the water policies, was dominated by agricultural lobbying groups (*ibid.*). As a result of these policies, water prices in the agricultural sector did not reflect the scarcity of the resource.

Instead, the agricultural sector was highly subsidised, and crops with a high water consumption rate such as cotton, citrus and flowers became profitable. Not only are these crops not adapted to the arid climatic conditions, but also they require high quantitative and quality water (Lipchin 2007).

In his 1990 report, Israel's state comptroller heavily criticised the existing water policies. In addition, the 2000 report was very critical of government pollution prevention policies (Plaut 2000). In 1997, the Arlosoroff Commission recommended to reduce agricultural water uses and change the institutional structures dominated by the agricultural sector (Feitelson 2001, cited in Lipchin 2007). However, it took nearly another decade and a couple of years of severe drought brought on by the political lock-in until the propositions favouring agriculture were taken down. This policy shift was underpinned by breaking up Mekorot's monopoly as Israel's only water supplier (Feitelson 2013).

Since 2007, Israel has gradually implemented several measures to overcome the water crisis. The main measures were:

1. Reform of the water management institutions
2. Enlargement of the water supply by different technical solutions (construction of sea water desalination plants on the Mediterranean coast, upgrading existing and construction of new wastewater recycling plants, use of the effluents for irrigation)
3. Introduction of economic incentives (water prices) to manage demand
4. A campaign to raise public awareness.

16.3   Reform of the Water Management Institutions

In 2007, the Israel Water Authority was created to replace the former Water Commission. The "Governmental Authority for Water and Sewage" (Israel Water Authority) gradually gathered all regulatory bodies concerning water issues under its authority. The main objective of the 2007 reform was to

> [...] enable the Authority implementing an integrative management of the whole 'water chain' and to transfer authorities from the political level of several ministers to one professional board. (Israel Water Authority 2012a, p. 1)

This has come into reality and today the Israel Water Authority is in charge of the management and regulation of the national water sector according to the water law.

The responsibilities of the Israel Water Authority include the preservation and regulation of the water resources. This includes the production (from seawater), the supply and use of water, the design and implementation of water supply schemes and the allocation and price setting for all water sectors (private households, agricultural and industrial) (OECD 2011). The Israel Water Authority intervenes in the sphere of withdrawal from the natural sources. Withdrawal without a permit is not possible. The Israel Water Authority issues permits at the beginning of each year according to hydrological considerations. Therefore, these permits differ from year to year (Kislev 2011). The users are required to present their requirements for the next year to the Israel Water Authority.

The Israel Water Authority Council has the primary responsibility for taking the water-related decisions. The council consists of eight elected members: the head of the Israel Water Authority as chairperson of the Israel Water Authority Council; one representative each from the Ministry of Agriculture and Rural Development, the Ministry of Environmental Protection, the Ministry of Interior and Management of Local Authorities, the Ministry of Finance, and the Ministry of Infrastructure; and two delegates of the public. The public representatives are appointed by the Minister of National Infrastructure. Kislev (2011, p. 136) criticised that the public representatives "cannot be taken as a realisation of public involvement".

With the creation of the Israel Water Authority, operations for managing waste and wastewater services were gradually transferred to water and sewerage corporations according to the Water and Sewage Association Law of 2001. This transfer of responsibility from the municipalities to the corporations was supposed to boost the construction of wastewater treatment plants. Before the reform, the financial resources required for the construction of wastewater treatment plants were not made sufficiently available to the municipalities (Hophmayer-Tokich 2010). As part of the reform, all revenues from water tariffs have now to be reinvested in water infrastructure (OECD 2011). The Israel Water Authority lists 55 corporations, which provide services for 6.2 million citizens in 147 local authorities (out of 187 local authorities bound by law). The law does not apply to the West Bank, regional councils and water associations. The largest corporation serves 415,000 residents in Tel Aviv; the smallest serves only 14,000 residents. The Israeli public criticises the number of the corporations as too many, and plans for merging corporations exist.

The government cut financial assistance and imposed restrictions on the use of water revenues for those municipalities that did not form corporations, while it provided financial assistance to those municipalities which formed corporations (Kan and Kislev 2011).

The partial privatisation of Mekorot and the shift to call for bids, as one of the results of the governmental market liberalisation efforts, opened up a new financing mechanism in Israel (public–private partnership, PPP) and started the era of large desalination projects.

One of the major achievements of the political reforms in the water sector was the introduction of the full cost recovery principle. In 2002 the agricultural ministry accepted that subsidies for the agricultural sector were shifted from water to land subsidies, which allowed raising prices for agricultural water uses (Menahem and

Gilad 2013). In 2006 the government signed an agreement with farmers' representatives according to which prices for agriculture would be set according to the costs of Mekorot for supplying water to agriculture (Kislev 2013). By 2015 the full cost recovery principle should be applied to agriculture.

As a result of the institutional reforms, Israel increased investment in water infrastructures, but compared to the energy sector, these increases have been far less significant (Israel Central Bureau of Statistics 2014).

16.4 Desalination

The 1997 master plan already foresaw the implementation of large-scale seawater desalination plants (Tenne et al 2013), but it took until 2000 – following a severe drought – for the Minister of Finance to reverse his position of not financing large-scale desalination plants and to issue a tender (Feitelson 2013). According to Feitelson, the issue of such a tender has been the turning point in decreasing the costs of desalination plants. But according to Becker (2010), desalination was still not the lowest cost alternative to close the water gap. Therefore cost reduction was one of the major concerns in the realisation of Israel's national planning programme for desalination. Besides the large-scale seawater desalination capacities, which today achieve an overall capacity of 577×10^6 m³/year, desalination capacities for brackish water exist for an additional 60×10^6 m³/year. This brackish water comes from groundwater wells (Tenne 2010).

In Israel desalination plants have to follow all public tender and bid procedures. With the exception of Ashkelon, which is operated by Mekorot, all seawater desalination plants are financed, constructed and operated by private companies (*ibid.*). But in fact, only two companies (Mekorot and IDE) exist on the Israeli market increasing the risk of possible monopolistic market structures. The desalination facilities in Israel are financed via a build–operate–transfer (BOT) contract. The concessionaires design, build and operate the plants over a period of 26.5 years after which the plant is transferred to state ownership (Spiritos and Lipchin 2013, p. 2014). During the time of operation, the companies receive the water revenues as private gains.

To reduce costs for the private investors of desalination plants, they are allowed to build power plants together with the desalination facilities, where the energy is not only used for the plants' own energy provision but sold into the national energy grid at their own profit (Tenne 2010).

Along the Mediterranean coast, Israel has constructed five seawater desalination plants (see Fig. 16.1). The Sorek plant is the newest and largest, with a desalination capacity of 150×10^6 m³/year. The long-term master plan foresees a seawater desalination capacity of up to 750×10^6 m³/year until 2020.

It was necessary to reshape the national water grid to integrate the new desalination plants in the water distribution system. The National Water Carrier transports water from north to south, but with the construction of the desalination plants, water

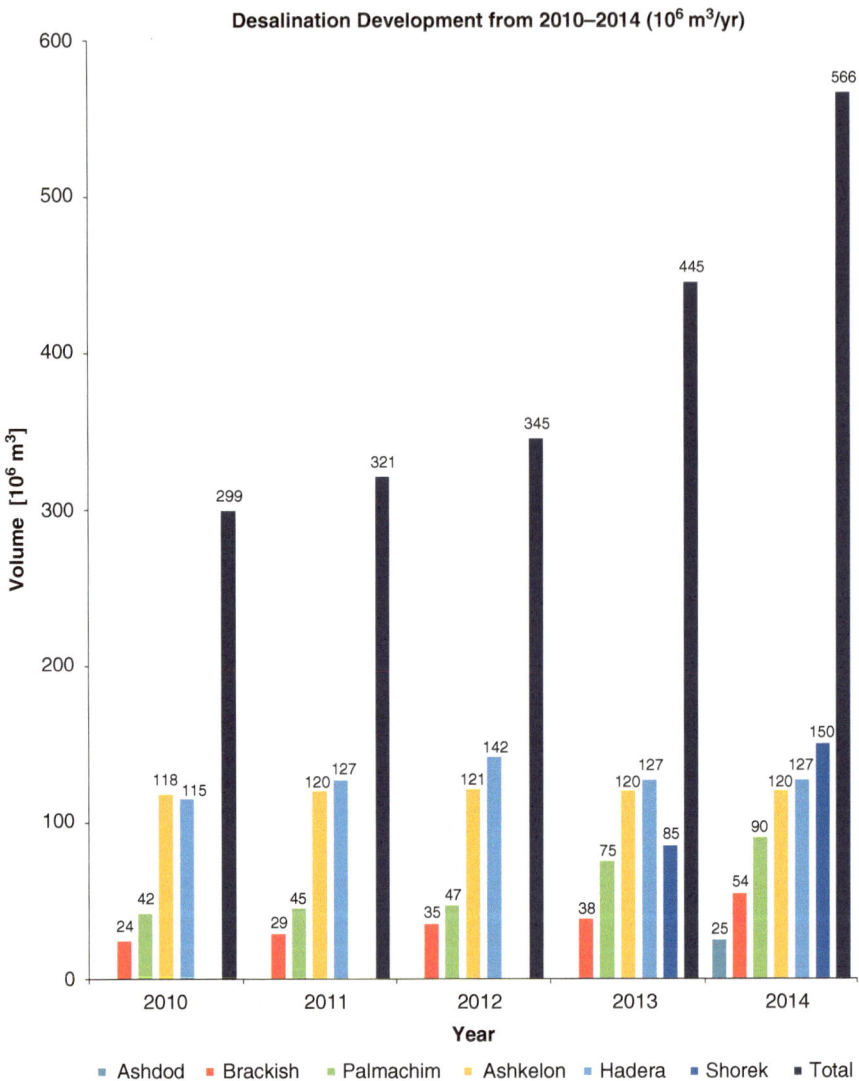

Fig. 16.1 Desalination development from 2010 to 2014 (10^6 m³/year) (Data source: Israel Water Authority (2014a))

had to be brought from the Mediterranean shoreline to the municipal centres. New lines from west to east were connected to the national water grid. With the extension of the drinking water network, the irrigation water network (recycled wastewater) had to be extended too (see Fig. 16.2).

The costs for desalinated water are highly dependent on energy costs, which make up more than 50 % of total costs. Consequently, energy saving is promoted in Israel for the desalination plants. As Israel discovered large natural gas reserves in

Fig. 16.2 Israel national water system (Source: Israel Water Authority)

2010, natural gas power generation is favoured for the desalination plants, while the national power system is still based on coal.

The prices for desalinated water are composed of a fixed price component for capital costs and a variable price component for variable costs, mainly for electricity prices (Lokiec 2011). The variable price is paid according to the quantity of water actually delivered, while the fixed price is paid according to the facilities' capacity. The prices of desalinated water from Israel's seawater desalination facilities are among the lowest worldwide. The national average is at USD 0.65; Sorek desalination plant is at USD 0.52 (Tenne 2010).

Cohen (2011) carried out a survey on the environmental impact of seawater desalination in Israel. He concluded that the Ashkelon, Palmachim and Hadera plants have no significant impact on the ecosystems of the receiving environments. One of the limiting factors for the construction of new seawater desalination plants in Israel is the limited space on Israel's shorelines. Therefore, planned construction of future seawater desalination plants will be offshore (Tenne 2014, personal communication).

The water from the desalination plants has improved the water supply quality both for private households and agricultural uses. Hardness and salinity were reduced. This has had positive effects on lifetime of heaters and other household devices, and the productivity of agriculture was increased due to lowered salinity levels in the water supplied to agriculture (Tenne et al. 2013).

The water from the desalination plants is introduced into the National Water Carrier (see Fig. 16.3). The desalination plants must operate continuously in order to get the fixed cost component financed through water sales (Dreizin et al. 2008). Changes in water demand are therefore adjusted by inputs from the various "natural" water supply sources (e.g. groundwater). As a result of the water privatisation policy, the management of the supply is more oriented towards the goal of reducing costs. This means that domestic users may be provided with water from other sources than desalinated water and that agriculture may also use water from desalination instead of using treated wastewater. This led to conflicts between the policy of the Israel Water Authority with its objective of saving water and taking water from wastewater reuse and the actual supply management of the desalination plants and Mekorot (Tenne et al. 2013) that is oriented towards using the full amount of desalinated water available.

16.5   Extension of Wastewater Treatment Plants

In 2011 96 % of the Israeli population was connected to an urban wastewater treatment system (Fig. 16.4). Most of the wastewater treatment plants are characterised by either secondary or tertiary treatment level. The water from tertiary treatment is unrestricted for irrigation uses. With the enlargement of the treatment capacities, new stringent quality parameters were introduced.

Fig. 16.3 National Water Carrier of Israel (Photo: Mekorot)

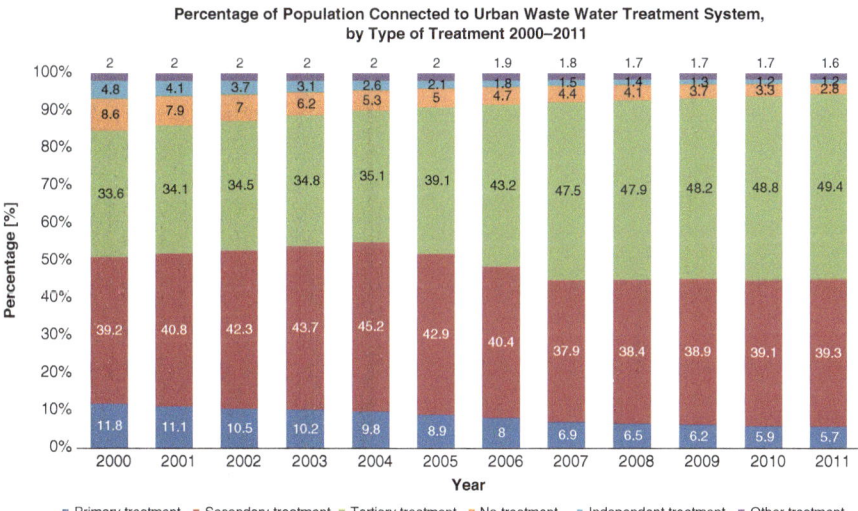

Fig. 16.4 Population connected to an urban wastewater treatment system, by type of treatment and level of treatment (Source: Israel Central Bureau of Statistics (2013a, p. 38))

From 2002 to 2013, Israel invested nearly 2.8 billion New Israel shekel (NIS) (approx. USD 734 million) in wastewater infrastructure and wastewater reclamation plants. Another 1.5 billion NIS (USD 393 million) will be invested in the near future. Not only were 100 water treatment plants built or upgraded, but a separate wastewater pipeline network, including a 90 km pipeline to the Negev, had to be built for the country (Israel Water Authority 2014a). In 2012, agriculture used 429×10^6 m^3 of recycled wastewater (*ibid.*). For 2014, it was required that 50 % of the agricultural water uses should come from recycled wastewater; according to the National Water Plan, this share is supposed to rise to more than 90 % by 2050 (Israel Water Authority 2012b). The rest will flow back to nature.

Stringent effluent quality standards came into effect in 2010. The regulations set maximum levels for dissolved and suspended elements and compounds and for 36 other parameters regarding effluents (European Environment Agency 2014). They came into effect to minimise the risk of soil and groundwater contamination by the use of wastewater for irrigation.

Apparently there is a conflict of interest between agriculture and environmental protection about the use of treated wastewater effluents. In the summer of 2014, the Israel Water Authority did not permit the use of treated wastewater for the rehabilitation of the Kishon River in the Haifa area. The Ministry of Environmental Protection had to suspend further river rehabilitation plans because of the decision of the Israel Water Authority (Israel Ministry of Environmental Protection 2014). This example might also illustrate a lack of coherence between the river rehabilitation plans and the national water management plans issued by the Israel Water Authority.

Environmentalists have criticised the partial dumping of the remaining sludge into the Mediterranean. The sludge from the Shafdan sewage plant contributes at least 75 % and up to 98 % of the total load to the Israeli marine environment. It is planned that this source will be terminated by the end of 2015 (European Environment Agency 2014).

16.6   Economic Instruments and Incentives

As a consequence of the restructuring of the water management institutions, the Israel Water Authority was able to considerably increase the prices for domestic water uses. The Israel Water Authority signed an agreement with the agricultural representatives that in 2015 the full cost recovery for agricultural water uses should be introduced. So far low water prices for the agricultural sector were cross-subsidised by higher water prices for domestic uses.

The water prices in Israel are set as a part of the general water law. The Israel Water Authority is responsible for establishing the water prices, and it does so according to the rules set by the Ministry of Finance, the Prime Minister's Office and the Ministry of National Infrastructures, Energy and Water Resources (Water.org.il 2014).

The current model for establishing the water prices is based on three main principles (Israel Water Authority 2014a):

Table 16.1 Water consumption distribution

	Low block rate (%)	High block rate (%)	Price a (%)	Price b (%)	Price c (%)	Gardening price (%)
2007	–	–	46.4	16.0	19.1	18.4
2008	–	–	48.8	17.2	18.6	15.4
2009	–	–	56.8	18.5	19.1	5.6
2010	57.1	42.9	–	–	–	–
2011	65.2	34.8	–	–	–	–
2012	71.7	27.7	–	–	–	–

Data source: Israel Water Authority (2014b)
Note: Between 2009 and 2010, there was a change in the water tariffs

1. Equal treatment: The price of water from all water corporations should be equal for all end consumers (except for industrial consumers – they have different prices due to different sewage components).
2. Block tariffs: Water prices for domestic uses are staggered in block tariffs. The first block goes up to 3.5 m³ per capita and month. The approved quantity for a single member household is fixed to 7 m³. According to Kislev (2011), only in few cases for small households with high consumption is a higher block rate likely to be applied. Table 16.1 shows the share of consumers paying different tariffs. The figures for 2010 to 2012 clearly indicate that water-saving measures for domestic consumption had been effective – the amount of households charged with a high block rate decreased considerably from 42.9 to 27.7 %. One reason is that private household can receive state support for the installation of water-saving devices. The reduction of freshwater consumption for domestic purposes was accompanied by the increase of treated wastewater uses for gardening, as quotas of water for watering private gardens had been cancelled (Kan and Kislev 2011). Also municipalities increasingly use recycled wastewater for the irrigation of public gardens and parks (*ibid.*).
3. Full cost recovery: The prices of water are set according to the costs of providing water for the municipalities, including coverage of all expenses (operating and maintenance costs, interest for capital costs for financing water-related infrastructures, calculatory costs).

The tariffs for domestic consumption are cost recovery rates.[1] These rates have two components (Fig. 16.6):

---

[1] It should be mentioned here that the cost recovery principle in Israel is based on a "company perspective". This means that the costs mirror all the necessary resources to safeguard the existence of a company – the company assets (operation and maintenance costs, capital costs for foreign capital, calculatory costs for company's own capital). This company-oriented concept of cost calculation goes beyond the "refinancing" perspective of cost recovery, which refers to the refinancing of company's expenditures (this does not include, e.g. calculatory costs). However, it does not include so-called social costs that mainly refer to the costs of environmental degradation; see Hansjürgens (1997). In Israel, this concept has led to increasing investments in water-related infrastructures, as the companies can receive money from the capital market.

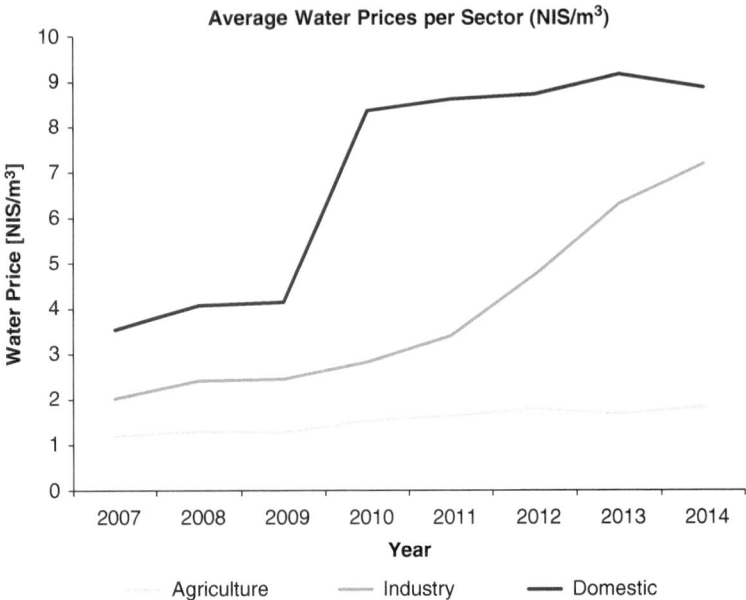

Fig. 16.5 Average water prices for agricultural, industrial and domestic water uses from 2007 to 2014 (NIS/m³). One New Israeli shekel (NIS) is equivalent to USD 0.26 (value as of November 2014) (Data source: (Israel Water Authority 2014b, c), Israel Central Bureau of Statistics (2012))

(a) "Costs for purchasing water from Mekorot at the municipality gate and costs for seawater desalination, wastewater treatment and subsidies".
(b) Costs for water distribution, sewage collection and wastewater treatment within the municipalities.

For a closer look on the tariff structure, we refer to Kislev (2011).

Water supply for agriculture is organised by an administrative allotment of quotas. Trading of the quotas among the agricultural companies is permitted under restricted conditions.

Figure 16.5 shows the evolution of the average water prices (NIS) for the agricultural, industrial and domestic sector from 2007 to 2014. Water prices for the domestic sector increased from 2007 to 2014 by 151 %, while in 2014, for the first time, the prices had been lowered by approx. 3 %. Prices for the agricultural sector increased between 2007 and 2014 by approx. 52 % and for the industrial sector approx. 254 %. Compared with other sectors, such as the energy sector, price rises had been less significant. The prices for energy augmented from 2011 to 2013 up to 31 %, compared with 6.3 % for price increases in the water sectors for the same period (Israel Water Authority 2014a). The share of household expenditures on water services and related services compared with total consumption expenditures is about 1.1 %, compared to an average 2.4 % in the EU (Israel Water Authority 2014a).

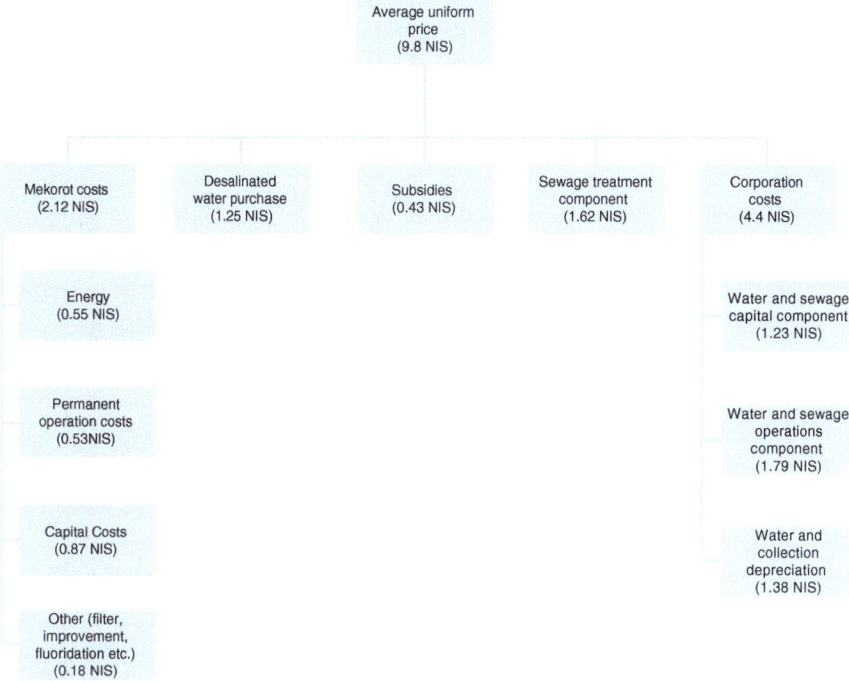

Fig. 16.6 Main components in the average uniform price of water for 2014 (NIS/m³) (Data source: (Israel Water Authority 2014b, c))

For comparison, the consumer price index for the period between 2007 and 2014 is 20.9 % (2006 = 100) according to the data provided by the Israel Central Bureau of Statistics (CBS).

Certain population groups such as disabled people, victims of hostilities and people who are qualified to receive social welfare transfers receive governmental support. This support allows these groups of people to double their amount of water used; however, they still have to pay the price for the low block tariff – instead of 3.5 m³ per capita and month, these individuals can use 7 m³ per capita a month for the low block tariff rate. The Knesset Finance Committee decides which population groups are eligible to receive this benefit, while the eligibility is decided by the Social Security (Bituach Leumi). The support itself comes from the local water corporation retroactively and is shown on the water invoice.

Figure 16.6 shows the components of the average uniform price. The corporation costs component (45 %) makes up the highest share, followed by the costs for Mekorot, which represents the costs of provisioning drinking water. The sewage treatment component is about 16 %, and the component for desalinated water is about 13 %. 4 % goes to subsidies. From 2007 to 2014, the costs for water provision were nearly constant.

The decisive point here is that water prices ease the financing of MWEPs and, thus, cause path dependencies for expanding water-related infrastructures. Once

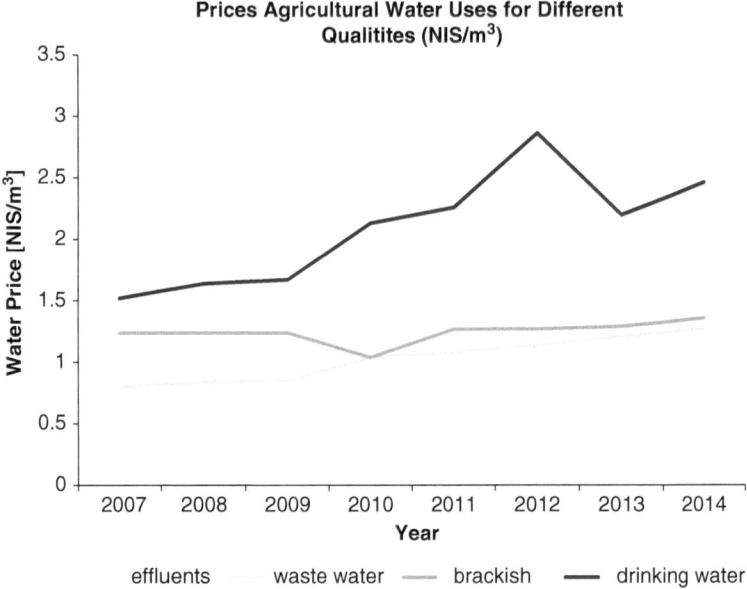

Fig. 16.7 Prices for agricultural water uses according to different qualities (NIS/m^3). Prices were approximated (Data source: (Israel Water Authority 2014b, c), Israel Central Bureau of Statistics (2012))

water-pricing schemes have been implemented, there is a strong push, not only for reinvestments but also for extensions and "modernisations" of such infrastructures.

The price structure for water for agriculture differs according to the quality of the water used (Fig. 16.7). It is separated into drinking water quality, water coming from wastewater treatment plants, effluents and brackish water. The price for drinking water is set according to the costs of production, but does not include the sewage treatment component, subsidies and costs for the purchase of desalinated water. Therefore there is a cross-subsidy from the domestic to the agricultural water sector. Compared with the price increases for the industrial and the domestic sector, the price increases for the agricultural sector were quite moderate. Within the agricultural sector, the share of energy prices caused the highest increases for agricultural inputs; water price increases played a much less important role (Fig. 16.8).

Besides water pricing instruments, which are related to end users, water levies for extraction are fixed in order to charge the value of water at the source. Those levies apply mainly to those who do not purchase their water from Mekorot but extract it by themselves. The levies are directly paid to the State Treasury. Changes in levies are suggested by the Israel Water Authority and subject to approval by the Knesset Finance Committee (Kislev 2011). The levy is set as a purely fiscal burden as it does not consider the characteristics of the services supplied by the ecosystem (Kedmi 2005). However, the calculation for the levies differs between different purposes: water for agriculture and water for domestic and industrial uses. The allocation is determined for each producer licence (Israel Water Authority). The

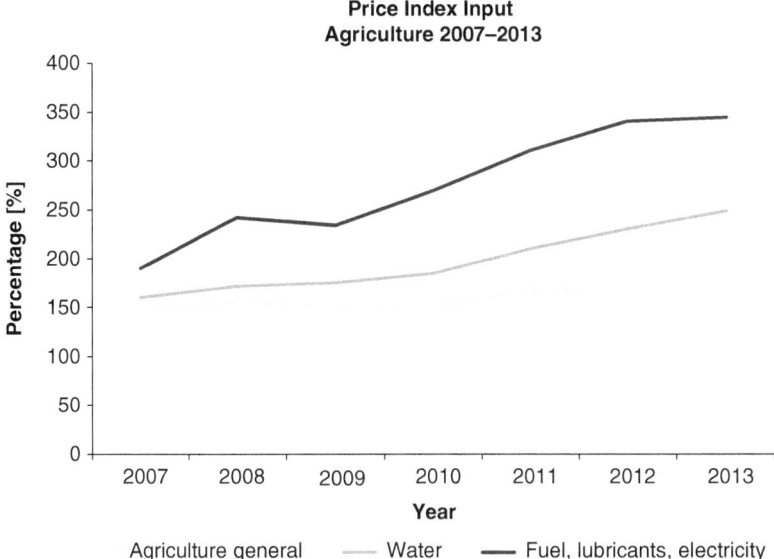

Fig. 16.8 Price index of input in agriculture from 2007 to 2013 (base year 2000 = 100) (Data source: Israel Central Bureau of Statistics 2013b)

levy level has increased steadily over the previous years, although not necessarily in accordance with the Israeli multi-annual plan (Kedmi 2005). The rates of the levies are determined on the basis of a number of parameters relating to the rate of water abstraction, the water production source, type and quality of water, hydrological situation, the purpose of the production and the quantity and type of the water produced (Israel Water Authority).

The levy system is divided into three regions: the country system area (in fact, those areas connected to the National Water Carrier), areas disconnected from the country system, and the Lake Tiberias area (*ibid.*).

Currently the levies for agricultural drilling within the country system are set at 50 % of the regular rates. In addition, agriculture is exempted from VAT (*ibid.*). Here we notice again that the agricultural level is benefitting from different (implicit) subsidies.

There is also an ongoing discussion in Israel whether the potassium companies at the Dead Sea should continue to be exempted from the extraction levy. With 300×10^6 m³ water extraction per year, the companies significantly contribute to the decline of the Dead Sea (TAHAL and GSI 2011). The authors of the Study of Alternatives (Allan et al. 2014, p. 74) proposed a levy of USD 0.1/m³, under the condition that the income generated from the planed hydropower production (exploiting the elevation difference between the Red Sea and the Dead Sea) is included into the calculations.

But the technical feasibility of hydropower production based on saline water or brine was never investigated in the World Bank's study programme and was not part of the terms of reference for the different studies. There are sound reservations concerning realisation costs and lifetime of such a project: the generation of energy via salt water, or even brine will be achievable only with considerable technical effort in the use of corrosion-resistant materials. Maintenance and lifetime expectations will be not comparable to normal hydropower plants. It is estimated that the costs of the pipeline according to the plans of the water swap will be approx. USD 300 to USD 400 million (Udasin 2013). The construction of the pipeline will take 3–4 years and will cross the Jordanian part of the Arava Valley. The conveyance of the brine from the Red Sea to the Dead Sea would therefore raise the price of water by at least USD $0.25/m^3$ (on the basis of an amortisation period of the pipeline of 30 years). With respect to the polluter pays principle, a levy for the potash companies should at least be equivalent to the additional costs for water for conveying brine to the Dead Sea to compensate for uses upstream and from the Dead Sea Works themselves. For comparison, the costs of water for the restoration of the Lower Jordan River are about USD $0.10/m^3$ (Allan et al. 2012, p. 30).

16.7   Awareness-Raising Measures

In 2008 the Israel government and the Israel Water Authority started a comprehensive campaign to inform and raise awareness on reducing domestic water consumption. The installation of water-saving devices in private households and gardens and metering for the irrigation in private and public gardens were financially supported.

Another important measure to raise awareness is the online publication of water use data and of data on the groundwater levels in the different basins. The levels of Lake Tiberias are reported on a monthly basis and can be viewed on an online graph. Its level is an important communication instrument as droughts and water shortages can be visualised directly with the passing of the critical lake level red line. Also the level of the Dead Sea is published on a monthly basis, making extreme water stress in the region visible.

Furthermore, water prices are published on the Israel Water Authority website. All water data is publicly available and presented in a transparent manner.

16.8   Discussion of Water Management Impacts

Israel is a striking example of how long it may take (nearly 20 years) to effectively accomplish a shift in water management practices – even though the development in water stress in the region requires an even faster pace of change. Israel is likewise an example that land and water management are deeply rooted in the nation's values, which make it psychologically difficult to introduce changes, as one has to give up long-standing traditions and values. With regard to agriculture, Israel is far from

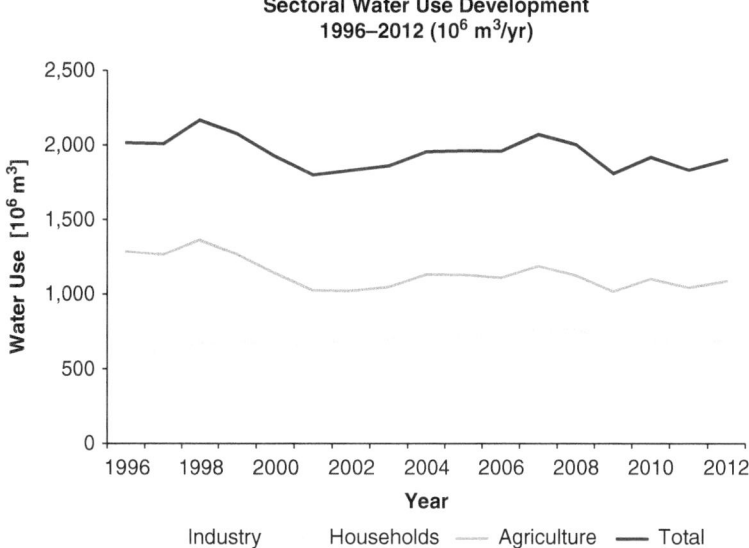

Fig. 16.9 Development of water uses for different sectors from 1996 to 2012 (10^6 m³) (Data source: Israel Water Authority (2011), Israel Water Authority (2014))

giving up its "agricultural Zionistic ideal" despite the negligible economic importance of the agricultural sector which counts for only 1.4 % of Israel's GDP.

Figure 16.9 reveals that changes in total water uses are mainly determined by decreases in agricultural water uses. Even though population increased considerably, domestic and industrial water uses remained nearly the same. However, this also means that per capita domestic water consumption decreased. The figure also indicates that the cuts in the water allocation for the agricultural sector in the late 1990s had a greater impact on the reduction of the overall water consumption by the agricultural sector than the agricultural pricing policies after the water reforms starting in 2007.

This also applies for the use of drinking water in the agricultural sector, as a detailed analysis of the agricultural water consumption reveals (Fig. 16.10). Drinking water was mainly replaced by the use of treated sewage water. But still, with more than 400×10^6 m³ of drinking water consumed in 2012, the agricultural sector uses a substantial amount of freshwater for irrigation purposes. This use goes along with prices, which only reflect the costs of the provision of water. However, the costs of seawater desalination and other investments costs, especially costs of the wastewater treatment infrastructures and the extension of the national water grid system, are still not included. If the cost recovery principle had been comprehensively applied to the agricultural sector, the sector would have been forced to pay water prices equivalent to those for private households. Thus the present pricing system only sets limited incentives for the agricultural sector in order to reduce its overall water uses.

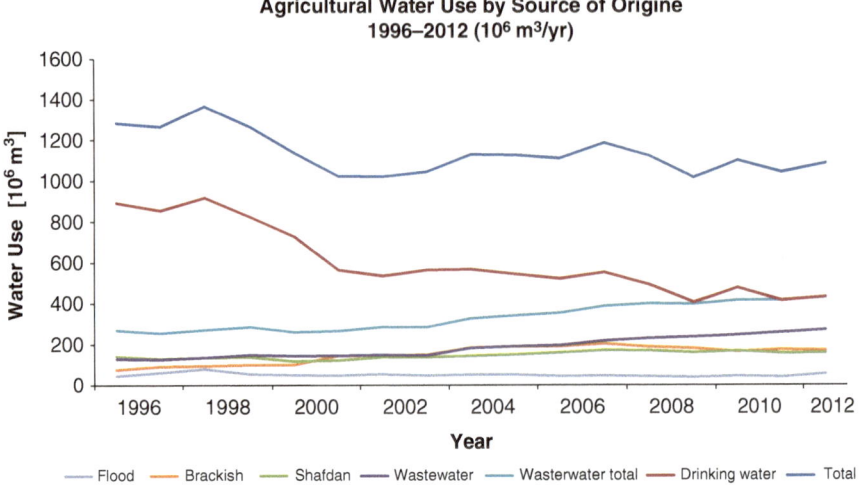

Fig. 16.10 Agricultural water uses divided into origin of source from 1996 to 2012 (10^6 m^3) (Data source: Israel Water Authority (2011), Israel Water Authority (2014a))

In some respect, Israel's water policy faces a dilemma: with the reduction of the household consumption of drinking water and the increased use of wastewater for public gardening (which also leads to a substitution of drinking water), there is less wastewater available, forcing the agricultural sector to use drinking water as a substitute to upholding its production. Practices in Israel's irrigation are of the highest state of knowledge, and their advanced techniques serve as a role model for the world. Although Israel is still undertaking major efforts in furthering increases in irrigation efficiency, its additional water-saving potential is limited due to the high standards already achieved. The only way to get out of this dilemma is either to reduce agriculture in some parts of the country or to convert the existing water-intensive agricultural practices to dry land agriculture. Becker et al. (2013) have produced evidence that it can be economically beneficial to give up agriculture for the rehabilitation of the Lower Jordan River.

Other regions in Israel with a high potential for recreation activities and rehabilitation of their ecosystems might also benefit from this. Israel is a country with a high population density, especially in the greater Tel Aviv area. Parks and pristine landscapes are a necessity for urbanised areas not only for their effects on the microclimate but also for the health and the well-being of the population. These ecosystem services might be of equal importance to the economic benefits obtained from agriculture.

Seawater desalination helped to sustain Lake Tiberias, but present allocations to the environment (80×10^6 m^3/year) are not sufficient to successfully rehabilitate the Jordan River and the Dead Sea. If in the future all treated wastewater will be used for agriculture, conflicts will increase between agricultural use and the water needs of the landscape and the environment. The Kishon example shows that the rehabilitation of rivers is far from being given priority within the Israel Water Authority. Israel's water management is highly centralised and linked to its scientific technocratic embedment. At present water planning mostly originates from this background, one of the reasons why water strategies have been criticised openly by the public and in politics. The civil society with its nature protection organisations and consumer protection groups is currently not represented in the water commission. A better representation of these groups might not make the decision-making process easier or shorter but in the end might lead to a better understanding and higher consensus within the Israeli society.

Pricing policies and levies to address the scarcity of natural resources as an incentive for water-saving measures are especially subject to differing interests and the volatile nature of political processes.

Many municipalities refuse to participate in the overall water allocations system. They expect a loss of independence and reduced potential for cross-subsidising other sectors (Kan and Kislev 2011). Improving civil society involvement and participation in water management planning processes could also help to reprioritise the objectives of water allocation and water pricing and increase support for unpopular measures within society.

Israel suffers not only from water shortages but also from increased income disparities and low welfare expenditures (Dahan and Hazan 2014). The outbreak of the social protests in summer 2011 is a reflection of these disparities. But water pricing would be the wrong instrument to mitigate these disparities. Low-income groups are supported by specific welfare programmes instead of lower water prices.

The restructuring process within Israel's water management has still not come to an end with regard to water corporations. Two specific problems are linked to this issue: the first is that there are far too many water companies in Israel, and the second is that under the uniform tariff structure system, companies who perform well have to support the ones who perform less well (Kan and Kislev 2011). This is due to water corporations with high water losses and high losses in water charges (because of compliance problems in water price payments) paying lower prices to Mekorot. To reemphasise – according to the equity principle, despite end-users' respective water resource use, all water prices are identical – subsequently this results in different costs in different regions. The government and Israel Water Authority need to find a way out of this dilemma (*ibid.*).

16.9   Can Israel Serve as a Role Model? 10 Lessons to Be Learnt

Israel has a very specific water management and the specificity of the availability of its water resources, technical facilities and governance structures limits the transferability of the Israeli experiences to other countries. Nevertheless, some lessons can be drawn from the Israeli case example. The first seven lessons are general in nature, while the last three refer to water pricing:

1. Innovative solutions to improve water supply and high technical standards are an important prerequisite to reduce water stress and improve the water situation. In Israel, major innovative technical solutions include new and innovative types of irrigation systems in the agricultural sector, highly innovative desalination systems, advanced wastewater treatment facilities for the reuse of water and the building of the corresponding infrastructure systems (e.g. water grid systems).

2. An important prerequisite for building up water-related infrastructures are sufficient financial resources. In Israel these resources have been made available since the foundation of the State of Israel. One important reason is that the agricultural sector received highest priority within the political discussions. Even though the significance of agriculture decreased in budgetary terms (in terms of the share of agriculture in the Israeli GDP), the agricultural sector is still seen as very important, and therefore, considerable public expenditure is devoted to this sector. A second reason for successfully financing the water sector lies in the raising of revenues from the users of water resources.

3. Nevertheless, it is not sufficient to just focus on water supply and the development and improvement of water infrastructure systems (and on the financing of these infrastructures). The improvement of existing and development of new water supply systems is a necessity, but not sufficient enough to improve the overall water situation. Technical supply measures have to be accompanied by management structures that complement technical developments, such as appropriate forms of governance and institutional reform. In Israel this has been achieved through comprehensive organisational reforms (the centralisation of powers) and the introduction of water prices intended to incentivise users to use water resources more effectively.

4. Strong institutions and the rule of law are preconditions for the successful implementation of water policy reforms, which are often neglected when policy makers and decision makers simply focus on technical solutions alone. If adequate institutional structures do not exist, policy reforms will fail. Even if technical solutions deliver success in the short run, links with institutional structures are decisive for long-term sustainability. Institution building and strengthening (with respect to water authorities, division of responsibilities between actors, regulation, etc.) are therefore necessary preconditions before any other reforms (such as pricing) can be introduced. Weak state institutions impact on other reforms in water policies.

5. Time is needed to successfully develop infrastructure systems, not only for technical reasons (long construction periods for large infrastructures) but mainly because they have to go hand in hand with institutional reform. As economic and societal rules are deeply embedded in society's legal system and culture, economic and societal changes have to be made gradually and with care. Even within highly centralised organised water management institutions as in Israel, the reshaping of the institutions (founding water authorities, defining and allocating

competences, defining regulatory rules, etc.) and the introduction of new steering mechanisms (such as full cost recovery principle or water quotas for the agricultural sector) take many years or even decades. Therefore, there is the danger of seeing one of these elements (e.g. the cost recovery principle) as the "magic pill" that will solve all "ills" in water management alone.

6. Israel is an outstanding example of science, practice and government interacting closely together. Scientists play an important role in the water sector and in agricultural organisations. Especially in agriculture, this is one of the major reasons why technological pace has been so rapid. To increase efficiency, in particular for irrigation, this close system of interactions and relations between science and practice can truly serve as role models in other countries.

7. Israel is the only country in the region where water management and water policy are presented transparently. Anyone who would like information on water uses and water prices can obtain it. The credibility of water policies is enhanced by such an open and transparent information policy, even though there might be space for improvement, especially with respect to the participation of the civil society in the formulation of the water policy.

8. Water prices and levies play an important role in Israeli water policy. If properly designed, they can set incentives for water savings and an efficient use of water. A water price reform can therefore be an important supplement for water management strategies. However, in order to be successful in incentivising end-users' behaviour, water prices and levies should address the full cost of water (cost recovery principle) and the fair distribution of water prices among users.

9. With respect to cost recovery and allocative efficiency, it is important that all costs are included in the water price and that all sectors that use water are in fact obliged to contribute to the water pricing scheme. Cost recovery should not only be applied to a share of the user groups. However, in many countries, this will cause problems: the agricultural sector is seen as fundamental for the welfare of countries, leading to a deep-rooted need in societies to protect their agricultural sector and an automatic rejection of burdens such as water prices or levies.

10. Social equity may also be seen as a major obstacle to introducing the cost-recovery principle on the level of domestic end users. Even in economically prosperous states such as Israel, water price policies are not disconnected from the overall social debate. The greater the income disparities, the more difficult will be the introduction of the cost recovery principle. Uniform tariff structures might be fair to the consumers but impede the efficiency of the local water utilities. This might only be a question for states with highly centralised water management institutions such as in Israel, but in many other regions under water stress, such centralised institutions are in place. Specific solutions have to be found to increase efficiency of the water utilities under such conditions.

References

Aaronsohn R (1995) The beginnings of modern Jewish agriculture in Palestine: "Indigenous" versus "Imported.". Agric Hist 69:438–453

Allan JA, Malkawi AIH, Tsur Y (2014) Red Sea – Dead Sea water conveyance study program. Study of alternatives. Final report. Executive summary and main report. World Bank Publications, Washington, DC

Allan JA, Malkawi AIH, Tsur Y (2012) Red Sea – Dead Sea water conveyance study program. Study of alternatives. Preliminary draft report. World Bank Publications, Washington, DC

Amnesty International (2009) Troubled waters Palestinians denied fair access. Israel – occupied Palestinian Territories. Amnesty International Publications, London

Becker N (2010) Desalination and alternative water-shortage mitigation options in Israel: a comparative cost analysis. J Water Resour Prot 02:1042–1056. doi:10.4236/jwarp.2010.212124

Becker N, Helgeson J, Katz D (2013) Once there was a river: a benefit-cost analysis of rehabilitation of the Jordan River. Reg Environ Chang 14:1303–1314. doi:10.1007/s10113-013-0578-4

Cohen Y (2011) Seawater desalination and the environment – the Israeli experience. Paper presented at the 12th Israel desalination society annual conference, Haifa, 14–15 December 2011

Dahan M, Hazan M (2014) Priorities in the government budget. Israel Econ Rev 11:1–33

Dreizin Y, Tenne A, Hoffman D (2008) Integrating large scale seawater desalination plants within Israel's water supply system. Desalination 220(1):132–149

European Environment Agency (2014) Horizon 2020 Mediterranean report. Annex 3: Israel. Publications Office of the European Union, Luxembourg

Feitelson E (2013) The four eras of Israeli water policies. In: Becker N (ed) Water policy in Israel. Context, issues and options, 4th edn. Springer, Dordrecht/Heidelberg/London/New York, pp 15–32

Feitelson E, Fischhendler I, Kay P (2007) Role of a central administrator in managing water resources: the case of the Israeli Water Commissioner. Water Resour Res 43:1–11. doi:10.102 9/2007WR005922

Hansjürgens B (1997) Gebührenkalkulation auf Basis volkswirtschaftlicher Kosten? AnwendungsproblemeundLösungsmöglichkeite.In:Difu(ed)ArchivfürKommunalwissenschaften, vol 36. Kohlhammer, Stuttgart, pp 233–253

Hophmayer-Tokich S (2010) The evolution of national wastewater management regimes – the case of Israel. Water 2:439–460. doi:10.3390/w2030439

Israel Central Bureau of Statistics (2014) Gross domestic capital formation in infrastructure, by industry. Table 8A. In: National accounts 2006–2012. http://www1.cbs.gov.il/publications14/1553/pdf/t08a.pdf. Accessed 17 Feb 2015

Israel Central Bureau of Statistics (2013a) Sustainable development indicators in Israel 2011. Israel Central Bureau of Statistics (CBS), Jerusalem

Israel Central Bureau of Statistics (2013b) Input, output and domestic product in agriculture, Table 19.14. In: CBS, Statistical abstract of Israel 2013. http://www.cbs.gov.il/reader/shnaton/templ_shnaton_e.html?num_tab=st19_14&CYear=2013. Accessed 11 Feb 2015

Israel Central Bureau of Statistics (2012) Satellite account of water in Israel 2007–2008. Israel Central Bureau of Statistics (CBS), Jerusalem

Israel Ministry for Environmental Protection (2014) Kishon rehabilitation project frozen due to inaction by water authority. http://www.sviva.gov.il/English/ResourcesandServices/NewsAndEvents/NewsAndMessageDover/Pages/2014/6June/Kishon-Rehabilitation-Project-Frozen.aspx. Accessed 10 Oct 2014

Israel Water Authority (2012a) Israel water authority – towards WWF VI. http://www.water.gov.il/Hebrew/ProfessionalInfoAndData/2012/01-Towards-WWF-VI.pdf. Accessed 10 Sep 2014

Israel Water Authority (2012b) Long-term master plan for the national water sector part A – policy document version 4. http://www.water.gov.il/Hebrew/Planning-and-Development/Planning/MasterPlan/DocLib4/MasterPlan-en-v.4.pdf. Accessed 17 Feb 2015

Israel Water Authority (2014a) Water Authority Report 2013. Israel Water Authority, Jerusalem (in Hebrew). http://www.water.gov.il/Hebrew/about-reshut-hamaim/DocLib/water-authority-report-2013.pdf

Israel Water Authority (2014b) Prices archives (in Hebrew). http://www.water.gov.il/Hebrew/Rates/Pages/prices-archive.aspx. Accessed 11 Dec 2014

Israel Water Authority (2014c) Production levies (in Hebrew). http://www.water.gov.il/hebrew/rates/pages/producer-levies.aspx?p=print. Accessed 11 Dec 2014b

Israel Water Authority (2014d) Allocation, consumption and production data (in Hebrew). http://www.water.gov.il/Hebrew/ProfessionalInfoAndData/Allocation-Consumption-and-production/Pages/Local-authorities-data.aspx. Accessed 11 Dec 2014

Israel Water Authority (2011) Water consumption by sectors: 1996–2011 (in Hebrew). http://www.water.gov.il/Hebrew/ProfessionalInfoAndData/Allocation-Consumption-andproduction/.

Kan I, Kislev Y (2011) Urban water price setting under central administration. European Commission, Brussels

Kark R (1983) Millenarism and agricultural settlement in the Holy Land in the nineteenth century. J Hist Geogr 9:47–62. doi:10.1016/0305-7488(83)90141-X

Katz S, Ben-David J (1975) Scientific research and agricultural innovation in Israel. Minerva 13:152–182. doi:10.1007/BF01097793

Kedmi N (2005) Integrated water resources management in Israel. http://www.unece.org/fileadmin/DAM/env/water/meetings/payment_ecosystems/Discpapers/Israel.pdf. Accessed 11 Dec 2014

Kislev Y (2011) The water economy of Israel. Taub Center for Social Policy Studies in Israel, Jerusalem

Kislev Y (2013) Water in agriculture. In: Becker N (ed) Water policy in Israel. Context, issues and options. Springer, Dordrecht/Heidelberg/London/New York, pp 51–64

Lipchin C (2007) Water, agriculture and zionism: exploring the interface between policy and ideology. In: Lipchin C (ed) Integrated water resources management and security in the Middle East. Springer, Dordrecht, pp 251–267

Lokiec F (2011) Sorek 150 million m³/year seawater desalination facility build. Operate and Transfer (BOT) Project. Paper presented at the IDA World Congress, Perth Convention and Exhibition Centre (PCEC), Perth, 4–9 September 2011

Menahem G, Gilad S (2013) Israel's water policy 1980s–2000: advocacy, coalitions, policy stalemate, and policy change. In: Becker N (ed) Water policy in Israel. Context, issues and options. Springer, Dordrecht/Heidelberg/London/New York, pp 33–50

OECD (2011) OECD environmental performance reviews: Israel 2011. OECD Publishing, Paris

Plaut S (2000) Water policy in Israel. Policy Stud 47:1–26

Spiritos E, Lipchin C (2013) Desalination in Israel. In: Becker N (ed) Water policy in Israel. Context, issues and options. Springer, Dordrecht/Heidelberg/London/New York, pp 101–124

TAHAL, GSI (2011) Dead Sea study (final report). http://siteresources.worldbank.org/INTREDSEADEADSEA/Resources/Dead_Sea_Study_Final_August_2011.pdf. Accessed 14 Jan 2015

Tenne A (2010) Seawater desalination in Israel: planning, coping with difficulties, and economic aspects of long-term risks. Israel Water Authority, Tel Aviv

Tenne A, Hoffman D, Levi E (2013) Quantifying the actual benefits of large scale seawater desalination in Israel. Desalin Water Treat 51:26–37

Udasin S (2013) Israel, Jordan, PA sign trilateral agreements to "swap" and share water. In: The Jerusalem Post. http://www.jpost.com/Enviro-Tech/Exclusive-Israel-Jordan-PA-to-sign-trilateral-water-swap-sales-agreements-334505. Accessed 16 Mar 2015

Water.org.il (2014) Water prices (in Hebrew). www.water.org.il. Accessed 1 Oct 2014

Part V
Outlook and Options for Action

Chapter 17
Lessons Learnt, Open Research Questions and Recommendations

Christine Bismuth, Bernd Hansjürgens, Timothy Moss,
Sebastian Hoechstetter, Klement Tockner, Valerie Yorke,
Hermann Kreutzmann, Petra Dobner, Shavkat Kenjabaev,
Reinhard F. Hüttl, Oliver Bens, Rolf Emmermann, Hans-Georg Frede,
Gerhard Glatzel, Hermann H. Hahn, Bernd Hillemeier,
Hans-Joachim Kümpel, Axel Meyer, Helmar Schubert, Herbert Sukopp,
and Ugur Yaramanci

Abstract This chapter represents the summary of the common analysis within the Interdisciplinary Research Group *Society – Water – Technology*. Lessons learnt, research gaps and recommendations are presented as the outcome of the analysis of the two case studies Fergana Valley and Lower Jordan Basin and as a conclusion from the cross-analysis based on the evaluation framework and the considerations outlined in Chap. 3 (Bismuth et al., Research in two cases studies: (1) Irrigation and land use in the Fergana Valley and (2) Water management in the Lower Jordan Valley. In: Huettl RF,

C. Bismuth (✉)
Interdisciplinary Research Group Society - Water - Technology,
Berlin-Brandenburg Academy of Sciences and Humanities,
Jägerstraße 22/23, 10117 Berlin, Germany

Helmholtz Centre Potsdam   GFZ German Research Centre for Geosciences,
Telegrafenberg, 14473 Potsdam, Germany

Enquiries should be directed to the spokeperson of the Interdisciplinary Research Group
R.F.Hüttl mail: huettl@gfz-potsdam.de
e-mail: bismuth@gfz-potsdam.de

B. Hansjürgens
Department of Economics, Helmholtz Centre for Environmental
Research – UFZ, Permoserstraße 13, 04318 Leipzig, Germany

T. Moss
Leibniz Institute for Regional Development and Structural Planning (IRS),
Flakenstraße 28-31, 15537 Erkner, Germany

S. Hoechstetter
Helmholtz Centre Potsdam - GFZ German Research Centre for Geosciences,
Telegrafenberg, 14473 Potsdam , Germany

R.F. Hüttl et al. (eds.), *Society - Water - Technology*, Water Resources
Development and Management, DOI 10.1007/978-3-319-18971-0_17

Bens O, Bismuth C, Hoechstetter S (eds) Society water technology: a critical appraisal of major water engineering projects. Springer, Dordrecht, 2015, in this volume).

Keywords Recommendations • Lessons learnt • Research gaps • Major water engineering projects • Complex and coupled systems • Options for action

K. Tockner
Leibniz-Institute of Freshwater Ecology and Inland Fisheries (IGB),
Müggelseedamm 310, 12587 Berlin, Germany

Department of Biology, Chemistry and Pharmacy, Freie Universität Berlin,
Altensteinstraße 6, 14195 Berlin, Germany

V. Yorke
NCCR Trade Regulation/World Trade Institute, University of Bern,
Hallerstr. 6, 3012 Bern, Switzerland

H. Kreutzmann
Department of Geography, Freie Universität Berlin, Malteser Str. 74, 12249 Berlin, Germany

P. Dobner
Institute of Political Science and Japanese Studies, Martin-Luther-Universität
Halle-Wittenberg, Emil-Abderhalden-Str. 7, 06099 Halle/Saale, Germany

S. Kenjabaev
Scientific-Information Center of the Interstate Coordination Water Commission (SIC ICWC),
h.11, Karasu-4, Tashkent 100187, Uzbekistan

R.F. Hüttl • O. Bens • R. Emmermann
Helmholtz Centre Potsdam - GFZ German Research Centre for Geosciences,
Telegrafenberg, 14473 Potsdam, Germany
e-mail: huettl@gfz-potsdam.de

H.-G. Frede
Institute of Landscape Ecology and Resource Management (ILR), Justus-Liebig-Universität
Gießen, Heinrich-Buff-Ring 26-32, 35392 Gießen, Germany

G. Glatzel
Universität für Bodenkultur, Wien BOKU, Vienna, Austria

H.H. Hahn
Heidelberg Academy of Sciences and Humanities, HAW,
Karlstraße 4, 69117 Heidelberg, Germany

B. Hillemeier
Technische Universität Berlin, Berlin, Germany

H.-J. Kümpel
Federal Institute for Geosciences and Natural Resources (BGR),
Stilleweg 2, 30655 Hannover, Germany

A. Meyer
Universität Konstanz, Universitätsstr. 10, 78457 Konstanz, Germany

H. Schubert
Karlsruhe Institute for Technology (KIT), Kaiserstr. 12, 76131 Karlsruhe, Germany

H. Sukopp
Technische Universität Berlin, Rüdesheimer Platz 10, 14197 Berlin, Germany

U. Yaramanci
Leibniz Institute for Applied Geophysics LIAG, Stilleweg 2, 30655 Hannover, Germany

17.1   Introduction

The chapters of this book have given an insight into various aspects of major water engineering projects (MWEPs) and the influence of water technologies on societies and natural resources. The key question asked was to which extent MWEPs may support the efficient and sustainable management of water and land resources. Directly linked to this overarching question are reflections on the extent to which such projects generate serious environmental, economic, and social changes, on how MWEPs contribute to the creation of path dependencies and on how such dependencies can be resolved (Moss and Dobner 2015, in this volume). The political component of MWEPs, particularly considerations of the problems related to transboundary water management, has been another important focus.

We have tried to address these questions from two general perspectives: The main aim for the Fergana Valley was to analyse the current and future impacts of the decisions made in the past. For the Lower Jordan Basin and the Red Sea–Dead Sea (RSDS) Conveyance Project, the main goal was to assess future options for action and to determine the conditions for their implementation.

The results from the two case studies do neither allow to draw general conclusions about the complexity of MWEPs nor to simply transfer the results obtained from these two regions to MWEPs globally. Nevertheless, we were able to put the two case studies into a broader context by a comprehensive literature search and to put emphasis on the general political, social and ecological processes associated with such projects. It allowed to identify transferable results and to formulate principal lessons learned from the analysis of these case studies.

For the cross-analysis of the case studies, we also refer to the evaluation framework and the considerations outlined in Chap. 3 (Bismuth et al. 2015a, in this volume). Comparing the two case studies enabled to draw general conclusions and recommendations from four different points of view: (1) from the perspective of international and transboundary water management, (2) from the perspective of the observed planning processes, (3) from institutional and governance-based viewpoints, and finally (4) from the perspective of "coupled socioecological systems".

Before we present this overarching analysis, we want to highlight and summarise the most important findings from both the Fergana Valley and the Lower Jordan Basin.

17.2   The Fergana Valley

As in other regions of Central Asia, the legacy of Russian colonisation and the Soviet Union has had a significant influence on the current water and irrigation management system in the Fergana Valley. This also applies to the ways decisions are made and the present political systems, economies and societies in the region are shaped. MWEPs, i.e. irrigation infrastructures, have led to path dependencies. This legacy determines the range of action available to the Central Asian countries and shapes the quality and nature of their mutual relations to a large extent. Both the formal and informal political, societal and administrative structures that exist currently can be regarded as a result of past decisions, and the irrigations systems inherited define the options in agricultural cultivation. The transition from former Soviet structures to a market-oriented economy has been slow, and the spirit of the past system controls many decision-making regulations (Kreutzmann 2015, in this volume). Irrigation cycles and drainage systems are oriented towards the cultivation of a limited number of crops, while political guidelines have created a parallel system of "state crops" and "secondary crops", offering only a limited degree of freedom to farmers.

The role of farmers in Uzbekistan – particularly with regard to informal institutions or their stake in the water users associations (WUAs) and the newly formed water users groups (WUGs) – is a key component to understanding the complexity of the socio-agricultural system (Moss and Hamidov 2015, in this volume). The WUAs, for example, are caught between two powerful state-driven institutional regimes: agriculture and irrigation. They lack a supportive institutional environment as well as the necessary resources and the financial and personal capacities needed to efficiently run and maintain the present infrastructure. In order to increase their effectiveness, it would be necessary to analyse existing funding sources for irrigation services and to ascertain how revenues from the agro-hydrological systems could be distributed more equally to enhance the capacities of the farmers and the WUAs. At the same time farmers should be given more discretionary powers to enable them to decide independently on resource use (which crops to plant, how much water to use, when to use water, etc.) which could be achieved by increasing their participatory power in the WUAs and WUGs (Hansjürgens 2015, in this volume).

17.3   The Lower Jordan Basin

One of the arguments brought up in favour of the RSDS Conveyance Project is that the project could be a vector for peace in the region. Therefore, our focus has been on the relations between the riparian states of the Lower Jordan Basin and on the institutional settings (legal provisions, power constellations, informal norms, etc.). The relations between the riparian states are central for medium- and long-term water security (Yorke 2015, in this volume, pp. 227–251). The asymmetry of powers and the deeply rooted mistrust between the parties involved, a result of the Israeli occupation of the West Bank and also of institutional fragmentation within the Palestine Territories and the Kingdom of Jordan, weakens the capacities of actors concerned to develop comprehensive strategies for sharing management of the water resources available (Bismuth 2015, in this volume).

One of the intriguing aspects of the RSDS Project is that alternatives were given serious attention, albeit belatedly. This suggests a learning process going beyond the conventional single-project debate. But the political background and the power asymmetries in the region impede the attainment of nonconventional, less technically orientated measures to prevent further decline of the Dead Sea. Other aspects, such as the technical feasibility to produce energy with saline water, have not been taken into account from the beginning.

Particularly in Jordan, the interests of vested groups work against the implementation of transformative political and administrative reforms (Yorke 2013). International donors can even be regarded as being complicit in these outcomes, since their funding of aid projects and budgets also underpinned state largesse from which these groups benefitted in the past (Yorke 2015, in this volume, pp. 227–251). Therefore, the future role of the international donor community should be evaluated critically, and their policies should match a common development strategy with projects and funds being fairly distributed among riparian states. Specific competences and duties should be assigned to each riparian according to their capacities, thus avoiding inconsistencies and enhancing efficiency.

Jordan in principle has a strong incentive to reap the benefits of improved regional cooperation. But a platform for such regional cooperation in the region is missing. It could be the role of international donor organisations particularly in the light of the planned RSDS Conveyance Project to provide such a platform for mutual exchange as a means of trust building, to enhance transparency of the decision-making process and to provide reliable databases about water resource availability and water distribution. At present, any advances in this direction have been blocked by the failing Israeli-Palestinian "peace process".

The influence of an "ideological narrative" rather than economic considerations on agricultural planning is particularly evident in the Middle East. In some parts of the Jordan Basin, giving up agricultural cultivation entirely or switching back to dryland agriculture in favour of ecological rehabilitation projects could even create win-win situations, especially where the contribution of agriculture to the state economy is minor. Research in enhancing productivity of dryland agriculture should therefore be intensified (Bismuth et al. 2015b, in this volume). Also, the role of "virtual water trade" in the resource strategies of the countries and its effect on the economies, revenues and trade balances still need further research. The same applies to the effects of global food shortages and price hikes.

17.4 Cross-Analysis of the Case Studies

Despite significant differences and varying framework conditions between the two case studies, we believe that it is exactly due to these variations in water management and water governance structures that allow us to define overarching trends regarding the management of MWEPs, to formulate open research questions and to derive options for action.

Even though cultures, societies, historical backgrounds and political and economic systems differ considerably between the two case studies, they show some striking parallels:

In both regions, MWEPs have been regarded as essential elements for economic development. We realise that it has not necessarily been technical failure that has led to the observed ecological and social problems in the respective countries, but rather the inadequate or even absent interaction between technical specifications and societal processes. Apart from infrastructure and human and institutional resources, MWEPs also require arrangements which do not serve only selected groups of users or beneficiaries. It needs political processes to introduce, coordinate and establish adequate governance rules in order to run these social-technical-environmental systems effectively (Dobner and Frede 2015, in this volume).

Our analyses of the situation in both regions have shown that transformation processes in water management have been slow and insufficient, even under generally favourable conditions. Transformation of water management institutions does not hold pace with the societal and environmental changes. As a consequence, the Aral Sea has already been given up, and the Dead Sea is threatened by the same fate.

Decision-making has been ad hoc and partial, too often guided by merely technocratic approaches and aimed at overcoming either economic or resource-based deficiencies. Decisions have often served the interests of influential interest groups and have supported neo-patrimonial structures (Sehring 2009; Yorke 2013). The processes have been organised in centralised and highly hierarchical structures dominated by a "hydro-scientific elite". This corresponds with the observations described by Molle et al. (2009) in their study on hydraulic bureaucracies and missions.

According to the four different perspectives on MWEPs, as described above, we can formulate the following key recommendations for action:

1. Strengthening coordination and cooperation among riparian states
2. Enhancing planning processes and decision-support tools
3. Improving responsiveness and responsibility of institutions and governance structures for MWEPs
4. Taking into account systemic feedbacks and contingent contexts of coupled systems

17.4.1 Strengthening International Coordination and Cooperation Around MWEPs

An analysis of international agreements concerning the management of shared water resources in the two case studies has revealed that they are inflexible concerning changing environments and societal developments, generally lacking a basin-wide approach. Explicit rules for cooperation, adaptation and conflict mitigation are missing, which is a key reason why they fall short.

In addition, the failure of transboundary initiatives is the consequence of an unequal distribution of costs and benefits among riparian states as well as among the various stakeholders within each country. Such asymmetrical relations may considerably hinder the development of transboundary initiatives and agreements. Indeed, one of the fundamental principles of international water law, the equitable utilisation of resources, is frequently violated.

Lim (2014) postulates the identification of multi-resource linkages in the negotiation of international water agreements as a means to increase the number of potential win-win solutions. This so-called "shared benefit model" promotes the identification of benefits of transboundary cooperation that accrue across sectors (e.g. the water sector, agricultural sector, trade sector).

To optimise the management of MWEPs in a transboundary context, it is necessary to improve the coordination and cooperation among riparian states. Important guidelines on how to improve coordination and cooperation in transboundary water management can be found in the so-called "Berlin Rules" developed by the International Law Association (International Law Association 2004). The authors of these rules set out criteria for determining an equitable and reasonable use of water and also for the consideration of the physical and geographical characteristics of the watercourse. While the acceptance of these rules might not be binding for riparian states when bargaining about MWEPs' benefits and cost sharing, international donors, who are in many cases involved in MWEPs, could place more emphasis on the implementation of the Berlin Rules.

The overarching goal of improved cooperation and coordination can be achieved by the following activities:

- Integrating transparent monitoring, evaluation and conflict resolution mechanisms into transboundary agreements
- Harmonising different national rules within a transboundary management area
- Integrating resource allocation methods, which are adaptive to changing environments and societies, as well to new predicaments
- Identifying benefits and costs of transboundary cooperation and trying to establish win-win solutions

It is important to understand that general objectives need to be broken down into practical, reliable and concrete milestones. Research efforts are needed both for the development of adaptive strategies (goals, procedures, strategies) and for the transformation of institutions, preferably achievable in a step-by-step approach. This has to go hand in hand with the development of monitoring and evaluation criteria as they play a critical role in adaptive management structures determining whether standards have been met or more interventions are needed (Lim 2014). A comprehensive analysis of present transboundary treaties and conventions as well as their success factors and failures would be a first step towards the development of adaptive strategies (see, e.g. Dombrowsky 2007).

17.4.2   Enhancing Planning Processes and Evaluation Tools for MWEPs

In the past, the benefits of MWEPs have been mostly overestimated, while the costs have been underestimated (Flyvbjerg 2007, 2012; Ansar et al. 2014). A comprehensive and integrative planning approach may help to improve the sustainability of water resource management. It incorporates the search for alternatives, the adjustment of the plans to technical innovations and the application of transparent auditing and planning procedures. MWEPs as technical structures are always simultaneously embedded into contextual societal and economic frameworks. The planning process should include options for dealing with situations in which – due to uncertain costs and/or benefits – high burdens on a nation's state budget can emerge. The use of economic cost-benefit analysis (CBA) or multi-criteria analysis (MCA) based on a comprehensive understanding of values (e.g. the total economic value, TEV, framework) may provide overarching analytical tools to assess changes in values due to large-scale investments (Young and Loomis 2014). Involved parties should ensure that they have sufficient resources to finance the project, either within the present generation or with acceptable burdens on future generations.

Regarding the planning process of MWEPs, it is also important to evaluate not only the financial but also the legal, institutional and societal capacities right from the outset. Legal instruments, formal and informal institutions and implementation strategies (e.g. the existence of an effective revenue system, the inhabitants' tax mentality, the effectiveness of governance structures) should match available resources at each level of authority and address the needs of the stakeholders (Lim 2014). Designing regulation and a decision-making process should take into account specific locations of capture and weak capacity and to improve the existing infrastructure (Faure et al. 2010).

We highlight the following essentials that a planning process for MWEPs should include:

- Developing alternatives, including the "no-action" alternative and exit strategies
- Drawing up a flawless set of rules for assessing and evaluating MWEPs and a transparent planning process
- Using appropriate decision-support tools such as CBA and MCA within the planning and evaluation process
- Analysing the economic, institutional and societal capacities to assure the sustainability of the project
- Clearly distinguishing between the planning and implementation process in order to minimise the risk of appropriation by vested interests
- Independent controlling accompanying the planning process
- Comparing MWEPs with successful reference systems

Even though planners do have a number of instruments available, there still appears to be insufficient information and knowledge on the complexity and interdependency of ecosystems and the values and services they generate for certain vulnerable groups (Hansjürgens et al. 2015, in this volume). In particular, we lack a comprehensive approach to cost-benefit analysis that includes an integrated instrument assessing all stakeholders and all values affected. In an ideal case, such an instrument would be able to integrate the knowledge of planners and of natural, technical and social sciences and of humanities and societal stakeholders in one integrative process of decision-making. Research on cross-sectorial planning – i.e. ways of connecting water management planning, infrastructure planning and urban planning – is highly desirable in this context.

17.4.3 Improving Responsiveness and Responsibility of Institutions and Governance Structures for MWEPs

Many formal institutions in fragile or failing states (e.g. legislation) lack effective backing and therefore remain ineffective. By contrast, non-codified customs and practices of established elites can remain powerful behind the scene. Building capacities is an important strategy to address weak formal institutions. Institutions should be in accordance with the capacity of the governmental and the economic systems (Lim 2014; Dobner and Frede 2015, in this volume). Designing regulations and a decision-making process should take into account the capacities and the powers of the addressed institutions and improve and optimise the existing infrastructure (Faure et al. 2010).

In the context of fragile or failing states, it is particularly important to also consider the existing institutional strengths that are available. In the past some of the institutional reforms towards liberalisation and decentralisation have undermined already existing, although weak, institutions. In this regard, international donor policy has a specific obligation. Better coordination and harmonisation of donor policy with overarching objectives and international regulations are needed.

Transforming water management institutions is a time-consuming and complex process. The introduction of economic instruments such as water pricing alone will not be sufficient for a successful transformation (Hansjürgens 2015, in this volume). Efficiency and economic performance are linked to the existence of inclusive institutions (Acemoglu and Robinson 2012). Economic instruments have to be designed properly. They have to consider aspects of calculation and distribution of costs among users, metering and monitoring water use and strong governance structures including social participation and social equitability. Transparency, trust and public participation in decision-making procedures from the beginning are important assets of inclusive institutions. Admission to financing and banking systems and access to markets and market information, specifically for farmers, should also be considered within the transformation process. Institutional reform can thus support economic and social development by creating opportunities for rural populations.

The formation of inclusive institutions requires commitment both from the authorities and the civil society. Such institutions also need sufficient financial resources and capacities to be able to play a beneficial role in water management. Structures (e.g. water users associations) have to be adapted to the specific local context. A polycentric model (the "bazaar"), in which various organisational forms coexist, may prove more suitable than a hierarchical one, particularly in situations characterised by little reliable data, highly variable water supply and demand and under-resourced regulatory agencies (Lankford and Hepworth 2010).

Institutional learning has to be backed by close cooperation between research and education. Technical schools, experimental farms and training courses for practitioners can be named as important instruments for transferring knowledge within institutional learning processes. Scientific and educational institutions themselves have to adapt to new requirements arising from increasing complexity of coupled systems. New communication instruments need to be developed to enhance the citizen-science dialogue.

The following activities can be regarded as critical in designing and adapting institutional arrangements:

- Fostering institutional development in accordance with governmental, societal and economical capacities
- Harmonising and coordinating donor policy with overarching objectives and international regulations
- Strengthening inclusive institutions that take all stakeholder interests into account
- Supporting participation processes, transparency and trust-building measures
- Applying the "polluter pay's principle" as far as it is economically and politically feasible
- Linking research and education with practice

National institutional reform and adaptation can only be effective in the context of supportive international development policy. In the past, mainly international donor organisations such as the European Development Bank or the World Bank have granted loans for investments to MWEPs. Recently, we are observing the emergence of other actors, such as certain countries that finance infrastructures (e.g. China) or private financing institutions independent from international conventions, regulation schemes and standards. Decision-makers mostly welcome these new investors without assessing in depth the societal, economical or environmental consequences of the planned investments. This development is a new challenge not only for the societies in the concerned regions but also for the international community as such.

17.4.4   MWEPs as Coupled Technical-Social-Environmental Systems

Technical solutions such as MWEPs are attractive for decision-makers because they promise high gains, which can be harvested in manageable timescales. Benefits can be expected within a short period after the construction phase is completed. This is in contrast to institutional transformation processes, which usually require longer planning, longer implementation and longer evaluation phases. Commonly overlooked when planning MWEPs are not only often "neglected values" like ecosystem and societal impacts but also their systemic rebound effects, their irreversibility and their effects on coupled systems. Coupled systems are characterised by high-complexity and self-organising structures leading to emergent phenomena. Such self-organising structures are illustrated by new system properties, which cannot be understood from the properties of the single component (Helbing 2014). Not only in socio-economic systems but also in ecosystems, complexity leads to emerging systemic risks. Especially in conditions of uncertainty, incomplete information or even a basic lack of knowledge, one has to be aware of the fragility of complex (eco)systems. Furthermore, MWEPs create path dependencies, which limit the available range of choices for future generations, and therefore pose ethical questions. On the other hand we can use the self-organising, adaptive nature of the coupled systems to reach favourable system behaviours, which are robust to external disruptions and align to changing conditions (Helbing 2010).

From the perspective of coupled systems, the planning, regulation and use of MWEPs should respect the following:

- Considering not only aspects of uncertainty (where probabilities can be defined), but a lack of knowledge
- Applying the precautionary principle
- Developing alternatives especially in the case of contingent events
- Using modelling tools (e.g. agent-based modelling) and analytical tools (e.g. network analysis) that explicitly address complexity
- Establishing favourable system attributes, which are robust to external disruptions and adaptive to changing conditions

Further research is needed to develop methods, models, tools and decision-support systems for coupled systems, which have stochastic characteristics. We underline the necessity for an interdisciplinary research approach to answer the pressing questions of a complex world.

17.4.5   Concluding Remarks

MWEPs are characterised by properties that exceed those of "normal" infrastructure projects not only quantitatively (by their mere size) but also qualitatively (by adding entirely new dimensions to an infrastructure project). MWEPs are complex investments that encompass not only technical but also societal, ecological and institutional dimensions. This complexity requires an understanding of the collective dynamics driving it. The answer cannot be technological approaches alone. Many crises occur because we are unable to understand the dynamics of such systems. There is a gap between problems and solutions. It is our task as researchers to reduce this gap, to develop new tools to address the complexity of systems and to provide knowledge for decision-makers regarding the behaviour and management of complex systems. But we have also to reflect whether we use the right approaches to understand these interrelated complexities, whether we derive fitting recommendations or action and whether we communicate them accordingly. In a citizen-science dialogue, we might find the first promising answers – but we should go further. The role of research is to raise understanding of complex processes, to query underlying assumptions, to uncover inconsistencies, to raise alternative options, to map out potential futures and to broaden perspectives. In these ways researchers would not just be providing new (specialist) knowledge but engaging with political debates and seeking ways of informing these.

We have to invest not only in the development of new technologies, which can reconcile environmental necessities with our human needs, but also in the capacities of our societies and our institutions to deal with growing complexities. We can no longer ignore the social dimensions of our decisions nor the increasing disparities between our societies. We need adaptive and resilient societies, based on transparency and humanistic values. In planning MWEPs we have to start now, as their lifespan far exceeds the lifetime of single human generations.

References

Acemoglu D, Robinson J (2012) Why nations fail: FBBVA lecture
Ansar A, Flyvbjerg B, Budzier A, Lunn D (2014) Should we build more large dams ? The actual costs of hydropower megaproject development. Energy Policy 69:1–14
Bismuth C (2015) Cooperation and power asymmetries in the water management of the Lower Jordan Valley – the situation today and the path that has led there. In: Huettl RF, Bens O, Bismuth C, Hoechstetter S (eds) Society – water – technology: a critical appraisal of major water engineering projects. Springer, Dordrecht, pp 189–204
Bismuth C, Hoechstetter S, Bens O (2015a) Research in two cases studies: (1) Irrigation and land use in the Fergana Valley and (2) Water management in the Lower Jordan Valley. In: Huettl RF,

Bens O, Bismuth C, Hoechstetter S (eds) Society water technology: a critical appraisal of major water engineering projects. Springer, Dordrecht, pp 89–98

Bismuth C, Hansjürgens B, Yaari I (2015b) Technologies, incentives and cost recovery: is there an Israeli role model? In: Huettl RF, Bens O, Bismuth C, Hoechstetter S (eds) Society – water – technology: a critical appraisal of major water engineering projects. Springer, Dordrecht, pp 253–275

Dobner P, Frede H-G (2015) Water governance: a systemic approach. In: Huettl RF, Bens O, Bismuth C, Hoechstetter S (eds) Society – water – technology: a critical appraisal of major water engineering projects. Springer, Dordrecht, pp 79–87

Dombrowsky I (2007) Conflict, cooperation and institutions in international water management – an economic analysis. Edward Elgar, Cheltenham

Faure J-M, Goodwin MEA, Weber F (2010) Bucking the Kuznets curve: designing effective environmental regulation in developing countries. Virginia J Int Law 51:95–156

Flyvbjerg B (2007) Policy and planning for large-infrastructure projects: problems, causes, cures. Environ Plan B Plan Design 34:578–597. doi:10.1068/b32111

Flyvbjerg B (2012) Why mass media matter to planning research: the case of megaprojects. J Plan Educ Res 32:169–181. doi:10.1177/0739456X12441950

Moss T, Hamidov, A (2015) Where water meets agriculture: the ambivalent role of the water users associations (WUAs). In: Huettl RF, Bens O, Bismuth C, Hoechstetter S (eds) Society – water – technology: a critical appraisal of major water engineering projects. Springer, Dordrecht, pp 149–167

Hansjürgens B (2015) Theory, market and the state: agricultural reforms in post socialist Uzbekistan between economic incentives and institutional obstacles. In: Huettl RF, Bens O, Bismuth C, Hoechstetter S (eds) Society – water – technology: a critical appraisal of major water engineering projects. Springer Berlin, Dordrecht, pp 169–186

Hansjürgens B, Droste N, Tockner K (2015) Neglected values of major water engineering projects: ecosystem services, social impacts and economic valuation. In: Huettl RF, Bens O, Bismuth C, Hoechstetter S (eds) Society – water – technology: a critical appraisal of major water engineering projects. Springer, Dordrecht, pp 65–78

Helbing D (2014) Complexity time bomb – when systems get out of control. Chapter 2 of Digital Society, forthcoming. http://papers.ssrn.com/sol3/papers.cfm?abstract_id=2502559. Accessed 10 Apr 2015

Helbing D (2010) Systemic risks in society and economics. International Risk Governance Council (irgc). http://irgc.org/IMG/pdf/Systemic_Risks_Helbing2.pdf. Accessed 10 Apr 2015

International Law Association (2004) Water resources law. Berlin conference. http://international-waterlaw.org/documents/intldocs/ILA_Berlin_Rules-2004.pdf. Accessed 10 Apr 2015

Kreutzmann H (2015) From upscaling to rescaling – The Fergana Basin's transformation from Tsarist irrigation to water management for an independent Uzbekistan. In: Huettl RF, Bens O, Bismuth C, Hoechstetter S (eds) Society – water – technology: a critical appraisal of major water engineering projects. Springer, Dordrecht, pp 113–127

Lankford B, Hepworth N (2010) The cathedral and the bazaar: monocentric and polycentric river basin management. Water Alternat 3:82–101

Lim M (2014) Is water different from biodiversity? Governance criteria for the effective management of transboundary resources. Rev Eur Comp Int Environ Law 23:96–110. doi:10.1111/reel.12072

Molle F, Mollinga PP, Wester P (2009) Hydraulic Bureaucracies and the hydraulic mission: flows of water, flows of power. Water Alternat 2:328–349

Moss T, Dobner P (2015) Between multiple transformations and systemic path dependencies. In: Huettl RF, Bens O, Bismuth C, Hoechstetter S (eds) Society – water – technology: a critical appraisal of major water engineering projects. Springer, Dordrecht, pp 101–111

Sehring J (2009) The politics of water institutional reform in neopatrimonial states. Springer, Wiesbaden

Yorke V (2013) Politics matter: Jordan's path to water security lies through political reforms and regional cooperation. NCRR Trade Regulation, University of Bern, Bern

Yorke V (2015) Jordan's shadow state and water management: prospects for water security will depend on politics and regional cooperation. In: Huettl RF, Bens O, Bismuth C, Hoechstetter S (eds) Society – water – technology: a critical appraisal of major water engineering projects. Springer, Dordrecht, pp 227–251

Young RA, Loomis JB (2014) Determining the economic value of water: concepts and methods. Taylor and Francis, New York

Index

A

Agricultural sector, 120, 142, 169, 171–173, 177–180, 182, 255, 256, 262, 264, 266, 269, 270, 272, 273, 285

Agricultural water use, 142, 255, 256, 262, 266, 269, 270

Agriculture, 4, 7, 27, 41, 48, 49, 69, 84, 92, 103–105, 107, 108, 113, 120, 124, 130, 133–135, 138, 141, 142, 144, 145, 149–166, 179, 180, 192, 202, 228, 230, 231, 232, 235, 236, 239, 243, 244, 254–257, 260, 262, 264, 266–270, 272, 273, 282, 283

Anthropocene, 48, 49

Aral sea, 11, 54, 55, 72, 91–93, 97, 102, 108, 114, 130, 284

B

Biodiversity, 14, 19, 25, 26, 36, 43, 48, 49, 58, 67, 69, 102

C

Cash crop, 38, 104, 114, 133, 134, 160, 164, 178, 180, 183, 184, 254

Central Asia, 5–8, 38, 54, 55, 61, 89, 90–93, 103, 106–108, 113, 116–122, 124, 125, 133, 142, 144, 150, 152, 159, 170, 171, 177–180, 184, 281

Christianity, 12, 13

Complex and coupled systems, 4

Complexity, 4, 34, 41, 80, 81–86, 94, 135, 281, 282, 287–290

Cooperation, 3, 6, 7, 24, 42, 83, 90, 94, 96, 106, 108, 109, 125, 142, 145, 152, 153, 157, 162, 166, 170, 173, 178, 189–202, 227–249, 283–285, 288

Costs, 14, 37, 50, 65, 103, 124, 130, 151, 172, 200, 205, 241, 257, 284

D

Dams, 4, 19, 21, 34–37, 40, 49, 50, 52, 55, 56–59, 65–70, 72, 92, 93, 137, 192, 197, 247

Dead sea, 5, 11, 40, 44, 84, 90, 94–97, 189–192, 196, 198–201, 205–224, 245, 247, 248, 254, 267, 268, 270, 281, 282, 284

Dead sea reclamation, 206, 207, 210, 212, 217–222, 224

Desalination, 7, 42, 62, 90, 94, 96, 192, 196, 201, 202, 216–220, 222–224, 230, 244–245, 247, 249, 255–260, 264, 269, 270, 272

Design elements, 171, 177

Distributional impacts, 74

E

Ecological impact, 72, 73, 102, 142–143

Ecological trade-offs, 71–73

Economic valuation, 65–76

Ecosystem services, 21, 43, 50, 58, 65–76, 90, 102, 270

Education, 8, 24, 36, 71, 106, 157, 162–164, 242, 254, 288

© The Author(s) 2016
R.F. Hüttl et al. (eds.), *Society - Water - Technology*, Water Resources
Development and Management, DOI 10.1007/978-3-319-18971-0